AN INTRODUCTION TO Crime Scene Investigation

Aric W. Dutelle, MFS
Department of Criminal Justice
University of Wisconsin–Platteville

JONES AND BARTLETT PUBLISHERS
Sudbury, Massachusetts
BOSTON TORONTO LONDON SINGAPORE

World Headquarters
Jones and Bartlett Publishers
40 Tall Pine Drive
Sudbury, MA 01776
978-443-5000
info@jbpub.com

Jones and Bartlett Publishers
Canada
6339 Ormindale Way
Mississauga, Ontario L5V 1J2
Canada

Jones and Bartlett Publishers
International
Barb House, Barb Mews
London W6 7PA
United Kingdom

Jones and Bartlett's books and products are available through most bookstores and online booksellers. To contact Jones and Bartlett Publishers directly, call 800-832-0034, fax 978-443-8000, or visit our website, www.jbpub.com.

Substantial discounts on bulk quantities of Jones and Bartlett's publications are available to corporations, professional associations, and other qualified organizations. For details and specific discount information, contact the special sales department at Jones and Bartlett via the above contact information or send an email to specialsales@jbpub.com.

Production Credits
Publisher, Higher Education: Cathleen Sether
Acquisitions Editor: Sean Connelly
Associate Editor: Megan R. Turner
Senior Production Editor: Renée Sekerak
Production Assistant: Jill Morton
Associate Marketing Manager: Jessica Cormier
Manufacturing and Inventory Control Supervisor: Amy Bacus
Text Design: Anne Spencer
Composition: Arlene Apone
Cover and Title Page Design: Scott Moden
Associate Photo Researcher: Jessica Elias
Cover Image: © Kevin Chesson/Dreamstime.com
Printing and Binding: Malloy Incorporated
Cover Printing: Malloy Incorporated

Library of Congress Cataloging-in-Publication Data
Dutelle, Aric W.
An introduction to crime scene investigation / by Aric W. Dutelle.
p. cm.
Includes bibliographical references and index.
ISBN-13: 978-0-7637-6241-4
ISBN-10: 0-7637-6241-5
1. Crime scenes. 2. Crime scene searches. 3. Criminal investigation. 4. Evidence, Criminal. I. Title.
HV8073.D875 2010
363.25—dc22

2009036755

6048

Printed in the United States of America
14 13 12 11 10 10 9 8 7 6 5 4 3 2 1

Dedication

This book is dedicated to the victims of crimes and to their families. I am sad to think that since the beginning of humanity there has been crime, and there likely always will be. However, I hope that the knowledge and insight of those who will learn from this text, and the professionalism and dedication of the contributors to it, will help to bring you closure, and to right the wrongs.

To my family who could not be around to see the result . . . I think you really would have enjoyed reading this.

Brief Contents

Contents

Preface

At present I am, as you know, fairly busy, but I propose to devote my declining years to the composition of a textbook which shall focus the whole art of detection into one volume.

Sir Arthur Conan Doyle (1859–1930)
Sherlock Holmes, in *The Adventure of the Abbey Grange*

In 2004 I was asked by the University of Wisconsin–Platteville to develop a program in forensic investigation. This opportunity intrigued me as it enabled me to attempt to fill a void that I had realized existed during my previous years of police and crime scene–related service. It was my experience as a new patrol officer that a large chasm existed about the topic of how to properly process a crime scene. The police academy provided a thorough education on search and seizure law, arrest control techniques, firearms, use of force, and driving tactics. However, the block of time devoted to the processing and documentation of crime scenes was typically less than one day, eight hours, in duration.

I thought that the academy lacked sufficient training on such a serious topic, and I was even more shocked when I was hired by a mid-size police department (100+ officers). During my six months of field training, not one hour was devoted to the topic of crime scene processing. We received a four-hour block given on police report writing, but there was no training on processing methodologies relating to fingerprint recovery, crime scene photography, sketching, and other processing-related topics. The disconcerting part of this observation was that each officer was issued, as part of his/her standard vehicle gear, a field latent processing kit.

Most people have the view (as a result of weekly primetime television) that crime scene personnel have received significant training in matters pertaining to crime scene processing methods. However, my first experience with dusting for fingerprints was not in a training session, but was instead at a crime scene. After completing my report and reconsidering the manner in which the investigation was conducted, I began to take notice of my knowledge- and training-related deficiencies. At no time during my academic, academy, or police department training was I instructed on the investigative process as it related to the documentation and collection of physical evidence. My department soon realized the problem as well, and was proactive in the development of a Crime Scene Unit. This was a developing trend around the United States and not an isolated event to my department. However, as the reader of this text will find, and as I discovered during my time in law enforcement, it is not simply those crime scene technicians or crime scene investigators who must be educated on the methodologies and theories relating to crime scene investigation. It is also necessary for police officers, police administration, medical personnel, and a great many others to understand such matters. However, because I was unable to acquire such education prior to my employment, it was through trial and error and the mentoring from others (who were hopefully properly trained on such matters) that I learned how to process crime scenes.

My frustrations did not end when I left full-time duties in law enforcement to accept the position of creating a forensic investigation program for the University of Wisconsin–Platteville. Although I managed to find myself in law enforcement without crime scene–related training, there was now a mass movement toward the development of forensic programs and academic resources related to crime scene processing. However, the development of such programs did not necessarily mean proper information was available to those attempting to educate and be educated on the topic.

This book was designed with that chasm in mind. I hope that the reader will find that this text offers a single source reference for the topic of crime scene investigation, which assumes no prior knowledge of the topic, and is assembled to be easy to follow and easy to instruct.

Introduction

During my six years of teaching at the university level, I was frustrated when trying to find a text on crime scene investigation that was worthy and applicable to both subject matter and audience. Several notable introductory texts on crime scene investigation exist; unfortunately, they are all written for those who are involved in some capacity in the field. There are over 400 academic programs within the United States (in undergraduate, graduate, and secondary education) that have a need for a text presenting a foundation in crime scene investigation. This need inspired me to cease use of texts that assumed too much for the level of my students, and to instead write one that is more appropriate. This text is written with the assumption that the reader is new to the field of crime scene investigation and crime scene processing.

An Introduction to Crime Scene Investigation is written to be a single-source reference for the investigative process as it relates to crime scene processing methods and procedures. The reader will find that the focus is on the day-to-day aspects of crime scene processing. It describes what the crime scene investigator does, details the steps in the process, and explains how to decide on the order of the methods. My hope is that this in-depth reference will help to eliminate the warped impressions of a crime scene investigator's duties created by modern forensic television and movie dramas, and will assist the reader to identify the various interrelated components of the investigative process. Such public misconceptions are addressed in Chapter 2, which deals with the so-called *CSI Effect*.

This text also provides the reader with a historical perspective to crime scene investigation. By understanding how we got to where we are today, the reader will have a better appreciation of, and an understanding for, the complexities and difficulties relating to the topic. If one is to be educated and to understand what crime scene investigation is, it is necessary to cover more than the simple "how-tos" relating to such matters. Therefore, in addition to the methods, motives, and motions needed to secure the crime scene, the text also covers an overview of the investigative process as well as the ethical considerations applying to such matters. Discussions about crime scene procedures, detailed figures, and real-life examples enhance the reader's understanding and demonstrate precisely how to apply the techniques and tools of the trade.

An Introduction to Crime Scene Investigation provides those in training and those attending community colleges, universities, and police academies with a true introductory text that is written for those with no assumed knowledge in the area and for those who are either not in or are new to the field. My hope is that this text will be considered an essential source of foundational development, and for the proper protocols used in a successful crime scene investigation.

This text was assembled using a pragmatic, critical, and multidisciplinary approach. This text recognizes that many of the concepts and methods relating to the field of crime scene investigation are likely to be unfamiliar. For this reason, special instructional devices and learning strategies are utilized throughout:

- ***Learning Objectives.*** The learning objectives are listed at the beginning of each chapter. Emphasis is placed on active learning rather than passive learning. It is hoped that the reader gains knowledge of how to apply the concepts and material, and not simply to retain the information temporarily with plans to regurgitate it on a test. The learning objectives concentrate on the acquisition of knowledge and foundations needed to understand, compare, contrast, define, explain, predict, estimate, evaluate, plan, and apply.
- ***Key Terms.*** If one is to study the field of crime scene investigation, it is necessary to become familiar with the terminology and vocabulary associated with it. Key terms are listed at the beginning of each chapter to key the reader to specific terms they should understand.
- ***Review Questions.*** Knowledge and skills must be reinforced. Review questions are provided for student self-study options and for use by instructors who are developing written assignments and examinations.
- ***"Ripped from the Headlines" Current Event Examples.*** In an effort to apply the theory and guidelines addressed within the book, I have given the reader examples of real-world incidents involving the content discussed in the chapter. It is hoped that this application to real-world situations will enable the reader to better grasp the concepts presented.

- ***"Case in Point" Examples.*** In addition to current events, there are case scenarios and summations to assist the reader in further comprehension of the presented material.
- ***"View from an Expert" Insights.*** As a way to further the real-world information that the reader is exposed to within the text, many chapters also include insights from a forensic expert in the field of study discussed in the chapters.
- ***Case Studies.*** At the conclusion of many of the chapters are case studies that challenge the reader to seek out additional information from outside sources to assist them with a more in-depth comprehension of the presented material.
- ***References.*** Each chapter is supplemented with references pertaining to the key areas addressed within the chapter.

A comprehensive student companion Web site, hosting downloadable Word searches, Flashcards, Glossary, Crosswords, and Case Study Weblinks, is available online at www.criminaljustice.jbpub.com/CrimeSceneInvestigation. PowerPoint presentations and a TestBank have also been created for instructor use.

Acknowledgments

There are a number of people who I would like to thank for their selfless assistance with this text, without whose assistance this work would fall incredibly short of hitting the mark.

I wish to thank:

Sandy Weiss, of Packard Engineering. Your insight into forensic photography and your friendship has meant a great deal. I am forever grateful.

Robert Ramotowski, of the United States Secret Service Forensic Division, for your contributions and professional insight throughout the writing process. Your friendship and professional advice over the years have been much appreciated.

Sarah Owen and Gary Lind, from the Northeastern Illinois Regional Crime Laboratory, for your professional opinions and contributions with regards to the collection and preservation of evidence.

Joseph LeFevre, for your gracious assistance with authoring Chapter 14 on arson and explosive evidence. Your experience relating to fire public safety information was greatly appreciated.

Carmine Artone, United States Secret Service (ret.), for your contributions and wealth of knowledge regarding the science of fingerprints.

Chief Randy Taylor, for your professional insight, support, and contributions. You are a good friend, and I am forever grateful.

Numerous students at the University of Wisconsin–Platteville assisted with contributions to this text: Alex Albright, Sarah Bedish, Larissa Larsen, Fall 2008 sections of Criminalistics, and many unnamed, but certainly appreciated students who have challenged me to come up with the answers. I consider it an honor to have been your mentor.

Elicia Bruchez and Erica Lawler, graphic art students at the University of Wisconsin–Platteville, who have exciting careers ahead of them. Your vision and professionalism have helped this text to leap off the pages and give the readers visual insight into the world of crime scene investigation. I could never thank you enough.

Vanessa Davis, the only reason that you are near the end is that I want my words of thanks reverberating throughout the entire text. Your assistance with the preparation of the ancillary materials, and reviewing and providing feedback on the entire text was absolutely wonderful. Your ability to see the material from both student and professional standpoints was tremendously appreciated and insightful. Trust me when I say that you will go on to be an incredible chemist and do some wonderful things.

I also wish to thank Megan Turner, Cathleen Sether, Sean Connelly, Jill Morton, Jessica Elias, Arlene Apone, Scarlett Stoppa, Shellie Newell, and the entire editorial, production, and marketing staff at Jones and Bartlett Publishers for giving me the opportunity, assisting me with the production, and helping to create what I hope will be a work that will fill a large void in professional education.

Lastly, I wish to thank my family for their love and support. This was a work that was produced at a difficult time, and I appreciate your understanding and support with the endeavor.

Reviewers

H. Dean Buttram, III, PhD
Jacksonville State University

Joseph LeFevre, MS
University Wisconsin–Platteville

Detective Sergeant Michael Payne
Racine Police Department

Terry L. Pippin
College of Southern Nevada

Jacqie Spradling
Chief Deputy District Attorney
Shawnee County District Attorney's Office

Michael R. Summers
Erie Community College North Campus

Michael T. Stevenson
University of Toledo

Gregory M. Vecchi, PhD
Nova Southeastern University

Sally Welch, PhD
Marygrove College

About the Author

Aric Dutelle has been involved in law enforcement for over a decade. During this time he has held positions as a police officer, deputy sheriff, crime scene technician, and reserve medico-legal investigator. He has a Master of Forensic Sciences (MFS) degree, is a POST-certified law enforcement officer, and a Wisconsin Law Enforcement Standards Board certified instructor in: Hazardous Materials, Interview and Interrogation, Physical Evidence Collection, Report Writing, Scene Management, Testifying in Court, and Traffic Crash Investigation. Since 2004, he has taught crime scene–related courses at the University of Wisconsin–Platteville. He is currently responsible for the Forensic Investigation Program within the Department of Criminal Justice.

In addition to his University obligations, since 2006, the author has been a forensic instructor for the U.S. Department of Justice's International Criminal Investigation Training Assistance Program (ICITAP), specializing in and providing training in crime scene processing methodologies and techniques around the globe. He continues to be actively involved in training, consulting, and assisting law enforcement agencies with criminal investigations and crime scene processing around the United States, and internationally.

Throughout these experiences, the author has worked on thousands of crime scenes including kidnappings, homicides, suicides, robberies, burglaries, sexual assaults, and drug trafficking–related cases.

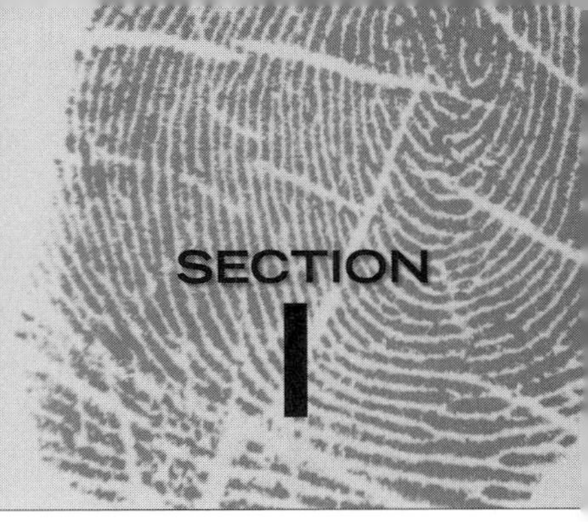

SECTION I

An Introduction to Crime Scene Investigation

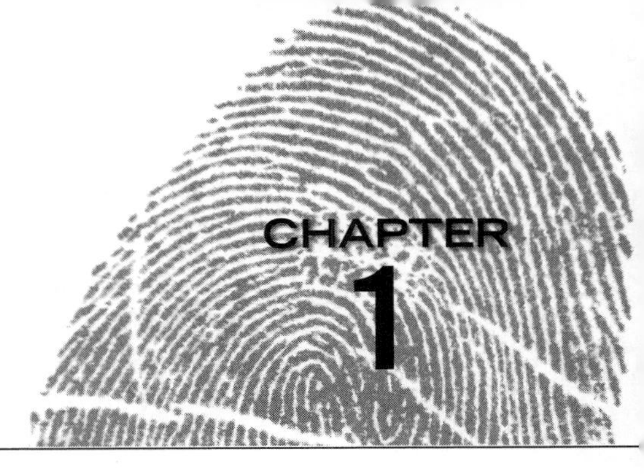

Truth is sought for its own sake. And those who are engaged upon the quest for anything for its own sake are not interested in other things. Finding the truth is difficult, and the road to it is rough.

Alhazen, 965–1039

LEARNING OBJECTIVES

- Understand the definition of crime scene investigation (CSI).
- Understand the objectives of CSI.
- Gain knowledge of the historical figures who have contributed to CSI.
- Understand the types of evidence associated with a criminal investigation.
- Understand the value of physical evidence.
- Define the scientific crime scene method.
- Know how to establish a proper chain of custody.

KEY TERMS

Chain of Custody
Circumstantial Evidence
Crime
Crime Scene
Crime Scene Investigation (CSI)
Criminalistics
Direct Evidence
Evidence
First Responder
Fruit of the Poisonous Tree
Investigate
Iterative Process
Locard's Exchange Principle
Real or Physical Evidence
Scientific Method
Scientific Investigation Method
Testimonial Evidence
Trier of Fact

What Is Crime Scene Investigation?

The word **investigate** means to make a systematic examination or to conduct an official inquiry. In law enforcement terms, this is used with reference to the investigation of a crime. **Crime** is defined as an act or the commission of an act that is forbidden by a public law and that makes the offender liable to punishment by that law. **Evidence** is anything that can help to prove or disprove that a crime was or was not committed, and by whom. The varieties of evidence are endless and are discussed later in this chapter.

Crime scene investigation (CSI) is the often used term relating to the process associated with the investigation of a criminal event. More specifically, it is the systematic process of documenting, collecting, preserving, and interpreting physical evidence associated with an alleged crime scene, in an effort to determine the truth relating to the event in question. Note that this does not define the purpose of a criminal investigation and thus the process of CSI as the determination of guilt or innocence. Rather, it is to document the truth. In the United States, the determination of guilt or innocence is left up to the **trier of fact**, who would be a judge or magistrate in a trial by the court, or a jury of one's peers in a trial by jury. The study of crime scene investigation therefore concentrates on guidelines, concepts, and principles associated with aiding the trier of fact to determine the truth. To accomplish this task involves techniques associated with locating, documenting, collecting, and preserving physical evidence.

CSI is not necessarily the analysis of the physical evidence and does not pertain to the entire investigative process. CSI often is confused with **criminalistics**, the analysis of physical evidence. More specifically, CSI is the application of science within the enforcement of law. Criminalists (also called forensic scientists) use instrumentation and education within the natural sciences associated with the interpretive value of the physical evidence submitted. (This will be discussed more in depth in Chapter 7.) However, it is important to note at this point that CSI is associated with the on-scene techniques applied in an investigation, and typically does not involve analyzing the physical evidence. The foundation of CSI is in documentation.

■ Who Is Responsible for Crime Scene Investigation?

There are many components of a criminal investigation. All support one another and none can stand alone or is more important than any other. The first tier of the investigative process begins with the **first responder**. This is typically the police officer who is dispatched to the initial scene. Through the course of his or her response to and documentation of the event, if the officer believes that a crime was committed, then that officer must analyze the necessity to have that crime and associated crime scene further processed. Sometimes the crime and associated crime scene are of a minor nature and can be handled by the first responder. Other times they will necessitate the response of specialized personnel. (This process is discussed in depth in Chapter 5.)

Most police departments use uniformed or patrol officers as well as detectives or crime scene investigators to process a crime scene. On smaller scenes, sometimes the initial officer on scene will be responsible for photographing, fingerprinting, and documenting the crime scene. On larger scenes, a detective or investigator may be called in to assist. Some agencies have specialists who can be called upon to process a scene. These specialists may be either uniformed or civilian personnel.

There are five basic personnel components of a criminal investigation: the first responder, detective/investigator, crime scene investigator/technician, criminalist/forensic scientist, and the courts.

First Responder

Typically, the first responder is the patrol officer who responds to the initial complaint. His or her duty is to secure the scene and not to touch anything. The officer begins the initial stages of documentation through the securing of and preservation of the physical evidence.

Detective/Investigator

This is a specialized police officer who has significant experience in investigations. He or she is concerned with the entire criminal investigative process. This officer could be compared with a central processing unit (CPU) for a computer, where each part of a computer's peripheral devices has a specific function, but all of them rely on the CPU to interpret the information fed into it in order to output a product that is understood and applicable to the task at hand. The detective will process the information presented by the first responder and others, and decide whether or not additional investigative methods are warranted, if search/arrest warrants need be issued, and whether or not arrests can be made. This individual rarely collects physical evidence but is primarily responsible for the collection of testimonial evidence.

Crime Scene Investigator/Crime Scene Technician

This individual (who could be a civilian or a sworn police officer) has advanced training in the documentation, collection, and preservation of physical evidence. Additional skills might include technical or college training and education in photography, mapping, crime scene processing, search and seizure law, and criminal justice practices. While this person is responsible for processing the crime scene, he or she is rarely involved with the subsequent analysis of the collected evidence, and is not involved with interviewing or interrogating suspects or witnesses. The primary area of concern is the crime scene and associated physical evidence. This may include conducting examinations for fingerprints or performing presumptive tests for the presence of blood at crime scenes. After documenting, collecting, and properly preserving the evidence, the CSI investigator/technician transports the evidence to the evidence booking location, where it is booked in (recorded) and held until such time that it is needed for analysis or checked out to be transported to court.

Criminalist/Forensic Scientist

This individual is responsible for the analysis of submitted items of evidence. A criminalist uses the latest in scientific instrumentation and knowledge to assist with interpreting the value of the evidence submitted. Typically he or she will have a four-year degree in chemistry, biology, or another applied science. It is not the duty of the criminalist to determine guilt or innocence, or to otherwise investigate the crime. His or her only obligation is the proper interpretation of the submitted evidence, and providing a report on such matters to the detective for final determination on the investigative value of the evidence.

The Courts

This step includes the prosecution, defense, judge/magistrate, and jury (if applicable). It is left up to this step to determine whether the previous components of the investigative process have resulted in evidence showing that a crime was committed or not and, if so, than it shows by whom as well as defines the warranted punishment. At this step, it is pointed out whether there has been a breakdown or fault within the investigative process that would warrant retraining, education, or implementation of policies or laws to reduce future problems.

Objectives of Crime Scene Investigation

Although different personnel may be involved in processing a crime scene, their objectives remain the same:

- Determine whether a crime has been committed. If it is determined that there is no crime involved, or if the issue is one for the civil courts, law enforcement personnel have no responsibility. If uncertain, contact the district attorney's office.
- If a crime has been committed, determine whether it was committed within the investigator's jurisdiction.
- Discovery and documentation of all facts pertaining to the complaint in question.
- Identify and eliminate suspects as a result of collected physical and testimonial evidence.
- Locate and apprehend the perpetrator. Sometimes this means issuing an arrest warrant if the suspect is unable to be located.

- Throughout the entire process, maintain a proper chain of custody to ensure that collected evidence is admissible in court.
- Effective testimony as a witness to the collected evidence within court.

While the aforementioned are objectives of CSI, at all phases of the process, the obligation of the individuals involved is to the truth. All personnel involved in processing and investigatory efforts must be conscious of not simply collecting evidence to prove what they believe happened. Also, investigators should approach the crime scene investigation as if it will be their only opportunity to preserve and recover these physical clues.

History of Crime Scene Investigation

When considering the history of CSI, it is important to define its parameters. As mentioned earlier, CSI is not truly the forensic laboratory analysis portion of the investigation, but pertains to the processing methodology related directly to the scene of a crime. As such, the history of CSI spans thousands of years. Appendix A lists the dates of historical significance as they relate to crime scene investigations and the sometimes interrelated field of forensic science. Suffice it to say that since mankind first walked upon the Earth, there have been cases of foul play. The following investigators are but a few of the individuals who have furthered the process that has come to be known as CSI.

Song Ci (1186–1249)

The Chinese death investigator Song Ci wrote a book entitled *Xi Yuan Ji Lu,* (translated as *Collected Cases of Injustice Rectified*), in which he discussed a number of murders. One such case took place in a village in which the victim had been slashed repeatedly. The local magistrate suspected that a sickle had been used, but repeated questioning of witnesses proved fruitless. Finally, the magistrate ordered all of the local men to assemble, each with his own sickle. It was a hot summer day and flies, attracted by the smell of blood, eventually gathered on a single sickle. Confronted with such evidence, the sickle's owner confessed to the murder. The book also offered advice on how to distinguish between a drowning (water in the lungs) and strangulation (broken neck cartilage), along with other evidence from examining corpses on determining if a death was caused by murder, suicide, or an accident (Gernet, 1962, p. 170).

Sir Robert Peel (1788–1850)

In 1829, Peel established and subsequently headed the Metropolitan Police Force for London, based at Scotland Yard. The 1,000-member force became known as "Bobbies." They proved to be very successful in reducing the crime rate. Peel also was responsible for defining ethical requirements of policing officers through what became known as "Peelian Principles." His most memorable principle is summarized by the concept that the police are themselves the public, and the public are themselves the police. This is the earliest recorded reference to the idea of community-oriented policing efforts (Becker, 2009, p. 21).

Auguste Ambroise Tardieu (1818–1879)

The preeminent forensic medical scientist of the mid-19th century, Tardieu participated in 5,238 cases as a forensic expert during his 23-year career. He wrote over a dozen volumes of forensic analyses, covering such topics as abortion, drowning, hanging, insanity, poisoning, and suffocation (Labbé, 2005, pp. 311–324). Tardieu wrote what may be the first book on the sexual abuse of children entitled, *Medical-Legal Studies of Sexual Assault*, which was published in six editions from 1857 to 1878. To this day, battered child syndrome is also known as "Tardieu's Syndrome."

Allan Pinkerton (1819–1884)

In 1849, Allen Pinkerton was appointed as Chicago's first detective by Mayor Levi D. Boone. This is the first known functional separation of investigations from patrol. Pinkerton is also known for creating the Pinkerton Detective Agency, which provided private service to railroad properties and other business enterprises.

Sir Arthur Conan Doyle (1859–1930)

While it might seem odd to list an author as a contributor towards advancing the work of crime scene investigation, few would argue the impact that Doyle had on popularizing the work associated with such matters. Doyle's fictional character, Sherlock Holmes, brought science to the process of investigation. In Doyle's first Sherlock Holmes novel, *A Study in Scarlet* (1887), Holmes used forensic investigation methods that were undiscovered at the time, but later would revolutionize crime scene work. One of these areas was the concept of *serology*, the study of blood and other bodily fluids. This foresight led many readers to become intrigued by the possibilities of the field of criminology, and undoubtedly was an impetus for many future scientists to explore its realm.

Hans Gross (1847–1915)

In 1893, Gross was responsible for authoring the first dissertation concerning the practical application of scientific disciplines to the field of criminal investigation. Gross spent much of his life as a public servant, serving as both a public prosecutor and judge in Graz, Austria. As such, he spent a great deal of time studying and developing principles relating to criminal investigation. His compilation of information became a classic book entitled, *Handbuch für Untersuchungsrichter als System der Kriminalistik* (*Handbook for Examining Magistrates as a System of Criminology*), which was later published in English under the title, *Criminal Investigation.* This is the first known text to propose that investigators could expect investigatory assistance from such scientific fields as microscopy, chemistry, zoology, botany, physics, fingerprinting, and anthropometry (Saferstein, 2004). Gross was a very strong advocate for the application of the scientific method within investigations, although he did not make any specific technical contributions regarding such work.

Edmond Locard (1877–1966)

Locard was instrumental in demonstrating how the principles articulated by Gross could be applied within the crime laboratory. His educational background was in medicine and law, and both were used extensively within his work. Locard is credited with starting the world's first crime laboratory, when in 1910 he persuaded the police department in Lyon, France to give him space to begin work on such matters. His work in criminalistics led him to develop what has come to be known as **Locard's exchange principle**. The essence of this theory was that whenever two objects come in contact with one another, there is a cross-transfer of evidence that occurs (Saferstein, 2004). Locard was specifically referring to dust particulates and trace materials, an area in which he performed significant research. According to the premise behind Locard's Principle, every crime scene could be connected to a criminal, witness, and victim, and every criminal, witness and victim could be connected to a crime scene. In addition to this concept, the theory has come to be the foundation behind many of the guidelines and steps taken at a crime scene to limit contamination.

Paul Leland Kirk (1902–1970)

A chemist and forensic scientist, Kirk is most known for his work on blood spatter evidence. He applied this expertise on bloodstain pattern analysis to the Sam Sheppard homicide case (from which the 1960s television program and 1993 Hollywood movie, "The Fugitive," were based). As a result of Kirk's contributions to the fields of forensic science and investigations, the highest honor that one can receive in the criminalistics section of the American Academy of Forensic Sciences carries his name.

Jobs in Crime Scene Investigations

The U.S. Department of Labor's *Occupational Outlook Handbook (OOH) 2008–2009* states that forensic science technicians are considered to be members of a high wage and high growth occupation. Jobs "are expected to increase much faster than the average. Employment growth in state and local government should be driven by the increasing application of forensic science to examine, solve, and prevent crime. Crime scene technicians who work for state and county crime labs should experience favorable employment prospects resulting from strong job growth" (U.S. D.O.L., 2008). OOH projects a 36.4% increase in employment opportunities for this field from the period of 2004 to 2014.

The demand for those with forensic investigations background continues to grow in the private sector as well. The insurance industry, private security, and retail loss prevention will demand employees with extensive investigative backgrounds. The OOH is projecting an 18% to 26% increase for private investigators, a 9% to 17% increase for surveillance officers, and similar growth of 9% to 17% for claims adjusters (U.S. D.O.L., 2008). This all equates to considerable job security for those already involved in the field of crime scene investigation, and job potential for those considering such employment.

Training and Education/Professional Development

Those interested in crime scene investigation will need to do more than simply hunker down and log endless hours of watching prime-time television. There are various levels of education that one can accumulate in the field of CSI. Some regional technical/vocational schools offer certificate programs and/or two year degrees relating to crime scene investigations. An increasing number of four-year colleges and universities are beginning to offer more comprehensive education than simply the criminal justice system to include in-depth studies relating to the specific sections and functions of crime scene investigation. Typical program offerings include: Introduction to Forensic Science, Introduction to Crime Scene Investigation, Evidence Collection and Preservation, Crime Scene Processing Techniques, Criminalistics,

ICITAP: The Global World of Crime Scene Investigation

According to the United States Department of Justice (2009), "the mission of the International Criminal Investigative Training Assistance Program (ICITAP) is to work with foreign governments to develop professional and transparent law enforcement institutions that protect human rights, combat corruption, and reduce the threat of transnational crime and terrorism." ICITAP is located within the Department of Justice's Criminal Division, and receives its primary funding from the State Department.

ICITAP was established in 1986 with the first mission being the training of Latin American police officers. Over the last two decades this mission has evolved into the ICITAP becoming a full-service criminal justice development agency that plays a vital role in international stability and rule of law, both of which are paramount to U.S. security in the post-September 11, 2001 world in which we live.

In 2008, ICITAP was involved in carrying out 1,117 training programs involving nearly 83,000 foreign participants.

ICITAP's Forensic Services program assists in development of crime laboratories that support criminal justice institutions, while also providing for technical assistance and training related to:

- DNA/Serology
- Engineering sciences
- Expert witness development
- Odontology
- Pathology and biology
- Physical anthropology
- Psychiatry and behavioral science
- Questioned documents
- Toxicology
- Forensic nursing
- Crime scene processing
- Forensic photography and digital imaging
- Computer evidence
- Forensic accounting

ICITAP recently has supported forensic developments in the following countries:

- Bulgaria
- Columbia
- Indonesia
- Iraq
- Kosovo
- Mozambique
- Senegal
- Tajikistan

(continues)

ICITAP: The Global World of Crime Scene Investigation (continued)

- Tanzania
- Thailand
- Turkmenistan
- Uganda
- Ukraine
- Usbekistan

Source: Information retrieved from U.S. Department of Justice, International Criminal Investigative Training Assistance Program (ICITAP). Retrieved July 13, 2009 from http://www.usdoj.gov/criminal/icitap/programs/forensics/.

Investigative Photography, Fingerprint Classification and Development, and Criminal Investigation.

Most departments today prefer—if not require—some type of college degree. It is advisable to contact the departments or agencies where one resides or will be residing to find out their particular requirements and duties. Regardless of whether one's major education is in general studies, criminal justice, or forensics, augmenting those studies with minor courses in basic computer training, drafting, and photography will increase one's value to future employers. Any curriculum designed for crime scene investigation in criminal justice classes at most colleges will cover more general studies and not involve specifics.

After being selected for employment, most departments will have a probationary period where the employee will go through a training period (on-the-job training [OJT]) and be assigned to a field training officer. Most of the experience about how to do the job will be gained in this phase of employment. Most departments also offer their employees opportunities for post-employment or in-service training to further their employees' skills and development. Most of the continuing educational classes for the crime scene investigator are specifically geared to crime scene response, evidence collection, forensic photography, fingerprint technology, and homicide and death scene investigation.

If one was interested in seeking a job in evidence recovery, it is helpful to spend a weekend at an auto body shop gaining the experience removing a door panel, the seat, carpet, head liner, or a light assembly. Often the investigator will have to remove a door or a section of wall from a structure.

Visiting the morgue or a local trauma center is another way to gain experience. This is a custom-tailored job not suited for everyone. What a person can do to a fellow human is not always a pretty sight. If someone is not able to stomach a busy weekend night in the local trauma center or morgue, then he or she surely will be unable to stomach some of the mutilation and/or uncommon sights that are part of the job.

To be successful in the field of crime scene investigation, an individual must dedicate him or herself to a career of continuing education and professional development. There are a number of professional accreditations that crime scene investigators can pursue in order to meet basic and advanced standards within their field. Competent crime scene investigators will acquire continuing education and training through regular attendance at conferences and training seminars as well as by seeking advanced education within their areas of interest and specialty.

The International Association for Identification (IAI) offers numerous programs and levels of certifications within those programs to encourage professionals to continue their training and education, while also ensuring common training and standards of investigative strategies. The standards for certification are high, and thus achievement of certification serves to underscore the competence of the individual to perform their job-related duties, which assists in reducing or eliminating scrutiny of their qualifications within the court. Appendix F contains a list of organizations that are associated with various aspects of crime scene investigation.

What Is a Crime Scene?

A working definition of a **crime scene** (both primary and secondary) is anywhere evidence may be located that will help explain the events (Ragle, 2002). The crime scene is thought of as the location at which the crime was committed. However, a single crime may have numerous associated locations. Crime scene personnel should consider the first scene where evidence is located as the primary scene, even if it is not the most significant, and any subsequent locations as secondary scenes (Ragle, 2002).

Every crime scene is unique and each is dynamic. Although a standardized set of procedures must be followed at every location by the crime scene investigator, experience and observation will assist in developing a strategic operational plan for processing the area.

Types of Evidence

To learn about the proper manner in which to document, collect, and preserve evidence, it is wise to recognize the two types of evidence associated with criminal investigations: testimonial evidence and real or physical evidence.

Testimonial Evidence

Testimonial evidence consists of vocal statements that most commonly are made by a person who is under oath, typically in response to questioning. However, this type of evidence also can be made by witnesses, victims, or suspects during the course of the investigation, while not under oath.

While there is a certain amount of testimonial evidence that may occur at the crime scene (witness statements, victim statements, and spontaneous utterances by suspects), typically this is not the subject of collection efforts associated with crime scene work. Therefore, such matters are not addressed within the confines of this text. However, it should be noted that, from an investigatory standpoint, all testimonial evidence is considered to be lies until corroborated by other physical evidence or testimonial evidence. That is to say, the easiest way to tell when a person is lying is if his or her lips are moving.

Real or Physical Evidence

Real or physical evidence includes any type of evidence with an objective existence, that is, anything with size, shape, and dimension. This type of evidence can take any form. Gases, fingerprints, glass, paint, hair, blood, soil, and drugs are all examples of physical evidence. The variety is infinite.

The area of concern with respect to crime scene processing is with relation to physical evidence. Some texts break evidence into further categories of "direct" and "circumstantial" evidence, which simply sub-categorize the testimonial and physical evidence. This is because **direct evidence** is that which proves a fact without the necessity of an inference or a presumption. This can be true of either physical or testimonial evidence. **Circumstantial evidence**, on the other hand, involves a series of facts that, although not the fact at issue, tend, through inference, to prove a fact at issue. Again, at times this can be the case for either testimonial or physical evidence. Therefore, this book recognizes two chief categories of evidence: either testimonial (not emphasized herein) or real/physical evidence.

Although some types of physical evidence may come from a single source (i.e., fingerprints, DNA), most physical evidence may only be associated with a single class or group (i.e., fibers, hairs, clothing, etc.). Most physical evidence cannot definitively connect a suspect to a crime, as can fingerprint evidence. This should not diminish the usefulness of that evidence (i.e., corroboration, place at scene, interrogation, etc.).

Value of Physical Evidence

It Can Prove That a Crime Has Been Committed or Establish Key Elements of a Crime

Example

An individual contacted on a traffic stop is found to have a large amount of ingredients commonly associated with the manufacture of methamphetamine. Based on this and other intelligence data, a search warrant is performed on the individual's home and an active methamphetamine laboratory is located.

It Can Establish the Identity of Persons Associated with the Crime

Example

Fingerprints and DNA are specific to individuals and, when located at a crime scene, help to associate the individual with having been at the scene of the crime.

It Can Place the Suspect in Contact with the Victim or with the Crime Scene

Example

Officers are dispatched to a series of burglar alarms at a retail outlet. When they arrive, they discover several businesses that have been broken into by gaining access through destruction of the windowed storefronts. An officer locates an individual walking in the area and, upon further investigation, the individual is found to have glass particles covering his clothing. Later analysis of the trace material would find that the glass was consistent with the glass windows that had been broken to gain access to the businesses.

It Can Exonerate the Innocent

Example

The Innocence Project was founded in 1992 by Barry C. Scheck and Peter J. Neufeld, to assist prisoners who could be proven innocent through DNA testing. At this writing, 240 people in the United States have been exonerated by DNA testing, including 17 who served time on death row. True suspects and/or perpetrators have been identified in 104 of the DNA exoneration cases. These people served an average of 12 years in prison before exoneration and release. Exonerations have been won in 34 states. The leading causes of wrongful convictions in these cases were: mistaken eyewitness testimony, lab error, false confessions, and incriminating statements. The proper DNA analysis of the evidence resulted in the true story being told (Innocence, 2008).

It Can Corroborate the Victim's Testimony

Example

A victim of sexual assault claimed that she had been raped by her neighbor. There was no initial evidence of semen present during the medical examination, and the victim claimed that this was because the suspect had worn a condom. However, the victim mentioned that the suspect had placed his mouth on her breast during the assault. The neighbor claimed to have never had intimate contact with the victim and said that he had never been inside her apartment. During a search of the victim's bedroom, hair was located that was consistent with the neighbor's. This evidence proved that the neighbor was being dishonest, but did not show that a sexual assault had occurred. Swabs collected from the victim's breasts were compared against saliva samples (called *buccal swabs*) taken from the suspect, and were found to be consistent in all respects with the biological profile of the suspect. Further forensic analysis of samples taken during the medical exam showed the presence of spermicidal lubricant commonly associated with condoms, located within the vagina of the victim. This information corroborated that what the victim had said was true.

A Suspect Confronted with Physical Evidence May Make Admissions or Even Confess

Example

A suspect of a series of construction site vandalisms was confronted with video surveillance tapes that had been set up to record the ongoing vandalisms. The suspect immediately admitted to the crimes and claimed to have been a disgruntled former employee.

Court Decisions Have Made Physical Evidence More Important

Example

Miranda v. Arizona, 384 U.S. 436 (1966) as well as a number of other high court decisions within the United States have limited the authority of law enforcement to rely on statements and custodial confessions. This has resulted in a paradigm shift of attention being placed on physical evidence as fundamental to proving a case.

Juries Expect Physical Evidence

Due to the vast amount of influence on the jury pool by prime time forensic dramas, jurors expect litigations to involve physical evidence. Juries also expect the science associated with the analysis and testimony regarding such evidence to be infallible.

Negative Evidence (Absence of Physical Evidence) Also May Provide Useful Information (Such as Fraudulent Reporting?)

Example

An owner of a local electronics store claimed that a burglary had occurred. Subsequent investigation revealed no sign of forced entry. There was no video surveillance evidence because the video system was not operating. The owner claimed that the video system had gone down from a power outage during the night and probably missed capturing the incident. A check of power records revealed no such outage, and when pressed on the inconsistencies the owner admitted to staging the incident due to financial difficulties. The lack of evidence prohibited the owner from committing the intended crime of insurance fraud.

■ Scientific Crime Scene Investigation

Most texts about crime scene investigation limit the investigator's responsibilities to the documentation of the crime scene and the subsequent collection of the associated physical evidence. While this is mostly true, and most responsibilities and actions are mechanical in nature, the reasoning behind this processing methodology is actually to ensure the admissibility of the evidence within a court of law. It must be remembered that the fundamental purpose behind a criminal investigation is the determination of the truth, and successful investigation of a criminal episode. Because of this, the crime scene investigator has to take such mechanical steps a stage further.

A crime scene investigator must implement a scientific approach to CSI. This process not only includes the mechanical aspects of scene security, documentation of the crime scene, and the collection and preservation of the physical evidence. It also demands and expects more dynamic approaches, such as scene survey, scene definition, and analysis, development of the link between physical evidence and persons, and the reconstruction of the crime scene. These more dynamic aspects of CSI play an extremely important role in the identification of the suspect(s) and the solution of the crime (Lee, 2003). The process of CSI is therefore scientific. This process adheres to the scientific method, which means that the investigation of the crime scene is a systematic and methodical approach.

The **scientific method** utilizes principles and procedures in the systematic pursuit of knowledge involving the recognition and formulation of a problem, the collection of data through observation and experiment, and the formulation and testing of a defined hypothesis. **Figure 1.1** shows the steps involved in the scientific method. If a crime scene investigator is to incorporate this methodology into CSI, the associated **scientific investigation method** could be viewed as shown in **Figure 1.2**.

A process such as the scientific investigation method that involves backing up and repeating is referred to as an **iterative process**. This means that the investigator repeats the process throughout, and continually re-checks and analyzes that they have properly conducted the scene processing. Re-checks are continued until the results are

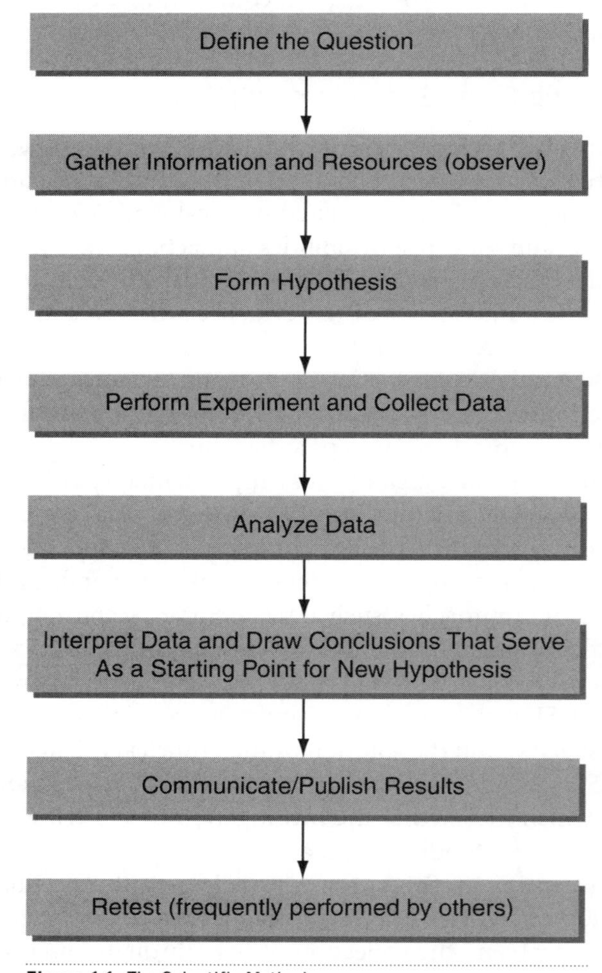

Figure 1.1 The Scientific Method.

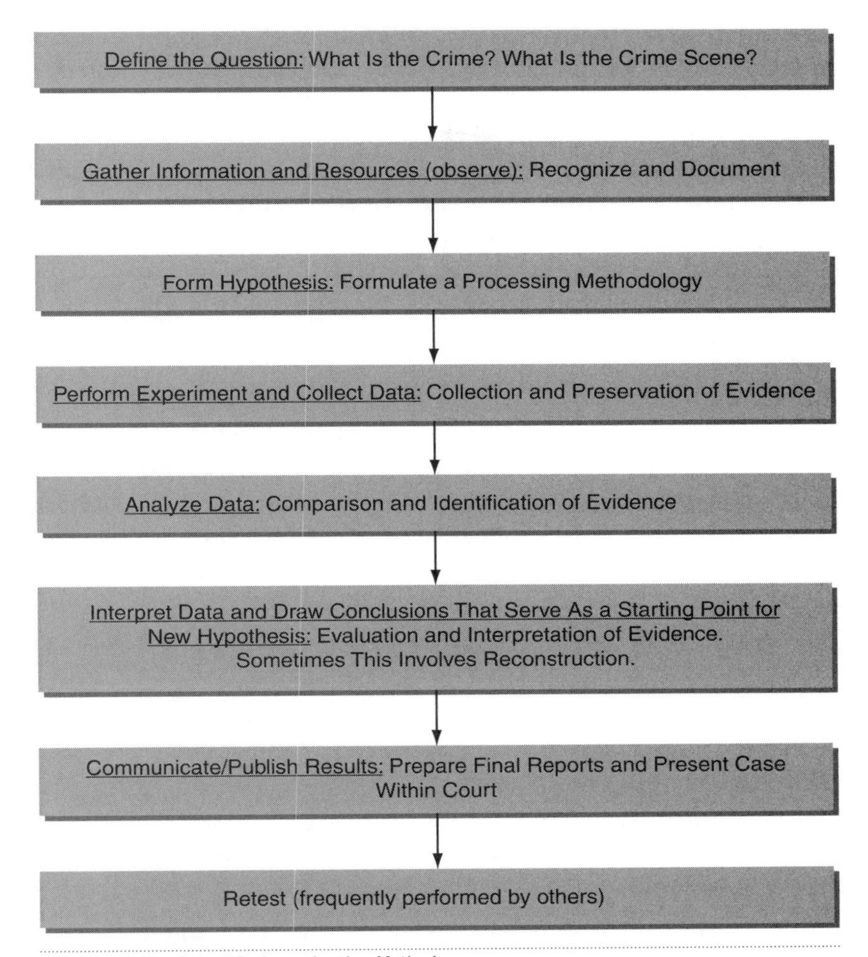

Figure 1.2 The Scientific Investigation Method.

negative, meaning that nothing further is required, and nothing has been overlooked. As with the scientific method, however, all steps within the scientific investigation method are subject to peer review.

Collection and Preservation of Physical Evidence

The need for proper recognition, collection, and preservation of physical evidence is apparent to all who are involved within the criminal justice system, and to those who regularly watch prime time TV dramas associated with such matters. Physical evidence can directly or indirectly lead to the solution of a crime. Charging and prosecutorial decisions are often affected by the quality of the physical evidence associated with the case, and the manner in which it was collected.

Because the laws and legal precedents concerning collection of physical evidence are subject to change, it is impossible to give specific, up-to-date information on acceptable procedures. It is, however, possible to provide guidelines. Foremost, it is of the utmost importance that extreme care be taken to use only approved and legal methods when collecting and packaging evidence (Wisconsin DOJ, 2005). If the individual involved in processing efforts is unsure of these procedures, close communication should be maintained between the prosecutor's office, the crime lab, and those persons responsible for the collection of physical evidence. This will ensure the evidence will not only be collected properly, but also be admissible within a court of law.

With regards to the admissibility and forensic usefulness of evidence, there are two main areas of concern relating to the collection and preservation of physical evidence. These include both legal and scientific considerations. Legally, before any evidence is seized, the investigator must decide if:

- Case law is pertinent. Has there been a case in which a decision was rendered relating to the case at hand?
- A search warrant is required. Is it necessary for the court to issue an order and permission for law enforcement to search an area for physical evidence?
- A court order is required. Is it necessary for the court to issue an order that requires a suspect/defendant to submit an evidence sample (DNA, hand writing, hair, etc.)?

If the investigator does not have legal grounds to be within the area at which the evidence is located, then the evidence cannot be collected. If evidence is found to be illegally collected (whether intentionally or unintentionally), then that evidence will be found to be inadmissible in court. This is commonly referred to as the **fruit of the poisonous tree doctrine** (Fisher, 2004). To avoid this and other pitfalls relating to the seizure of evidence, investigators should consult with the prosecutor's office and stay current with case law and court decisions.

With regards to scientific considerations in CSI, the "validity of information derived from examination of the physical evidence depends entirely upon the care with which the evidence has been protected from contamination" (Wisconsin DOJ, 2003, p. 16). More to the point, if the evidence is found to have been improperly collected, handled, or stored, its value may be destroyed and no amount of forensic laboratory work will be of assistance in restoring it. Therefore, it is imperative that items of evidence be collected, handled, and stored in a way that will ensure their integrity. In doing so, the likelihood is increased that useful information can be extracted by examination and that the item will be considered admissible within subsequent court proceedings (Wisconsin DOJ, 2003).

Scientific considerations include the following:

- Physical evidence should be handled as little as possible.
- Items should be packaged separately in individual containers.
- Known or control samples are needed for comparative laboratory analysis. (For example, if an item is submitted to the lab for DNA analysis, it would be necessary to also include a reference sample taken from the individual from whom the evidence is to be compared.)
- As a general guideline, paper is the preferred packaging material, because it allows the evidence to breathe and not to mildew or degrade, while not allowing cross transfer of evidence to occur with nearby objects.
- Sealable, airtight containers are necessary for chemical evidence and evidence associated with accelerants (arson) to prohibit the gaseous evidence from escaping the container and thus impossible to be analyzed.
- At all phases of collection, efforts should be made to avoid contamination of the evidence. This can include cross-contamination from incorrect collection and packaging methods as well as contamination directly from the individual conducting the collecting and packaging.

Areas associated with the correct collection and preservation methods of physical evidence will be covered in more depth throughout the course of this book. This is simply an overview of the scientific considerations that one must consider when collecting and packaging evidence.

■ Legal Duty to Preserve Evidence

In addition to the obvious importance of collection and preservation of evidence from an investigative standpoint, crime scene personnel have a legal duty to collect and preserve physical evidence associated with a crime. In fact, courts across the country have enacted legislation imposing both civil and criminal penalties for failure to properly preserve evidence.

In addition to the District of Columbia, 23 states have created legislation that compels the preservation of evidence: Arkansas, California, Colorado, Connecticut, Florida, Georgia, Hawaii, Illinois, Kentucky, Louisiana, Maryland, Michigan, Minnesota, Missouri, Nebraska, New Hampshire, New Mexico, North Carolina, Oklahoma, Rhode Island, Texas, Virginia, and Wisconsin. To date, only 25 states do not have statutes requiring the preservation of evidence. However, that certainly does not suggest that such preservation is not warranted or departmentally required within those jurisdictions lacking state legislation. Many departments have individual policies governing preservation and retention of evidence, along with strict sanctions for those failing to comply.

Requirements around the preservation of evidence are usually embedded in DNA testing access statutes. In 2004, Congress passed the Justice for All Act (H.R. 5107), which provides financial incentives for states to preserve evidence and withholds those same monies for states that do not adequately preserve evidence. Preserved evidence can help solve closed cases—and exonerate the innocent. Preserving biological evidence from crime scenes is critically important because DNA can provide the best evidence of innocence—or guilt—upon review of a case. None of the nation's more than 200 DNA exonerations would have been possible had the biological evidence not been avail-

CASE EXAMPLE

Robin Lovitt, Virginia Death Row Inmate

Robin Lovitt was convicted of the capital murder and robbery of a pool hall employee in Arlington, Virginia and was sentenced to death in early 2000. When Lovitt sought to appeal the decision, it came to light that the evidence associated with his case had been destroyed. Despite being reminded that Virginia law required the preservation of evidence from the case, a court clerk nonetheless discarded the murder weapon: a blood-stained pair of scissors. The DNA testing available at the time of the trial could only conclusively tie the blood on the weapon to the victim and not to anyone else. More sophisticated DNA testing is now available, but the evidence—which could have proven guilt or innocence and/or informed the appropriateness of the death penalty—is not. The Supreme Court declined to address this issue, and Lovitt was ultimately scheduled to become the 1,000th person executed since capital punishment resumed in 1977. The wrongful destruction of the evidence that could have conclusively proven innocence or guilt denied a conclusive answer. Recognizing the ambiguity caused by the destruction of evidence, Virginia Governor Mark Warner commuted Lovitt's sentence to life in prison (Innocence, 2008).

able to test. Had the evidence been destroyed, tainted, contaminated, mislabeled, or otherwise corrupted, the innocence of these individuals would never have come to light.

What and How Much to Collect?

One question a crime scene investigator asks of him or herself is, "How much should I collect?" Because it is often impractical, and sometimes impossible, to return to a crime scene at a later time to collect additional evidence, it is suggested that the crime scene investigator ensure a thorough and systematic processing of the crime scene to be certain that all necessary evidence, in the correct quantity, has been collected. As a general rule, more is better than less. Judgment with regards to the amount of material to collect is largely a matter of experience.

One must remember that regardless of the amount that is collected, department policy and sometimes statutes will require that the items be preserved for a considerable amount of time, if not indefinitely. This can create significant storage issues within an evidence and property room setting. Therefore, crime scene personnel should consider the proper collection method that will ensure the proper amount of evidence has been collected that will result in the least amount of storage-related issues. For example, if a crime scene investigator is confronted with a bloody king-sized mattress at a homicide scene, would it be best to collect the entire mattress, or simply to photograph the complete mattress and then cut out a sufficient amount of the bloody mattress to submit as evidence? Due to statute of limitations regarding homicide-related evidence, such evidence may be required to be kept indefinitely. Which takes up more room? The bloody mattress material, which could fit within a medium-sized paper bag, or the entire king-sized mattress? The evidence property room custodian will be much happier with the crime scene investigator who chooses to collect the sample rather than the entire mattress, because it will take up considerably less space and will have the same forensic analysis potential.

As illustrated above, it is imperative that the crime scene investigator recognize what is evidence and what is not. Also, it is important for the investigator to have a working knowledge of the forensic value of the evidence located and collected. If all of the natural and commercial objects located within the vicinity of a crime scene were collected and submitted as evidence, the deluge of evidence would quickly incapacitate any department's evidence and property facilities, in addition to further inundating the forensic laboratories charged with analyzing the material. Therefore, evidence can achieve its optimum value only when its collection is performed with a selectivity that is governed by the collector's knowledge of the crime laboratory's capabilities and limitations, and the forensic value of the evidence collected. Such matters will be the topic of discussion throughout this book.

Chain of Custody

The Court will require proof that evidence collected during the course of an investigation and the evidence ultimately submitted to the court are one and the same. In order to accomplish this, those involved in the crime scene investigation must establish a secure **chain of custody**, sometimes referred to as a *chain of evidence*. This "chain" shows who had contact with the evidence, at what time, under what circumstances, and what changes, if any, were made to the evidence. What is needed to establish a chain of custody?

- Name or initials of the individual collecting the evidence and each person subsequently having custody of it.
- Dates the item was collected and transferred (as well as signatures of both parties, validating the item transfer).
- Agency, case number, and type of crime.
- Victim or suspect's name.
- A brief description of the item.

In addition to proper labeling and documentation, all collected evidence must be booked into, and subsequently stored, in a secured area prior to transportation to court. Any evidence reasonably assumed to have been tampered with by unauthorized persons because it was kept in an unsecured area may be inadmissible in court.

Teamwork

Successful CSI involves a team approach. A crime scene team is a group of professionals, each trained in specific disciplines. The common goal of the team is to locate and document all of the associated evidence at the scene. Of utmost importance, however, is that each member of the team is allowed to work the scene independently of the other's influence and to have the ability to challenge the other's findings. This system was purposefully designed so that no one person or entity can operate independently. No single element or person is more important than any other. Each has a vital role. There is no place within this process for those seeking fame or glory.

The team approach to CSI begins with planning. This includes a procedure that is agreed upon by all of the team members and agencies involved with the process. Training for each member is typically mandatory, but more important is the understanding of exactly what is expected of each individual at the scene and in the process.

Assemble any group of individuals and they will be able to investigate a crime scene, albeit not necessarily in a thorough and systematic manner. A successful investigation, and thus a methodical approach to CSI, is much more likely with a well-trained, experienced, and cohesive unit.

As it turns out, however, this is not necessarily the case in many jurisdictions. If a crime scene investigator does only what they are told to do, and collects only what the on-scene police investigator tells them to collect, then (this is by no means a systematic approach) it falls woefully short of the scientific method discussed previously.

Attributes of a Successful Crime Scene Investigator

While it is universally recognized that every person is unique and brings his or her own abilities and insights into any given situation, it has been found that there is a type of person who makes for a successful investigator. As with all information contained within this text, these are guidelines and by no means hard, fast rules. The following is offered as a framework for those taking a self-inventory, and those thinking about choosing individuals who would be successful at involvement within a crime scene investigation capacity.

Intuition

This attribute is not trainable. Someone either has this ability or does not. This component of intellect is what those in law enforcement term a "hunch" or a "gut feeling."

Eye for Detail

An individual's observation skills cannot be overlooked when discussing desirable attributes for a crime scene investigator. Someone who is detail oriented in both observation and notation will find him or herself to be very successful as a crime scene investigator.

Good Communication Skills

The ability to communicate is an extremely valued, and unfortunately rare, commodity. An individual who is interested in crime scene work should be adept at both oral and written forms of communication, as these are imperative for proper testimony and documentation regarding crime scene activities.

Knowledge of Methods for Locating and Preserving Evidence

If one is not familiar with the correct way to properly collect the various forms of evidence, or know where to locate the information that could aid in educating them on the correct manner, then the evidence risks being collected improperly and thus may be found inadmissible in court or unusable. This trait requires that the person be both scientific as well as process-minded, and have a working knowledge of pertinent case law.

Enjoyment of Continuing Education

Because crime scene work is dynamic, such things as technological innovations, case law, and human ingenuity changes it on an almost a daily basis. It is important for a crime scene investigator to have an affinity for continuing education. Without a drive to learn the latest strategies to achieve a successful gathering of appropriate evidence, the individual quickly will be collecting and processing materials in an archaic or perhaps even illegal manner.

■ Mistakes and Errors

Crime scene investigators are human. Mistakes will be made. The important thing is to limit these mistakes, learn from them, and not to repeat them. Repetition of mistakes is seen as incompetence. Critics will always "Monday-morning quarterback" the actions and decisions taken at a crime scene. This is why it is extremely important to adopt a systematic approach to crime scene processing so as to minimize the possibility for errors. Courtroom focus of forensic science is shifting from the lab to the crime scene. Defense attorneys are quick to learn that if they can show that the initial handling of the physical evidence at the crime scene was faulty, then the evidence may be kept out of the trial or at least be tarnished in the eyes of the jury.

Those entrusted with the investigation of criminal activity bear a heavy burden. Mistakes are often unforgiven; there are scores of examples where unintentional errors have been made that resulted in evidence being declared inadmissible by the courts. The best advice is to consider all of the ramifications before embarking on any action (Fisher, 2004).

■ Chapter Summary

Successful CSI begins at the crime scene. If the investigator cannot recognize physical evidence or cannot properly preserve it, no amount of sophisticated laboratory instrumentation or technical expertise can rectify the situation. The crime scene investigator must implement a thorough and systematic methodology towards the documentation and processing of a crime scene in order to ensure a successful outcome to the investigation. While involved in this process, a crime scene investigator must remember that the underlying obligation is not to either the prosecution or the defense, but rather to the truth. Adherence to a scientific, legal, and systematic approach, in an unbiased manner, will ensure that the chain of custody for collected evidence will be intact, and will result in the subsequent admissibility for such evidence.

Review Questions

1. Define crime scene investigation.
2. What are the differences between criminalists and crime scene investigators?
3. What is meant by fruit of the poison tree?
4. Define Locard's exchange principle.
5. List the objectives of CSI.
6. What are the differences between real evidence and testimonial evidence?
7. List the steps of the scientific method.
8. List the steps of the scientific investigation method.
9. What was Hans Gross' contribution to CSI?
10. How did Sir Arthur Conan Doyle impact modern day CSI?

Case Studies

1. Look up the case of *Silverthorne Lumber Co., Inc. v. United States*, 251 U.S. 385 (1920) on the student companion Web site (http://criminaljustice.jbpub.com/CrimeSceneInvestigation) and describe how the fruit of the poisonous tree doctrine was used.

2. After reviewing the case of Columbus County, on behalf of *Katie Brooks Plaintiff v. Marion A. Davis*, on the student companion Web site (http://criminaljustice.jbpub.com/CrimeSceneInvestigation), discuss the importance of a complete chain of custody.

References

Alhazen (Ibn Al-Haytham). *Critique of Ptolemy*, translated by S. Pines, *Actes X Congrès internationale d'histoire des sciences*, Vol I Ithaca (1962), as referenced on p. 139 of Sambursky S. (1974). *Physical thought from the presocratics to the quantum physicists.* New York: Pica Press.

Becker, R. F. (2009). *Criminal Investigation* (3rd ed.). Sudbury, MA: Jones and Bartlett Publishers.

Doyle, A. C. (1887). *A Study in Scarlet.* London, UK: Ward Lock & Co.

Fisher, B. A. J. (2004). *Techniques of crime scene investigation* (7th ed.). Boca Raton, FL: CRC Press.

Gernet, J. (1962). *Daily life in China on the eve of the Mongol invasion, 1250–1276* (p. 170). Stanford, CA: Stanford University Press.

Innocence Project. News and information. (n.d.). Retrieved July 14, 2009, from http://www.innocenceproject.org/news/Fact-Sheets.php

Labbé, J. (2005). Ambroise Tardieu: The man and his work on child maltreatment a century before Kempe. *Child Abuse & Neglect, 29,* 311–324.

Lee, H. (2003). *Henry Lee's crime scene handbook.* Burlington, MA: Academic Press.

Ragle, L. (2002). *Crime Scene.* New York: HarperCollins.

Saferstein, R. (2004). *Criminalistics: An introduction to forensic science* (8th ed.). Upper Saddle River, NJ: Pearson Prentice Hall.

US Department of Justice. International Criminal Investigative Training Program. (n.d.). Retrieved July 14, 2009, from http://www.usdoj.gov/criminal/icitap/programs/forensics/

US Department of Labor Bureau of Labor Statistics. (2008). *Occupational outlook handbook, 2008–09 edition.* Retrieved July 14, 2009, from http://www.bls.gov/oco/home.htm

US Supreme Court Media OYEZ. *Miranda v. Arizona.* Retrieved July 14, 2009, from http://www.oyez.org/cases/1960-1969/1965/1965_759/

Wisconsin Department of Justice. (2003). *Physical evidence handbook.* (7th ed.). Madison, WI: Wisconsin Department of Administration.

The CSI Effect

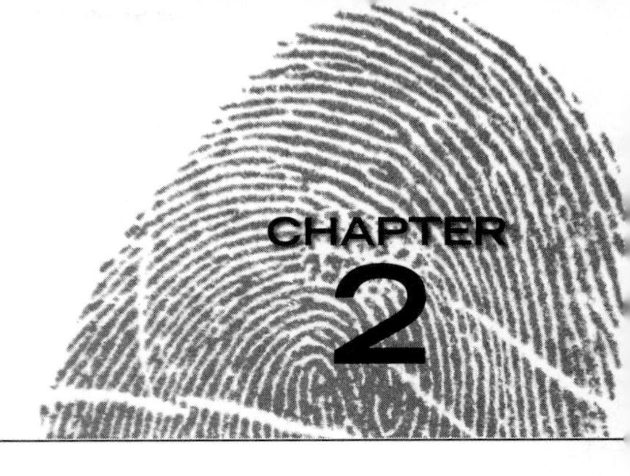

From 8 to 11 o'clock each night, television is one long lie . . .
David Rintels, Former President, Writers Guild of America

All television is educational. The only question is: what does it teach?
Nicholas Johnson, Commissioner, FCC

LEARNING OBJECTIVES

- Understand the "CSI effect."
- Understand how Hollywood's portrayal of CSI-related work has impacted the various levels of the criminal justice system.
- Define and apply voir dire.
- Know how to overcome the CSI effect.

KEY TERMS

American Academy of Forensic Sciences (AAFS)
CSI Effect
Forensic Science Educational Program Accreditation Committee (FEPAC)
Voir Dire

Introduction

A look at news headlines and other media sources quickly reveals how the media is impacting the area of CSI. This impact has permeated nearly every aspect of society. As an example of its widespread influence, even religion is not devoid of its touch. A recent trip by the author to the Atlanta area revealed a church marquee with the words, "Stop in and ask us about CSI." A brief visit with the pastor revealed that the church was capitalizing on the forensic acronym to educate their community about the "Christ Saves Initiative." Another connection to the religious and literary world is Ian Wilson's 2007 book entitled, "Murder at Golgotha: A scientific investigation into the last days of Jesus' life, His death, and His resurrection." But the list of media influences prevails especially in the criminal justice system, particularly with respect to the areas of criminal investigation, the court system, and U.S. education.

Target Stores to Sponsor Forensics Lab in National Law Enforcement Museum

Due to the large number and popularity of television programs such as *CSI: Crime Scene Investigation*, Americans have become fascinated with forensic science. Through support from retail giant Target Stores, the public will now have the opportunity to experience the process associated with the forensic analysis of physical evidence in a realistic forensics lab setting within the National Law Enforcement Museum.

The National Law Enforcement Museum is scheduled to open in downtown Washington, D.C., in 2011. When it opens its doors, the Target Forensics Lab will help visitors better understand the science behind forensic investigation and gain appreciation for the dedication of the criminal justice professionals who are responsible for investigation of criminal events.

"To most Americans, television presents crime solving as a simple, almost routine process that takes 60 minutes, minus the commercials. But in reality, crime scene investigation is an intricate, scientifically grounded process," said Craig W. Floyd, chairman and CEO of the National Law Enforcement Officers Memorial Fund. "Thanks to generous support from Target, we will be able to show Museum visitors what really goes into investigating crimes and how modern forensics labs are helping to bring criminals to justice and make our communities safer."

Exhibit features will include forensic stations that will give museum visitors hands-on experience with topics such as: fingerprinting, trace analysis, blood spatter analysis, DNA, toxicology, and impression evidence (firearms, toolmarks, footwear, etc.). It is also anticipated that visitors will have the ability to explore forensic accounting, entomology, and a realistic example of a medical examiner's office. Museum visitors will be given the chance to process evidence associated with one of four featured crimes within the exhibit. They will do this by collecting evidence and then analyzing it in the Target Forensics Lab before they identify a suspect.

For more information about the National Law Enforcement Museum, including a virtual tour, visit www.LawEnforcementMuseum.org.

Source: Information retrieved from http://www.policeone.com/community-policing/official-announcements/1701902-Target-to-Sponsor-Forensics-Lab-in-New-National-Law-Enforcement-Museum/.

Although many agree with this impact, it is not without controversy. Some believe that television programs and movies about CSI concepts are simply more Hollywood extravagance, giving too much credit to a media that could not possibly influence society to such a degree. While the concept of the so-called "CSI effect" is indeed controversial, it may be ignorant to assume that it does not exist in any manner. The degree to which it exists is the topic of this chapter. It is now left up to the reader to determine through his or her experience, training, knowledge, and the information contained within this book, whether or not the CSI effect is a reality.

■ Birth of a Problem

I remember the first time I heard it. It was near the first week of November 2000. I had been paged out of a deep sleep at 2:00 AM to process a motor vehicle that had been broken into and had its stereo stolen. As I worked groggily, taking photos and attempting to locate evidence that might aid the investigating officer in determining the identity of the offender, the victim watched intently. I was in the midst of attempting to dust for fingerprints when she spoke up. "That's not how they do it on TV." I was not sure that I had heard her correctly so I responded with "What?" "You know, on *CSI*. That's not how they do it. And aren't you going to vacuum for fibers?" There it was. Hollywood had impacted my job and made a 17-year-old high school student my new supervisor.

During the 1950s it was *Perry Mason*. The American public watched the fictional defense attorney on TV as Mason extracted last-minute confessions from the witness stand, thereby influencing public perceptions of the U.S. court system. From 1976 to 1983 television viewers were treated to the forensic wizardry of *Quincy M.E.* As a result of the show, the American public was lead to believe that forensic pathology could be used infallibly to solve criminal mysteries. Today there is yet another primetime drama that is both educating and influencing television viewers and, thus, the potential jury pool.

CSI: Crime Scene Investigation is a CBS television series depicting physical evidence being gathered, processed, and used to solve crimes. The characters portrayed in the series use a number of costly and sophisticated techniques to produce unquestionable results through the scientific analysis of evidence. *CSI* first aired on CBS on October 6, 2000 and the show celebrated its 100th episode on November 18, 2004, and had aired over 200 episodes as of this writing (imdb.com, 2009). According to recent data from the A.C. Nielsen Co. (2008), each week 60 million viewers tune in to watch as gorgeous investigators use techno-wizardry to uncover and analyze physical evidence, never failing in their efforts to solve the crime.

CSI's television cousin, *Forensic Files*, first aired on Court TV in October 2001. The program was touted as the "real life/real case" version of *CSI*. It was the first in a long line of spin-off forensic shows that have exploded from the *CSI* craze. CBS eventually capitalized on its popularity by producing two additional shows, *CSI Miami* and *CSI New York*.

Highly publicized trials such as those of Scott Peterson, Robert Blake, and O.J. Simpson also have drawn many people into forensics.

Basic-cable TV networks such as Court TV, Discovery Channel, USA Network, and A&E also carry many programs depicting forensic investigations of actual criminal cases. These include: *Forensic Files*, *Cold Case Files*, *Body of Evidence: From the Case Files of Dayle Hinman*, *NCIS*, and *American Justice*, among others.

Recent U.S. data show a drop in cinema audiences and a huge rise in TV viewing. Americans are now staying home to watch shows like *CSI*, which currently is the most influential show on world television (Goodman-Delahunty & Tait, 2006, p. 58). With all of these viewers, there is no doubt that some sort of change in society will result. The portrayals presented within these episodes can best be described as "infotainment," a highly stylized, edited, and formatted form of entertainment that is disguised as informative or realistic, but in fact is not scientifically correct and in some cases not grounded at all in reality.

Anytime a television show or movie reaches cult status, there are effects on society. Sometimes this is in regards to marketing or business. Other times it relates to morals, law, or politics. However, this infotainment may affect a person's view of reality. The *CSI* franchise is being pointed to as exerting an influence on all of these areas. This phenomenon has been coined as the CSI effect.

■ Defining the CSI Effect

The CSI effect has been defined in a number of ways. In this text, the author chooses to combine the varying theories into one definition, although it might seem to contradict itself. The **CSI effect** is a phenomenon whereby forensic drama television has created unreasonable expectations, thereby increasing the prosecution's burden of proof, while presenting an air of infallibility with regards to forensic science, while also increasing overall interest, and perceived understanding in the area of forensic science and crime scene work.

It is blamed for romanticizing forensic science as well as creating unreasonable expectations in the minds of jurors. Jurors have become conditioned to believe that every crime can be solved through forensic evidence, and that forensic evidence of guilt exists in every crime. Under this theory, jurors expect forensic evidence to be presented in every case, and require it before they will convict. Where forensic evidence does not exist, jurors may conclude that the evidence necessary to justify a guilty verdict does not exist, or interpret its absence as justifying acquittal. What many jurors do not understand is that even when available, forensic evidence is often contaminated and thus cannot be used or introduced at trial. In other instances, although powerful forensic tools such as DNA are available, they are neither the only nor the best method to prove guilt.

In a criminal case, the prosecution carries the burden of proving the defendant guilty "beyond a reasonable doubt." Courts have described "beyond a reasonable doubt" to mean that: (a) the evidence excludes to a moral certainty every hypothesis but guilt, (b) the inference of guilt is the only one that can be drawn from the facts; or (c) the evidence excludes every hypothesis of innocence. Where jurors refuse to convict without definitive forensic evidence, regardless of the strength of testimonial evidence, or unless the prosecution can exclude innocence via such scientific evidence, it increases the constitutional burden from "beyond a reasonable doubt" to "beyond any and all doubt" (Podlas, 2006, p. 436).

On the *CSI* program, science leads to a singular, objective correct answer. Yet, in real life, forensic conclusions are only as good as the technicians who retrieve the evidence, test it, and draw conclusions from it. Television goes a step further through its character portrayals. *CSI* quashes concerns of human error while heightening the expert status of crime scene technicians. The *CSI* team never possesses a personal motivation to engage in wrongdoing or to violate the (non-existent) standards of their profession. In the real world, however, forensic technicians often have come under scrutiny. Numerous forensic technicians, crime scene investigators, and crime-reconstruction experts have lied under oath, faked their credentials, and fabricated evidence (Podlas, 2006, p. 436). This will be discussed further in Chapter 3's discussion of ethics in CSI.

Some view this supposed CSI effect as an aberration and say that there is not as large an effect on the public as we have been led to believe. However, when one considers the impact on education, on the court system, on crime labs, and on law enforcement, this impact is undeniable.

■ Impact of the CSI Effect

On television, forensic evidence is nearly always correct and the detective is never wrong. Unfortunately, in real life this is not quite true. Many cases are not solved. Many cases do not have physical evidence associated with them.

Popular television shows have resulted in a phenomenon that drives jury verdicts across America. People who end up on a jury know (or think that they know) a great deal about forensic science and the kind of science necessary to solve crimes. Criminal practitioners are being confronted by a population that believes TV shows are realistic.

Prosecutors say juries expect scientific evidence in every case, even though the majority of criminal cases do not call for such evidence. Resourceful defense attorneys are using this to suggest to the jury that fingerprints or DNA evidence should have been introduced, and that failure to do so shows incompetence or an incomplete investigation, and is thus enough to create reasonable doubt. Also, on television, forensic evidence is nearly always correct, and this assumption pervades the jury room. Unfortunately, it is not quite true. This is a growing area of concern for those in the law enforcement, legal, and scientific communities, as well as the public as a whole.

However, not all of the impact has been negative. *CSI* and its spawn of dramas about crime have increased the public's awareness of forensics and thereby increased funding resources for the profession of forensic sciences. It has created a profound interest in forensic science and crime scene work. People now look forward to jury duty. Jurors, now equipped with a better sense of what a crime scene technician does, are better able to understand and follow expert witness testimony.

Elizabeth Devine, an original supervising producer on *CSI* who worked for 15 years as a forensic scientist in the Los Angeles area, also agrees that the CSI effect is positive. According to Devine, "Everyone's jumping on the downsides but we have girls now interested in science, and kids looking at alternative careers they never considered before. Sixty million people are watching a science show every week, millions more around the world and that has to be a good thing. *CSI* has also convinced people to put more funding into crime labs" (Goodman-Delahunty & Tait, 2006). In her defense, all she claimed she wanted to do was to make science sexy. She says there has always been crime-solving television and that her focus was figuring out how to make the analysis of evidence interesting (Fraser, 2007, p. 24).

Another positive aspect of the CSI effect is that the shows have made terms such as *DNA* and *gunshot residue* (GSR) household names. However, as a result, juries are calling for DNA and GSR every time. Juries have unrealistic expectations that simply cannot be met by existing technologies. While citizens or jurors might expect it, not every department or jurisdiction has the latest and greatest technology that is touted in the media.

■ Impact on the Courts

Since the advent of television, legal scholars and practitioners alike have contemplated the impact of law-oriented entertainment programming on the public. Even the Supreme Court and the American

VIEW FROM AN EXPERT

Media Impact on CSI over the Past Three Decades

Thirty years ago, crime scene investigations were conducted to fool victims into believing that officers were exhausting all resources to solve their case as much as actual evidence was being gathered. It was common to spread fingerprint dust mostly to appease the victims because no viable central database existed for identifying suspects. Today law enforcement operates in a much different environment.

In the last 15 years advancements in the areas of fingerprint technology, impression evidence processing and recovery, voice recognition and analysis, trace evidence identification, and DNA have enhanced law enforcement's capability to competently investigate crimes and deliver conviction verdicts for prosecutors as well as exonerate the innocent. Partial prints three decades ago would have been trashed. No method existed to cast an impression in the snow. And a hair follicle might help to identify only a person's hair color.

In most cities, crime scene investigators were the officer that [sic] arrived on scene and took the original report. Gradually, investigators began to arrive at crime scenes and after a decade of increased success and technology, the concept of a specialized investigator emerged. This was the creation of the crime scene investigator.

After a decade of amplified successes, Hollywood leaped on board. The sensationalization of Hollywood is entertaining, yet few crime scene investigators can meet the expectations of make-believe. This can make the crime scene investigator's job difficult when dealing with real victims that [sic] live in make-believe dimensions. Even the most advanced technology cannot compete with the imaginations of Hollywood. It constantly sets a pace for us to endeavor. This is not always a bad motivator. With the progressions of the last several years and the recent popularity of *CSI*-type shows, we are challenged to govern crime scene investigations like never before imagined.

The popularity of these shows has also helped in the education of victims. While many ideologies are ill-received, victims are still learning how to protect and preserve evidence. They've learned to leave crime scenes alone, contact law enforcement immediately, and maintain themselves as evidence when applicable by not bathing. They have even learned key elements to remember when witnessing a crime. These factors appear to be influential in the steady increase in the rate of criminal convictions over the past three decades.

Now is a great time to be involved in law enforcement. The public is holding us at a higher level of accountability and with a few exceptions we are able to meet these expectations. Technology and education has provided us with tools that give us a much greater chance at detecting and preserving evidence. While criminals are also becoming more educated, the public's education and enhanced self accountability is helping us narrow the gap of good versus evil. Just 10 years ago DNA was only a scientific term thought to be 30 years from providing evidentiary value. Today it is identifying suspects, victims, witnesses and innocent people daily.

What will tomorrow bring? It is rewarding to be a part of this profession that has grown from the flatfoot walking the beat to the crime scene professionals that Hollywood tries to replicate.

Randy S. Taylor, Chief of Police
Ellis Police Department, Ellis, Kansas

Bar Association have acknowledged that television impacts the public's perception of the legal system (Podlas, 2006).

Most people do not read statutory or scholarly legal resources; instead, they tend to learn about the law from secondary sources. Empirical evidence shows that most people learn about the law from media, and specifically from television. According to A.C. Nielsen, Co., 99% of Americans have at least one television set, and watch at least 28 hours of television programming per week. This results in television being, in essence, an institutionalized story-teller, explaining to the American populace how things work and what to do.

Over a decade ago, the forensic evidence used in the O.J. Simpson trial had most trial viewers perplexed. Today's viewers are a more educated lot, fueled by what they believe they have learned from television. Defense attorneys are seizing the opportunity to suggest to the jury that fingerprints or DNA evidence should have been introduced, and that failure to do so shows incompetence or an incomplete investigation, and thus is enough to create reasonable doubt. To combat this phenomenon, prosecutors have begun to use "negative witnesses" to explain to juries why forensic evidence might be absent or is not being introduced into court.

Influence on Jurors

Due to much of this false information pouring out of television infotainment, juries and potential jurors are being influenced inside and outside of the courtroom. The impact has been so dramatic on juries that the CSI effect tends to create unrealistic expectations. One example is that jurors believe if there is a DNA link then a suspect must be guilty, when in fact there might be a completely plausible explanation for the suspect's DNA being present (such as being a roommate, etc.).

Before a person becomes a jury member, they must go through **voir dire** (the preliminary examination of a witness or juror to determine his or her competency to give or hear evidence), where both the prosecutor and the defense are provided with an opportunity to question each of the potential jurors. If any of the potential jurors have any biased opinions about the case, they are eliminated from the jury.

Lawyers are now asking would-be jurors whether they watch any of these crime drama shows, and then changing their legal strategies depending on the answers. An example of this phenomenon is that prosecutors are now asking jurors whether any of them believe the state must produce scientific evidence in order to find the defendant guilty. Prosecutors may request certain scientific tests even if the jurors believe they are unnecessary. These prosecutors might even call experts to the stand to explain why evidence such as fingerprints or ballistics is not available (previously referred to as "negative witnesses").

People who overestimate the reality-basis of shows such as *CSI* may develop unreasonable expectations of actual forensic practitioners. Although the technologies lauded on these fictional programs are found in real crime labs, they often require much more time and deliver answers more equivocal in real life than depicted in television crime dramas. Analysts worry that people will come to believe that real forensic science has become as swift and certain as we have always wished justice to be. DNA evidence in particular is expected more and more by jurors, no matter if it is relevant or not found in a given case. Some potential jurors feel startled during voir dire when they are asked whether they are viewers of shows such as *CSI*.

Prosecutors Feeling the Pressure

Along with the juries being influenced, prosecutors also are feeling the pressure of the CSI effect. Before, they rarely had to worry much about making a sophisticated presentation when explaining their case, or about presenting some sort of evidence in every case. Now, however, this is changing because juries have higher expectations after watching crime drama shows.

Prosecutors fear that trial juries who watch *CSI*-type programs are dazzled by the millions of pounds worth of equipment, huge forensic laboratories, and unlimited resources, and think a case can be solved conclusively within the hour. They say that these types of shows can make it more difficult for them to win convictions in the large majority of cases in which scientific evidence is irrelevant or absent (Willing, 2004). Prosecutors now have to supply more forensic evidence because jurors are expecting it, having seen it on television (Dowler, Fleming, & Muzzatti, 2006). Many prosecutors believe that this is unfair, because in many cases certain evidence was not found or was not applicable to the particular case. Some prosecutors now seem to be apologizing at trial for not having "reliable evidence." They believe that they must do this to have a chance to prove their case without the use of evidence. Other prosecutors do not go to the extent of apologizing, but have changed the way they present their case to the jury. Prosecutors in some areas are now telling the jury in their opening statements what type of evidence they can expect to see during the trial, and reminding them that this is real life and not a television crime drama like *CSI*. Some prosecutors believe defense attorneys are taking advantage of situations when physical evidence is lacking or minimal, as an opportunity to create holes within the prosecution's case (DiPasquale, 2006).

Defense Attorneys and the CSI Effect

Defense attorneys have both positive and negative opinions about the CSI effect. Some defense lawyers say that *CSI* and similar shows make jurors rely too heavily on scientific findings and are unwilling to accept that those findings can be compromised by human or technical errors (Willing, 2004). They fear that jurors may be overly influenced by the presence of DNA evidence due to their perception of it being "infallible" and the "authority" of its scientific exponents, even if, in the lawyers' view, the DNA evidence does not directly bear on their client's guilt (Goodman-Delahunty & Tait, 2006).

There are those defense attorneys who embrace the *CSI* Effect phenomenon due to their clients being found "not guilty" as a result of insufficient evidence. This lack of evidence is often enough for juries to find "reasonable doubt," after having this seed of doubt planted by talented lawyers. Many of them do not see why all of the attention is being paid to such matters, and claim that they have not noticed any changes in patterns of jury verdicts. They argue that some forensic sciences, such as fingerprint science for example, have been around since the 1800s, so to say that forensic science is a new phenomenon that people were not aware of is very misleading. "Other defense lawyers say that jurors are not expecting anything different than in the past. Juries have always expected reliable evidence that corroborates eyewitness testimony" (DiPasquale, 2006, p. 2).

Impact on Law Enforcement

Police departments and crime laboratories have increased their efforts and spent much of their annual budgets to acquire modern technology and training to produce reliable results. Some departments are spending hundreds of thousands of dollars on DNA technology or trace evidence collection and analysis technology, while areas such as fingerprint or footwear evidence are being overlooked. This is financially burdensome for many departments. They find themselves in the position of believing they need to improve their forensic capabilities despite the fact that such areas as DNA and trace evidence are seldom—if ever—used in their jurisdiction.

Investigators, crime scene technicians, and officers are finding themselves collecting and booking more evidence than they did in the past. This is happening because they have found that failing to do so is pointed to as incompetence or inefficiency. However, the vast majority of these professionals know through their training and experience that the evidence they are collecting is not necessary to prove the elements of the crime, they are simply collecting it because of the results of the CSI effect.

■ Impact on Crime Labs

Crime labs also are feeling the overwhelming effects of this new phenomenon. Due to prosecutors' belief that if they present more evidence in more cases, then the courts will convict, the quantity of evidence that is being sent to these labs is increasing at an enormous rate. Investigating police are starting to put too much faith in the powers of forensic scientists, and sometimes expect miracles from their crime labs. They now "contain a scene" of murder for days, pouring over every inch while using a variety of chemicals to unveil evidence. They also take videos and hundreds of photos so they can be shown even years later to a jury, to help the jurors understand where and how someone died (Robertson, 2006). Police are now sending more samples to be analyzed because they think that forensic scientists can provide the answer to any case.

As the numbers of samples increase, so does the backlog of work for crime labs (Moscaritolo, 2005). These backlogs may be creating injustice in the criminal justice system by freeing the guilty and incarcerating the innocent. Crime labs now have to spend much more time analyzing different types of DNA. Files thick with evidence required before justices of the peace will issue arrest or search warrants, plus transcripts of police interviews with witnesses and suspects, often fill several cardboard cartons and accordion folders. In addition to unsolved cases, many of the brown cardboard boxes that fill endless shelves and rooms contain clues that can be revisited years later. It is also said that nothing is thrown out in a murder case (Robertson, 2006, p. 32). A result of this is shown by the Department of Laboratories and Research in Westchester, New York, whose division of forensic sciences benefited from a $9 million overhaul in 2007. The renovation of its 17-year-old DNA lab aimed to address the steady increase in demand for its services. The lab, which has about 100 employees, handles nearly 1,000 DNA tests, up from 400 just five years ago. Nearly half of these cases stem from rape and murder-related investigations (Kelly, 2007).

■ The CSI Effect on Education

Since *CSI* first aired in the year 2000, America has seen approximately a 332% increase in the number of programs featuring forensic science as a course of study in colleges and universities. There also has been an impact on early education, with courses of study involving forensic science showing up within America's secondary education systems.

Some police departments were critical of these education programs, saying that in an effort to attract a larger student population, universities were offering unsuitable courses that left graduates unprepared for real-world forensic work. Traditionally, the academic route followed by a would-be forensic scientist had been to pursue a primary (bachelor's) degree in a general-science subject such as chemistry or biology, followed by a suitable postgraduate course or some type of in-service training.

In 2003, the **American Academy of Forensic Sciences (AAFS)** assembled an outline for accreditation standards relating to forensic science education through its **Forensic Science Educational Program Accreditation Commission (FEPAC)**. These suggestions were based on recommendations from the U.S. National Institute of Justice (NIJ). This was an important step and the information was seen as being professionally accurate due to the history of AAFS, which has served the forensic community for over 60 years. The outlining of educational standards fits precisely with the organization's mission statement:

> The American Academy of Forensic Sciences is a multi-disciplinary professional organization that provides leadership to advance science and its application to the legal system. The objectives of the Academy are to promote education, foster research, improve practice, and encourage collaboration in the forensic sciences (AAFS, 2003).

Conforming to the organization's mission, FEPAC was created with the stated mission:

> to maintain and to enhance the quality of forensic science education through a formal evaluation and recognition of college-level academic programs. The primary function of the Commission is to develop and to maintain standards and to administer an accreditation program that recognizes and distinguishes high quality undergraduate and graduate forensic science programs (AAFS, 2003).

Statistics compiled by the AAFS show that in recent years, vocational and technical schools have begun adding forensic and practical investigations courses to their criminal justice programs. Colleges and universities are slowly beginning to follow suit. This trend is seen in **Table 2.1**. A number of programs at the University level dealt with forensic science prior to the *CSI* boom that erupted since 2000. However, all of these programs were in the area of forensic science sometimes referred to as "criminalistics." This education was based on the analysis of physical

TABLE 2.1 Comparison of the Growth of Forensic Science Programs in the United States and Other Countries

	United States			Other Countries		
Type of Program	**2000**	**2005**	**2009**	**2000**	**2005**	**2009**
Certification program	0	5	13	0	0	0
Associate program	0	6	18	0	0	0
Undergraduate program	22	63	106	9	19	31
Graduate program	19	36	56	9	14	20
Doctoral program	5	9	6	1	2	5
Total	46	119	199	19	35	56

Source: Data from the American Academy of Forensic Sciences.

evidence, typically from a chemistry, biology, and physics standpoint. It was heavily rooted in science and did not deal with actual investigative techniques. Such education was targeted towards the individual who desired to pursue employment in one of the over 300 forensic labs in the United States after graduation and become a forensic scientist or "criminalist." Prior to the year 2000, only 46 of these programs were available in the United States. Today there are almost 200 (AAFS, 2003). However, currently only about 10,000 forensic scientist laboratory positions are available within the United States. According to the AAFS, this same trend is evident worldwide, as seen in Table 2.1.

Students are being lured to these programs by the glamour of Hollywood forensic dramas without being informed that after four years of higher education, most likely they will be jobless, or at least unable to gain employment within their field of education. However, they certainly will have cool things they learned about CSI to tell their family and friends.

■ Influence on the Criminal Mind

The CSI effect may be altering how crimes are committed. Many criminal experts have noticed an increase in criminal cases in which suspects burn or otherwise tamper with evidence (e.g., using bleach to destroy DNA evidence) in an attempt to carefully clean the crime scene of trace evidence such as hairs and clothing fibers.

Certain viewers watch these shows as a training ground to learn not only educational facts about forensic science, but also how to cover up their own tracks pertaining to crimes they have committed in the past or might commit in the future. It appears that these shows are educating potential offenders at an alarmingly quick rate, and may even encourage them when they see in the media how simple it is to get away with crimes.

A well-known example of this phenomenon comes from a man charged in a double-homicide in Ohio. Jermaine "Maniac" McKinney, a 25-year-old *CSI* fan, allegedly broke into a house, killed a mother and daughter, and used bleach to remove their blood from his hands. He also allegedly covered the interior of a getaway car with blankets to avoid transferring blood. Prosecutors said McKinney burned the bodies, his clothing, and removed his cigarette butts, with his DNA, from the crime scene. His actions appeared to resemble many of the methods used and seen on episodes of *CSI*. Today, the use of bleach—which destroys DNA—is not unusual in a planned homicide. People seem to be getting more sophisticated with making sure they are not leaving trace evidence at crime scenes (Milicia, 2006). Many people fear the result of these shows, but are unable to decide on a way to eliminate this factor, besides having the shows taken off the air.

In April 2008, the Associated Press reported a story about a group of third graders in Georgia who had an assault attempt on their teacher foiled. These students plotted to attack their teacher and had brought a broken steak knife, handcuffs, duct tape, and other items for their planned assault. The nine children involved in the plot had gone so far as to assign children various tasks, including covering the windows and cleaning up after the assault. When interviewed, several of the students reportedly commented that the ideas were hatched as a result of "television" (Bynum, 2008).

■ Overcoming the CSI Myths

Many viewers are watching forensic drama television and taking their implied messages too literally. Among the growing group of international viewing public, the guiding principles of *CSI* shows that the premise—people lie but the crime labs do not lie—has been accepted as gospel truth when, in fact, it is quite far from the truth. Samples degrade, tests are inconclusive, and lab technicians make mistakes. Some of the science that is seen on *CSI* is state-of the-art but can only be performed by real lab technicians. However, many techniques used on these shows do not exist except conceptually; that is, they have yet to be invented. Too many people in the United States believe everything they see on television is true; unfortunately, however, many of the forensic sciences used on these crime dramas are pure fantasy. Because much of the technology that people see on fictional *CSI*-type programming does not exist, this fact is becoming a concern. One forensic scientist estimates that 40% of the so-called "science" on *CSI* does not exist, and most of the rest is performed in ways that crime lab personnel can only dream about (Schweitzer, 2007, p. 357).

Other examples of fictional technology include the lightning speed with which cases are solved; the computer software that from a partial tire print can tell the analyst the model and year of a suspect's car; fingerprints miraculously pieced together from a shattered glass; among many examples of simple misinformation (Appleby, 2006). Real technicians do not pour caulk into knife wounds to make a cast of the weapon. Caulk does not work in soft tissue. There are machines on the show *CSI* that can identify cologne from other scents on clothing. In reality these are still in the experimental phase, though this is unknown to viewers who are watching.

Forensic drama television such as *CSI* is convincing the television viewing audience that forensic science and crime scene work is not only science, but super science. The irony of this myth is that it is a departure from the current reality of dealing with the law and science. Currently, both the courts and scientists are beginning to realize that there is surprisingly little science in some of the areas referred to as "forensic science." After such court rulings as *Daubert v. Merrell Dow Pharmaceuticals, Inc.,* (505 U.S. 579, 1993), and *Kumho Tire Co. v. Carmichael,* (526 U.S. 137, 1999), courts have begun to scrutinize such evidence with more skepticism and have been surprised to discover the degree of scientific weakness of some forensic fields.

An example of one weakness is the use of neuro-linguistic programming (NLP), a pseudoscientific therapy developed in the 1970s that has since been largely scientifically unsupported. NLP touts that lies can be detected by the way a person's eyes shift when responding to a question, a strategy that has been used in primetime forensic dramas. In reality, evidence gained from an NLP session most likely would be dismissed by a judge due to its lack of credibility.

Lastly, on *CSI* the forensic scientist is a combination of a police officer, crime investigator, and forensic scientist, but in the real world, different professionals fulfill each role (Reville, 2006). *CSI* producers acknowledge that they take some liberties with facts and the capabilities of science, but they say it is necessary to keep their story lines moving (Willing, 2004). People are now asking whether crime drama shows should issue a type of warning at the beginning of episodes, stating that some of the technology used in the programs is fictional. They think that this might give viewers a better idea of what is real and what is not. Hopefully, doing so might help the current situation involving the influence of the CSI effect.

Overcoming the CSI Effect in Law Enforcement

To determine if a department or its personnel are being affected by the CSI effect consider the following questions. Are officers collecting more evidence at individual cases than they have in the past? Is this a result of additional training and experience or outside pressure and influence? Have citizen complaints regarding the manner in which criminal investigations are handled or processed increased since the year 2000? Has the department found itself spending more money on training and equipment relating to forensic investigations during the past four to five years? Have officers returned from court or the streets with stories or experiences where juries or citizens have expected or suggested that evidence should have been collected?

Some view this supposed CSI effect as an anomaly and say that there is not as large an effect as the public is being led to believe. However, when one considers the impact on education, the court system, crime labs, and on law enforcement as discussed in the previous sections, the impact is undeniable. So, how can this effect be countered?

Refresher Trainings

It has been suggested that departments conduct frequent refresher trainings to explain what is and is not needed with regards to preservation, documentation, and collection efforts. These trainings should cover the areas of evidence identification, documentation, collection, packaging, and preservation. They should concentrate on the procedures for processing a crime area and how to avoid its contamination. Educating personnel in the proper methodologies will ensure that the sworn officers, and not the citizenry whom they serve, are the experts.

Caseload

Departments should examine their current caseload. If the vast majority of their cases revolve around identity theft, fraud, and forgery, they might ask whether DNA technology is appropriate. For this jurisdiction, it would be wiser to train someone in handwriting analysis. Therefore, it is suggested that the department look at the demographic served and the case load to determine the most appropriate training and technology necessary to efficiently and effectively work the cases presented.

Informing the Public

Departments would be well served to implement the area of forensic investigation into their Citizen Police Academies to help inform and educate the public as to the realities of forensic science available in their area. While there is often a fine line between educating the public and teaching criminals, it is important for the citizens served to feel safe and to know that law enforcement is capable of conducting a thorough investigation of a crime. Oftentimes this means informing the public about what the department is not able to do themselves but can depend upon other agencies or jurisdictions for assistance.

Educated Personnel

Perhaps the most important way in which departments can ensure that they are not caving into the CSI effect is to hire educated and trained personnel who have a background in forensic investigation.

Popular forensic television dramas have brought the science of forensic investigation into American's homes. When presented as such, the drama is not completely realistic. Law enforcement is left to react and respond to misconceptions relating to the bombardment of Hollywood myths about CSI. Departments should attempt through education and hiring to deal with the pressures to ensure that they continue to conduct their criminal investigations effectively and efficiently.

Chapter Summary

The legal system within the United States demands proof beyond a reasonable doubt before the government is allowed to punish an alleged criminal. The criminal justice system must find ways to adapt to the increased expectations of those who are asked to determine guilt or innocence. If this is going to happen, it will take a major commitment to increasing law enforcement resources and will require equipping police and other investigating agencies with the most up-to-date forensic science equipment and technology. In addition, significant improvements will need to be made in the capacity of the nation's crime laboratories to reduce evidence backlogs and keep pace with increased demands for forensic analyses.

Courts must also find a way to address juror expectations. When scientific evidence is not relevant, prosecutors must find more convincing ways to explain the lack of relevance to jurors. Most importantly, prosecutors, defense lawyers, and judges should understand, anticipate, and address the fact that jurors enter the courtroom with potentially false preconceptions about the criminal justice system and the availability of scientific evidence (Shelton, 2007).

Successful CSI begins at the crime scene. If the investigator cannot recognize physical evidence or cannot properly preserve it, no amount of sophisticated laboratory instrumentation or technical expertise can rectify the situation. The crime scene investigator must implement a thorough and systematic methodology towards the documentation and processing of a crime scene in order to ensure a successful outcome to the investigation. While involved in this process, a crime scene investigator must remember that her or his underlying obligation is not to either the prosecution or the defense, but rather to the truth. Adherence to a scientific, legal, and systematic approach, in an unbiased manner, will ensure that the chain of custody for collected evidence will be intact, and will result in the subsequent admissibility for such evidence.

Review Questions

1. What is meant by the "CSI effect?"
2. How has modern forensic television drama affected the way in which police investigations are handled?
3. How has modern media impacted America's court system and processes?
4. What does voir dire mean?
5. What is the AAFS?
6. What is FEPAC and how does it relate to forensic education such as this text and perhaps even the course with which you are using this text?

Case Studies

1. Access the student companion Web site at http://www.criminaljustice.jbpub.com/CrimeSceneInvestigation/ and review the case of *King v. The State of New York*, #2007-009-171, Claim No. 88273, then consider that one result of the CSI effect is the belief that forensic investigators are infallible. Discuss the fiction of this belief.

2. Look up the case *Greer v. United States*, 697 A.2d 1207, 1210 (D.C. 1997), on the student companion Web site at http://www.criminaljustice.jbpub.com/CrimeSceneInvestigation/ and investigate how the following statement supports the existence of the CSI effect, "in assessing whether the government has met its burden of proving guilt beyond a reasonable doubt, the jury may properly consider not only the evidence presented but also the lack of any evidence that the government, in the particular circumstances of the case, might reasonably be expected to present. For this reason, defense counsel may appropriately comment in closing argument on the failure of the government to present corroborative physical evidence".

References

American Academy of Forensic Sciences (2003). Retrieved August 1, 2009, from http://www.aafs.org

Appleby, T. (2006, November 22). When guilt or innocence depends on fact or fiction; CSI and similar shows are blurring the lines between television and reality. *The Globe and Mail* [Canada].

Aronson, E. (2004). *The Social Animal* (9th ed., p. 87). New York, NY: Worth Publishers.

Bruckheimer J. (Executive Producer). *CSI: Crime Scene Investigation.* (2000–). Universal City, CA: Universal Studios, CBS Productions.

Bynum R. (2008, April 2). Experts dubious of Georgia 3rd-grader plot. Retrieved August 1, 2009, from http://www.usatoday.com/news/nation/2008-04-02-4018418455_x.htm

Daubert v. Merrell Dow Pharmaceuticals, Inc., (505 U.S. 579, 1993). Retrieved July 19, 2009, from http://caselaw.lp.findlaw.comscripts/getcase.pl?court=US&vol=509&invol=579

Deutsch, L. (2006, January 21). Justice going down the tube. *Courier Mail.*

Deutsch, L. (2006, January 16). Show me the DNA. *The Gazette* [Montreal].

DiPasquale, C. (2006). Beyond the Smoking Gun. *The Daily Record,* (September 8).

Dowler, K., Fleming, T., & Muzzatti, S. (2006). Constructing crime: Media, crime, and popular culture. *Canadian Journal of Criminology and Criminal Justice, 48(6).*

Dutelle, A. (2006). The CSI effect and your department. *Law & Order, 54(5).*

Fraser, J. (2007, January 20). Crime time. *Weekend Australian,* p. 24.

Goodman-Delahunty, J., & Tait, D. (2006). DNA and the changing face of justice. *The Australian Journal of Forensic Science, 38(2)*:58.

Greer v. United States, 697 A.2d 1207, 1210 (D.C. 1997).

Internet Movie Data Base (IMDB), (2009). Retrieved August 1, 2009, from http://www.imdb.com/title/tt0247082/

Johnson, N. (1971, September 10). Federal Communications Commission. *Life.* Retrieved November 10, 2009, from http://worldcongress.org/wcfupdate/Archive02/wcf_update_228.htm

Kelly, C. (2007, September 16). Crime lab gets a shot in the arm. *The New York Times.* Retrieved July 19, 2009, from http://www.nytimes.com/2007/09/16/nyregion/nyregionspecial2/16mainwe.html?_r=1

King v. The State of New York, #2007-009-171, Claim No. 88273.

Kumho Tire Co. v. Carmichael, (526 U.S. 137, 1999). Retrieved July 19, 2009, from http://www.oyez.org/cases/1990-1999/1998/1998_97_1709

Milicia, J. (2006, February 4). CSI breeds killers who stick to script. *The Advertiser.*

Moscaritolo, M. (2005, September 28). TV sleuths put law under the microscope. *The Advertiser.*

Nielsen Co., A. C. (2008), *Television Statistics*, compiled by TV-Free America. Retrieved August 1, 2009, from http://www.csun.edu/science/health/docs/tv&health.html

Podlas K. (2006, Winter) 'The *CSI* effect': Exposing the media myth. *Fordham Intellectual Property, Media, and Entertainment Law Journal* 16: 429–465.

Robertson, I. (2006, January 4). Courts feeling CSI effect; prosecutors say jurors want more forensic evidence than ever before because of hit television dramas. *The Toronto Sun.* Retrieved July 19, 2009, from http://www.thefreeradical.ca/Courts_feeling_CSI_effect.htm

Reville, W. (2006, November 22). 'CSI effect' tainting the real forensic evidence. *The Irish Times.*

Schweitzer, N. J., Saks, M. (2007, Spring). The CSI effect: Popular fiction about forensic science affects public expectations about real forensic science. *Jurimetrics*, 47, 357.

Shelton, D. E. (2007, March). The 'CSI Effect': Does it really exist? *National Institute of Justice, 259.*

Tibbetts, J. (2007, June 7). Crime dramas affect verdicts, study shows; psychology; non-forensic evidence increasingly seen as 'second-tier'. *National Post* [Canada].

Willing, R. (2004, August 5). 'CSI effect' had juries wanting more evidence. *USA Today.* Retrieved July 19, 2009, from http://www.usatoday.com/news/nation/2004-08-05-csi-effect_x.htm

Ethics in Crime Scene Investigation

Always do right—this will gratify some and astonish the rest.
Mark Twain (1835–1910)

You cannot make yourself feel something you do not feel, but you can make yourself do right in spite of your feelings.
Pearl S. Buck (1892–1973)

LEARNING OBJECTIVES

- Define and apply ethics in crime scene investigation.
- Understand a code of ethics and how it is incorporated into an agency.
- Understand the ethical concerns governing retention and disposition of crime scene artifacts.
- List and explain the history of important court decisions pertaining to expert witness testimony.
- Define and distinguish between an expert witness and a lay witness.
- Understand and research unethical practices involving crime scene work and forensic analysis, and provide arguments of why each is unethical.

KEY TERMS

Code of Ethics
Credentials
Daubert Standard
Ethics
Expert Witness
Federal Rules of Evidence (FRE)
Frye Test
Hearsay
Morals
Perjury
Values

What Are Ethics?

Theodore Roosevelt said, "To educate a man in mind but not in morals is to create a menace to society" (Lewis, 2009). It is for precisely this reason that the topic of ethics is discussed within this text as a vital component of CSI. Public servants, of whom a crime scene investigator is one, must not only do technical things correctly, they must do ethically correct things. Everyone encounters ethical dilemmas. The question is when and whether we are ready for them.

Before the topic of how ethics plays a role in CSI can be discussed, there first must be an analysis of the meaning associated with the term.

Ethics is the study of moral standards and how they affect conduct. The word stems from the Greek word *ethos*, which emphasizes the perfection of the individual and the community in which he or she is defined (Foster, 2003).

Often the term *ethics* is confused with the terms *values* and *morals*. These terms actually have quite different meanings and yet all are interrelated. **Values** are beliefs of an individual or group, for or against something in which there is some emotional investment. Typically, values are the rules by which an individual or group makes decisions as to what is right or wrong. **Morals**, on the other hand, can have a more widely reaching social acceptance than values. For example, a person is judged by societal standards of morality or immorality (right or wrong). Ethics are typically codified by a group of people or organization, and are not necessarily associated with societal norms. One can have professional ethics but seldom does one hear about having professional morals. Ethics are internally defined and self-imposed; morals relate to how individuals should behave but can be determined from personal conscience and/or common standards of justice so that the individual is encouraged to do good and be decent according to commonly accepted societal standards of justice. However, despite the differences in meaning between these three terms, this text intertwines them and encourages the reader to incorporate values and morals into their personal ethical decisions.

Revisiting the Basics

Nearly all people acknowledge the importance of ethics. However, unfortunately few really understand ethics as well as they think that they do or as well as they should. Ethics can be meaningfully discussed and applied only when it is fully understood. This understanding requires a periodic revisiting of the basics. So, what then is ethics about?

Ethics is about right and wrong. John Stuart Mill addressed this in his work *Utilitarianism* (1863) when he wrote, "We do not call anything wrong, unless we mean to imply that a person ought to be punished in some way or other for doing it; if not by law, by the opinion of his fellow creatures; if not by opinion, by the reproaches of his own conscience."

Ethics is about virtue and vice. "Vice, the opposite of virtue, shows us more clearly what virtue is. Justice becomes more obvious when we have injustice to compare it to" (Quintilian, *Institutio Oratoria*).

Ethics is about benefit and harm. "The two essential ingredients in the sentiment of justice are the desire to punish a person who has done harm, and the knowledge or belief that there is some definite individual or individuals to whom harm has been done" (Mill, 1863).

While ethics encompasses all of the above, it is more simply about principle, "fixed, universal rules of right conduct that are contingent on neither time nor culture nor circumstance" (Foster, 2003). And yet, it is also about character, "the traits, qualities, and established reputation that define who one is and what one stands for in the eyes of others" (Foster, 2003). Lastly, it is about example, that is, "an established pattern of conduct worthy of emulation" (Foster, 2003).

Ethics Involves What?

There is more to ethics than simply knowing its definition. It is just as important to know what is involved in its makeup. Ethics is the way values are practiced. As such, it is both a process of inquiry (deciding how to decide) and a code of conduct (a set of standards governing behavior).

To think well is to think critically. Critical thinking, the conscious use of reason, stands clearly apart from other ways of grasping truth or confronting choice, such as impulse or habit. Impulse is nothing more than an unreflective spontaneity, a mind on autopilot. On the other hand, habit is programmed repetition. Similar to muscle memory, it applies to behavior that is repetitive without thinking about it.

Therefore, "the object of critical thinking is to achieve a measure of objectivity to counteract or diminish the subjective bias that experience and socialization bestow on us all" (Foster, 2003). This is imperative because, "when we are dealing with matters of ethical concern, the well-being of someone or something beyond ourselves is always at stake" (Foster, 2003).

The Road to Unethical Behavior

Scandals can be prevented. They result from an evolution of predictable and preventable circumstances (Trautman, 2000). When an individual makes a conscious decision to commit a tiny indiscretion, it becomes exponentially easier for that individual to then rationalize his or her behavior and decisions and continue down the road of unethical decisions. This is sometimes what is referred to as "traversing down the slippery slope."

The most effective solutions to corruption must be instilled with straight-forwardness and honesty. The following suggestions may be implemented during a new employee's training in order to minimize the instances of corruption and unethical behavior within the scope of crime scene investigations:

1. Ensure thorough background investigations of all employees.
2. Ensure a high-quality field training program for new crime scene investigators.
3. Ensure consistent and fair accountability for employee actions.
4. Conduct effective ethics training.
5. Implement an effective employee intervention process.

Ethics in Crime Scene Investigation

An Internet search very quickly will drive home the importance of instilling a strong ethical foundation within the field of CSI. Ted Bundy's fingerprint chart ($15), death warrant ($15), and his last Christmas card ($1200) can be purchased on Web sites, as can animal bones and teeth from the Spahn Movie Ranch that are affixed to a Charles Manson letterhead and signed by the individual who obtained the items ($40). Items such as the clipped fingernails of serial killer Roy Norris have sold on eBay as well.

The majority of forensic specialists realize that it is illegal to remove objects from a crime scene without proper authorization. For ethically uncertain investigators, at least two states (Texas and California) now prohibit profiting from material taken from crime scenes. However, although it is legal in the majority of jurisdictions to turn around and sell crime scene-related evidence (including photographs and mug shots) for a profit, the integrity of the police force must be upheld at all times. Crime scene investigators and officers of the law thus must maintain the spirit of the law as well as the letter of the law. By preventing even the appearance of impropriety, the professional integrity of the police is maintained. Consider the damage a defense lawyer could do to an expert's reputation if it became known that the expert sold items obtained from former crime scenes.

While police and forensic specialists are ethically obliged to preserve the integrity of their investigation and their agency's reputations, the American Academy of Forensic Sciences (AAFS) provides no guidelines for crime scene ethics or the retention of items from former crime scenes. It is suggested that guidelines are "necessary to define acceptable behavior relating to removing, keeping or selling artifacts, souvenirs or teaching specimens from former crime scenes, where such activities are not illegal, to prevent potential conflicts of interest and appearance of impropriety" (Rogers, 2004). Once items have been legally obtained as part of a criminal investigation, their disposition is generally guided by local agency policies. Some items, by law, must be returned to the owner, while other items may be deemed contraband or too dangerous for release. If the owner cannot be located, the item may be classified as "found" property and thus subject to disposition options ranging from destruction, to auction, to conversion for educational and training purposes.

When the law is explicit, ethical decisions are not always required to guide behavior, but the variety of crime scene types and circumstances facing forensic investigators produces many ambiguous situations. Because her or his skills and knowledge of the criminal investigation may establish the innocence or guilt of a defendant, professional ethics and integrity are essential to a crime scene investigator's effort. Their education, training, experience, skills, and opinions on technical matters often are of considerable importance during a criminal investigation. For this reason, all forensic practitioners owe a duty to the truth. They may never be biased for or against a suspect in an investigation. The forensic practitioner's sole obligation is to serve the aims of justice. What is important is that the forensic practitioner conducts an investigation in a thorough, competent, unbiased manner. Forensic personnel have an obligation not to overstate or understated scientific findings. As experts in the criminal justice system, they are placed in positions of authority and responsibility.

To ensure ethical behavior, veracity of testimony, and professionalism among individuals engaged within the field of crime scene investigations, some departments and organizations have implemented a **code of ethics** that an employee must sign and agree to function by as terms of their employment or membership. This code lists what the department or organization believes are acceptable behaviors, professional expectations, and values to which their employees should adhere. Failure to comply with the code can result in dismissal or removal of membership and certification. An example of one of these codes of ethics for CSI personnel is on the homepage of the International Association of Identification (IAI, 2009). Examples of codes of ethics for various crime scene professionals are on that Web site as well.

Personnel involved with the processing of a crime scene have professional and moral obligations to work ethically and professionally. Legal, scientific, and ethical values can become tangled in the courtroom; however, the most important aspect of the trial is that the guilty are convicted and the innocent are exonerated (Fish, Miller, & Braswell, 2007). The obligation to the truth thus is paramount above all else.

The crime scene investigator must be professionally neutral to be effective. He or she must remain objective and all their actions must be unimpeachable as they follow the proscribed protocol at every crime scene. Acts of commission (intentional) and omission (unintentional) are not permitted and may in fact be criminal offenses in themselves.

No crime scene investigator wants to live with the possibility that a guilty person escapes prosecution or that an innocent person is punished based on his or her actions or inactions. Therefore, the investigator should do everything possible to preserve the chain of custody, take all necessary precautions to prevent cross-contamination or deterioration of physical evidence, and leave the forensic analysis up to the criminalists and courts. It is up to the judicial system—not the crime scene personnel—to weigh the evidence and come to a determination of guilt or innocence (Fish et al., 2007).

■ Expert Witnesses

Specially trained personnel such as crime scene investigators, forensic scientists, criminalists, forensic technicians, and identification technicians collect and examine physical evidence from crime scenes. These individuals are often called upon to testify as expert witnesses. An **expert witness** is someone who is called to answer questions on the stand in a court of law in order to provide specialized information relevant to the case being tried. Because scientific principles relating to physical evidence are often beyond the knowledge of lay people, courts permit persons with specialized training and skills to appear in court to explain and interpret scientific evidence to juries. A person can be considered an "expert" when they have sufficient skills, knowledge, or experience in their field to help the "trier of fact" to determine the truth. It therefore is the duty of the expert witness to educate the jury and provide testimony using terminology that is easily explainable and not misunderstood (Fish et al., 2007). Clarity, simplicity, and honesty are essential elements of expert witness testimony. It is suggested that supervisors review reports and conclusions with an unbiased attention to detail to ensure that the paperwork submitted is clear, concise, and accurate. Witnesses cannot deliberately omit relevant facts or encourage incorrect conclusions; these are distortions of the facts. Overstatements of the facts or a suggestion that an individual is guilty will cost an expert witness their integrity.

RIPPED FROM THE HEADLINES

Probe of Wisconsin Crime Lab Requested

On Wednesday, September 10, 2008, the *Wisconsin State Journal* (Hall, 2008) ran an article entitled, "Probe of Wisconsin Crime Lab Requested," regarding a requested investigation into possible ethical issues and employees' actions of the Wisconsin State Crime Lab system. The probe was requested by an attorney for the convicted murderer Steven Avery. In the 11-page complaint, the work of up to six state employees was called into question, specifically as it related to misconduct and ethical issues. The complaint alleged that not only had the employees been involved in misconduct and received discipline for their discretions, but that the incidents were investigated internally and not by an outside agency. Jerome Buting argued that an external and independent probe of the labs and incidents should occur because perpetrators of the crimes related to the mishandled evidence relating to the various situations may have eluded justice. The result of this activity could very well have been that innocent individuals may have been convicted with the aid of poorly executed and possibly unethical laboratory forensics.

The complaint specifically related to and was founded upon six disciplinary letters given to six forensic analysts employed by the State of Wisconsin. According to the letters:

- In 2002, an analyst falsely claimed to have conducted a fingerprint match, and had submitted falsified documentation to support it. The analyst received a written reprimand for this ethical violation.
- In 2004, an analyst in an unnamed department was fired after supervisors documented an extremely high error rate and a pattern in inattentiveness over a three year period.
- In 2006, a DNA analyst was suspended for two days for being intoxicated while on the job. The misconduct was reported to have occurred around the same time, and in the same laboratory, where evidence in the Steven Avery case (for which Jerome Buting served as defense counsel) was analyzed.
- In 2004, an analyst received a two-day suspension for performing an incorrect fingerprint identification that incorrectly eliminated a suspect in a fingerprint match. The same analyst had "false positive" fingerprint matches that were documented in two previous cases.
- In 2005, a fingerprint technician was suspended for three days for a series of incidents, which included taking fingerprint cards home.

The probe was requested and eventually conducted due to the possibility that serious negligence or misconduct may have compromised an unknown number of Wisconsin cases. The ethical decisions of several affected the work and lives of many.

History of Expert Testimony in U.S. Courts

But what exactly qualifies an individual as an expert? And when is such testimony admissible? There has been great debate and much litigation pertaining to what should be allowed as "expert testimony" and what should qualify an individual to be considered as an expert within court. To answer these questions, we must look back at a few historical rulings.

The Frye Test

In 1923, the D.C. Circuit court offered a ruling pertaining to the admissibility of testimony associated with a systolic blood pressure deception test, an early precursor to the modern polygraph. In this case, a Federal Court of Appeals ruling [*Frye v. U.S.*, 293 F. 1013 (D.C. Cir. 1923)] held that evidence could be admitted in court only if "the thing from which the deduction is made" is "sufficiently established to have gained general acceptance in the particular field in which it belongs." Because in 1923 such a blood pressure test was not widely known about or even accepted among scientists, the court ruled that testimony associated with such a process could not be used. Therefore, according to this ruling, if an expert is to testify on a specific matter or area of science, that matter or area must be "generally accepted" by the relevant scientific community. This has come to be known as the **Frye test.**

Federal Rules of Evidence

On January 2, 1975, the **Federal Rules of Evidence (FRE)** became federal law. These rules were developed to govern the introduction of evidence within proceedings, both civil and criminal, in U.S. federal courts. While they did not specifically apply to suits brought within state courts, the rules of many states have been closely modeled upon the provisions found within the FRE. The FRE is continually updated and continues to be the governing rules for federal courts. Of specific interest is rule 702 of the FRE that states:

> If scientific, technical, or other specialized knowledge will assist the trier of fact to understand the evidence or to determine a fact in issue, a witness qualified as an expert by knowledge, skill, experience, training, or education, may testify thereto in the form of an opinion or otherwise, if (1) the testimony is based upon sufficient facts or data, (2) the testimony is the product of reliable principles and methods, and (3) the witness has applied the principles and methods reliably to the facts of the case (Expert, 2008).

The underlying difference from the Frye standard, a ruling made over 50 years earlier, therefore was no incorporation of a "general acceptance standard." Instead, three provisions of the FRE governed admission of expert testimony in court.

- Scientific knowledge. Nothing is known with absolute certainty; however, the "knowledge" had to be arrived at by use of the scientific method.
- Assist the trier of fact. The scientific knowledge offered must be an aid in assisting either a jury or a judge in understanding the evidence or determining a fact in issue in the case.
- The judge makes the threshold determination. Such an assessment is meant to focus on methodology and principles, not the ultimate conclusions generated. It is left up to the judge to determine if the reasoning or methodology upon which the testimony is based, is properly applied to the facts in issue.

The Daubert Standard

In 1993 the U.S. Supreme Court made another important ruling regarding the admissibility of expert witnesses' testimony during federal legal proceedings. The case was *Daubert v. Merrell Dow Pharmaceuticals,* 509 U.S. 579 (1993).

In the Daubert ruling, the court held that federal trial judges are the "gatekeepers" of scientific evidence. Under the **Daubert standard**, trial judges must evaluate whether testimony is both "relevant" and "reliable." This resulted in a two-pronged test of admissibility.

1. Relevancy: whether or not the expert's testimony is relevant to the facts of the case and directly pertains to the matter in question.
2. Reliability: The expert must have derived their conclusions from the scientific method.

With Daubert, the Court supplanted the older Frye test's "general acceptance" standard. Because this applied to federal cases only and each state is permitted to create unique rules of evidence for use within its courts, the Frye test, or a variation thereof, may still be used in those states that have chosen not to adopt Daubert.

Kumho Tire Co. v. Carmichael

In July 1993, Patrick Carmichael was driving his minivan when the front tire blew out. This led to a vehicular accident resulting in one of the passengers in the vehicle dying and others being severely injured. The resulting lawsuit, filed by Carmichael, claimed that the tire was defective and the defect caused the accident. The Carmichael's case rested largely on testimony from a tire failure expert. However, under Daubert, certain factors contribute to the reliability, and hence the admissibility, of expert testimony, one of which is the validity of the expert's methods. In the Carmichael case, the district court found the tire expert's methods not to adhere to the scientifically valid standard, and thus excluded the testimony.

In this ruling, the court looked at both the Daubert standard and the FRE for its ruling. The court noted that the line between "scientific" and "technical" knowledge is not always discernible. As such, the Court held that the gatekeeping function set forth in the Daubert ruling applied to all expert testimony detailed in FRE 702, because there was no "convincing need" to draw a distinction between "scientific" and "technical" knowledge; both kinds of knowledge normally would be outside the grasp of the average juror.

■ Qualifying an Expert Witness

Usually when one talks about experience, the term **credentials** is used. Often this refers to a certificate, letter, experience, or anything that provides authentication for a claim or qualifies somebody to do something. However, as pertains to forensic and crime scene-related work, credentials as an expert will be established by the court through questioning pertaining to the witness' education, training, and experience. The ability and competence of the witness must be demonstrated through testimony relating to college degrees, continuing education, attendance at conferences, publications, ongoing research, and a variety of other possibilities that show rigor and knowledge within the area of expertise under consideration.

Unlike non-expert witnesses, once credentialing as an expert has been established, an expert witness can provide opinions based on the outcomes of the examinations and the significance of their findings. Non-experts who state opinions as part of their testimony will have such statements ruled as inadmissible due to them being classified as **hearsay**. Hearsay is unfounded information that is heard from other people. However, the court allows experts to state opinions due to their ability to assist the court in better comprehending the topic under consideration. In the case of an expert rendering an opinion on their findings, the opinion has a foundation in the expert's training and experience, and is not an arbitrary opinion with no factual relevance. It should be mentioned that simply because there is this ability to state an opinion, it does not mean that it is always a legal possibility. FRE Rule 703 provides for an explanation of the bases of opinion testimony relating to expert witnesses.

The facts or data in the particular case upon which an expert bases an opinion or inference may be those perceived by or made known to the expert at or before the hearing. If of a type reasonably relied upon by experts in the particular field in forming opinions or inferences upon the subject, the facts or data need not be admissible in evidence in order for the opinion or inference to be admitted. Facts or data that are otherwise inadmissible shall not be disclosed to the jury by the proponent of the opinion or inference unless the court determines that their probative value in assisting the jury to evaluate the expert's opinion substantially outweighs their prejudicial effect (Expert, 2008).

Although it is not possible for an expert to render an opinion with absolute certainty, as an advocate of truth, the expert must base opinions on a reasonable scientific certainty. An expert must be confident in their statements made within a court of law. If they are found to be contradictory, or if it is pointed out that the witness intentionally lied or misrepresented the facts, they could be charged with perjury. **Perjury** is the telling of a lie within a court of law by somebody who has taken an oath to tell the truth.

As an expert witness, a crime scene investigator must remember that her or his integrity and professionalism are open for inspection. Each must be familiar with the scope of her or his actions and knowledge, and know where their level of expertise ends. When subpoenaed to testify as an expert witness, the way others perceive the expert is more important than the way experts perceive themselves. Once credibility as an expert witness is compromised, it is nearly impossible to recover in court (Rogers, 2004).

Issues with Ethics in Forensics and Crime Scene Work

A search of recent cases involving mismanagement, improper documentation, unethical testimony, and improper analysis of physical evidence is bound to bring the searcher a plethora of cases associated with such matters. Some of the most notorious names associated with these types of cases center upon those involved with the physical analysis and subsequent testimony about the physical evidence in a case. Names such as Joyce Gilcrest and Fred Zane, and locations such as the Washington state and Tucson, Arizona crime labs are infamous because they made media headlines. But for each person and laboratory that made a headline, there are dozens that did not. With only a cursory amount of research, you will find that ethical transgressions occur at all steps in the evidentiary process including: crime scene security, physical evidence collection and documentation, physical evidence processing and analysis, testimony regarding all aforementioned phases, and final evidence disposition. Ethics—or lack thereof—permeate all areas of the criminal investigative process. They are not isolated to particular geographic regions or departments. Instead, ethics are a product of training and personal values that are present in any setting.

As one can see, ethical issues by no means are isolated to only large or small agencies. Ethical issues transcend departmental size, demographic makeup, and job title. Examples of unethical behavior can be pulled from all areas of the criminal investigative process and from all levels of rank.

An examination of unethical issues relating to crime scene work also shows a variety of motivations for committing such transgressions. Sometimes the motivation is greed, and other times it is power, status, or promotion. But often it is simply a case of the individual forgetting that their obligation is to the truth and not to one side or the other. Many times individuals believe as though they are on the prosecutor's team, and so failing to present testimony or physical evidence and analysis of evidence that might assist the prosecutorial team would be tantamount to letting their team down. As addressed previously, such instances can be avoided by a thorough background investigation before hiring an employee, and proper ethical training, correct management practices, and frequent refresher trainings after they have been hired.

Chapter Summary

Every time that a crime scene investigator responds to a crime scene, the potential for presenting the actions and observations documented at the scene within a courtroom exists. A crime scene investigator's reputation is based on the veracity of their work and the integrity of their actions. Their actions—or lack thereof—are the voice of the victim. While justice may be blind, investigators must present objective and unbiased testimony that clearly and accurately recreates the crime scene for the judge and jury. A crime scene investigator's actions should not detract from the credibility of the physical evidence. His or her behavior or testimony should not intentionally tarnish the voice of the evidence. As Paul Kirk (1953) said of a crime scene investigation,

> Wherever he steps, whatever he touches, whatever he leaves, even unconsciously, will serve as silent evidence against him. Not only his fingerprints or his footprints, but also his hair, the fibers from his clothes, the glass he breaks, the tool marks he leaves, the paint he scratches, the blood or semen that he deposits or collects-all of these and more bear mute

witness against him. This is evidence that does not forget. It is not confused by the excitement of the moment. It is not absent because human witnesses are. It is factual evidence. Physical evidence cannot be wrong: it cannot perjure itself; it cannot be wholly absent. Only its interpretation can err. Only human failure to find it, study and understand it can diminish its value.

Review Questions

1. Explain the term ethics.
2. What is a code of ethics?
3. What is meant by the term hearsay?
4. How is an expert witness different from a layperson within a court of law?
5. What does it take for a person to qualify as an expert witness?
6. Discuss the Frye test's direct impact on forensic testimony and evidence.
7. The Daubert standard is used by many courts today as a way to ensure the authenticity and veracity of testimony and forensic analysis. In this ruling, judges were given the role of ______________.
8. The Daubert standard resulted in a two-prong test of admissibility, which means that testimony must be both ______________ and ______________.
9. How did Rule 702 of the Federal Rules of Evidence (FRE) impact expert testimony? Does it apply to all courts?
10. ______________ is the telling of a lie within a court of law by somebody who has taken an oath to tell the truth.

Case Studies

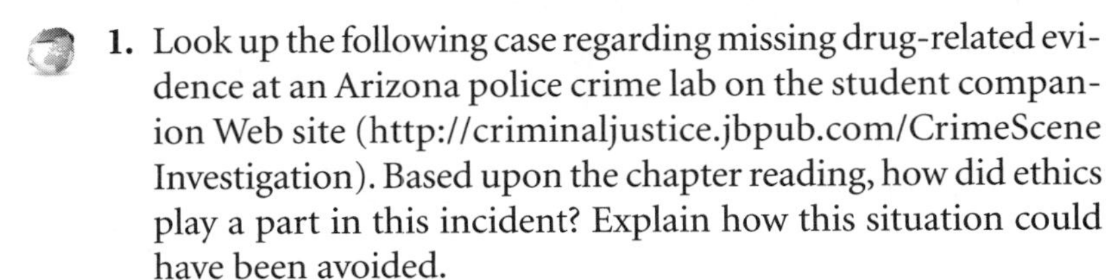

1. Look up the following case regarding missing drug-related evidence at an Arizona police crime lab on the student companion Web site (http://criminaljustice.jbpub.com/CrimeScene Investigation). Based upon the chapter reading, how did ethics play a part in this incident? Explain how this situation could have been avoided.

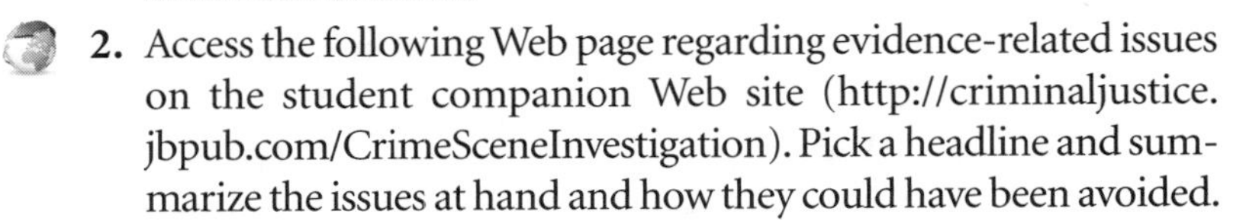

2. Access the following Web page regarding evidence-related issues on the student companion Web site (http://criminaljustice.jbpub.com/CrimeSceneInvestigation). Pick a headline and summarize the issues at hand and how they could have been avoided.

References

Daubert v. Merrell Dow Pharmaceuticals, 509 U.S. 579 (1993).

Expert Pages. (2008). *The leading directory of expert witnesses* (homepage). Retrieved on July 20, 2009, from http://www.expertpages.com

Fish, J. T., Miller, L. S., & Braswell M. C. (2007). *Crime scene investigation: An introduction*. New York: LexisNexis Anderson Publishing.

Fisher, B. A. J. (2004). *Techniques of crime scene investigation* (7th ed.). Boca Raton, FL: CRC Press.

Foster, G. D. (2003). Ethics: Time to revisit the basics. In J. P. West & E. M. Berman (Eds.), *The ethics edge, Second edition*. Washington, DC: International City/County Management Association. Originally published in *The Humanist*, *63*(2).

Frye v. U.S., 293 F. 1013 (D.C. Cir. 1923).

Hall, D. J. (2008, September 10). Probe of state crime lab requested. Analyst's faulty work cited. *Wisconsin State Journal.* Retrieved July 20, 2009, from http://www.buting.com/CM/Custom/ProbeofCrimeLabRequested.pdf

International Association for Identification. (2009). *Code of ethics for certified crime scene personnel*. Retrieved July 20, 2009, from http://theiai.org/certifications/crime_scene/ethics.php

Kirk, P. L. (1953). *Crime investigation; physical evidence and the police laboratory*. New York: Interscience.

Lewis, J. J. (2009). Ethics quotes. Retrieved July 20, 2009, from http://www.wisdomquotes.com/cat_ethics.html

Mill, J. S. (1863). *Utilitarianism*. Retrieved July 20, 2009, from http://www.utilitarianism.com/mill1.htm

Quintilian, M.F. *Instituto oratorio*. (posted 2006). Retrieved July 20, 2009, from http://penelope.uchicago.edu/Thayer/E/Roman/Texts/Quintilian/Institutio_Oratoria/home.html

Quotationspage.com (2007). Retrieved July 20, 2009, from http://www.quotationspage.com/quote/225.html

Rogers, T. (2004, March). Crime scene ethics: Souvenirs, teaching, materials, and artifacts. *Journal of Forensic Sciences*, American Academy of Forensic Science, Colorado Springs, CO, *49*(2).

Thinkexist.com (2009). Retrieved July 20, 2009, from http://www.thinkexist.com/quotes/pearl_s._buck/

Trautman, N. (2006). The corruption continuum: How law enforcement organizations become corrupt (avoiding public scandal and corruption). In J. P. West & E. M. Berman (Eds.), *The ethics edge, Second edition*. Washington, DC: International City/County Management Association. Originally published in *Public Management Magazine*, (June 1, 2000).

Duties of the First Responder to the Crime Scene

Crime is common. Logic is rare. Therefore it is upon the logic rather than upon the crime that you should dwell.

Sir Arthur Conan Doyle (1859–1930)
Sherlock Holmes, *The Adventure of the Copper Beeches*

LEARNING OBJECTIVES

- Understand the primary duties of a first responder to a crime scene.
- Identify the considerations involved with establishing a crime scene perimeter.
- Describe what is meant by transient evidence.
- Articulate when loss or destruction of evidence at a crime scene is acceptable.
- Explain multilevel containment.

KEY TERMS

Command Post
In Situ
Mechanical Loss
Multilevel Containment
Perimeter
Primary Scene
Secondary Scene
Transient Evidence

The First Responder

As discussed in Chapter 1, patrol officers, firefighters, and emergency medical personnel are usually first to arrive at the scene of a crime, and are thus referred to as *first responders*. Crime scenes are dynamic, rapidly evolving environments. Because of this, the initial actions—or inactions—of first responders can dictate the success or failure of investigative efforts.

First responders have two primary duties:

1. ***Preservation of life.*** The primary duty assigned to an officer responding to a crime scene is to render it safe and provide for emergency services for those in need.
2. ***Secure and preserve the crime scene and associated evidence.*** By effectively securing the crime scene, the first responder will prevent the destruction or diminished value of physical evidence.

Determining Jurisdiction

Prior to entering a crime scene, a first responder must ensure that he or she has the legal right to do so. As discussed in Chapter 1, a search warrant or other legal right may be necessary in order for a crime scene be properly entered and evidence obtained. Simply because an officer was dispatched to an area does not mean it was necessarily within the department's legal jurisdiction. An officer must be familiar with their assigned district, and if unsure of jurisdiction, such matters should be addressed before continuing with an investigation (Fish, Miller, & Braswell, 2007).

While city limits, county lines, or even nautical boundaries are invisible lines that do not impede an officer's ability to carry out his or her responsibilities, their legal significance is of paramount importance within the confines of a courtroom. There are numerous documented cases where jurisdictional issues or disagreements regarding which agency should be involved in conducting an investigation have lead to countless number of delays and the dismissal of charges because of technicalities not associated with the crime scene investigation itself. Such situations can be avoided by ensuring that the officer is within her or his jurisdictional domain and that any agencies that need be notified, have been notified before beginning a detailed crime scene investigation.

Approaching the Scene

A first responder must arrive at the scene safely if they are to have a chance to perform their necessary duties. It therefore is incumbent upon the officer to approach the scene in a controlled and safe manner. All movements should be prompt, deliberate, and calm. While it might appear to an outsider that police race with lights and sirens to each and every call, this is in fact a rare event. An emergency response (lights and sirens) is a very hazardous event that requires great skill and awareness. If an officer is injured, or injures another en route to a call, he or she defeats the purpose of the mission: to arrive safely to ensure the safety of whomever is calling for the police response.

Some police agencies have very specific policies for uniformed personnel concerning their duties and responsibilities at the crime scene. An officer should be familiar with these policies and execute their job to the best of their ability (Fisher, 2004). The duties of the first officer who appears at a crime scene are the same, regardless of his or her rank. The duties are also the same regardless of the type of crime scene.

Actions of the First Responder upon Arrival

Maximize the Safety of the Scene

The officer should remain observant for any persons, vehicles, possible evidence, and overall conditions of the scene. The officer's approach to the scene should be in such a manner that it will reduce the risk of harm to her- or himself, while maximizing the safety of potential victims, witnesses, and other persons in the area. If suspects or dangerous persons or situations are at the scene, those are the first priority.

Guide EMS to the Injured and Preserve the Scene

After controlling any suspects and/or any potential hazardous situations or persons on scene, the officer's next responsibility is to ensure that emergency medical services (EMS) are provided to injured persons. Officers should make efforts to guide medical personnel to the victim by an indirect pathway established to reduce scene alteration and contamination. The indirect pathway established by the first responder should be used by both arriving medical personnel and investigative personnel to reduce potential contamination and destruction of evidence. It is suggested that such a pathway, as well as the entrance and exit to a crime scene, be established that is not consistent with the most likely path and entrance/exit of the suspect(s). This will ensure diminished evidence destruction.

If there is potential evidence in the proximity of a victim, officers should point out such evidence to medical personnel, and request that they minimize contact with such evidence. An example would be to cut around bullet holes or stab marks in clothing. Officers who are not actively involved in lifesaving efforts should concentrate on documenting the movement of persons or items during the lifesaving efforts. However, if through the efforts of saving a life, evidence is lost, this is what is referred to as **mechanical loss**. This is an accepted loss of physical evidence and is easily articulated within court. If a life is saved and evidence is lost, that is acceptable. It is far better to have saved a life and lost evidence than it is to save evidence and lose a life. If a person dies, their visual and testimonial evidence dies with them. Gross incompetence or neglect are not acceptable reasons for losing or destroying evidence.

Protect the Integrity of the Crime Scene

Once these responsibilities have been fulfilled, concern is shifted towards the integrity of the scene. The scene must be protected until all physical evidence has been documented and collected. The first officer on the scene must attempt to "freeze" the scene as closely as possible to the condition in which it was found. This means that no one is allowed to needlessly move about or alter the scene. Persons present—victims or witnesses—must be isolated or removed from the scene so that they do not purposely or inadvertently alter or destroy evidence (Wisconsin DOJ, 2001). Suspects should be kept out of the scene in order to avoid scene contamination. If a suspect were to be brought back to a scene, and later DNA, fingerprints, hairs, and/or other evidence were collected in the course of the investigation, there would be an innocent explanation for their deposit.

Defining Crime Scene Scope and Establishing a Perimeter

Once emergency situations have been attended to, and witnesses and suspect(s) have been identified and removed from the scene, it is imperative that the first responder take actions to secure the identified crime scene to attempt to preserve the evidence present within. In order to do this, the officer must first identify the scope of the crime scene. Once the boundaries have been identified, a perimeter should be established. A **perimeter** is the outer confines of the crime scene. This involves some sort of delineation or physical boundary as to what area is considered inside of the supposed crime scene, and what is external to the area of investigation.

A perimeter can be established through utilizing a number of methods. The most common is stringing crime scene barrier tape around the defined area (**Figure 4.1**). For some night scenes, flares or traffic cones may be used to identify a crime scene area (traffic accident). Sometimes, marked patrol cars or personnel are used in the absence of tape. In other situations, securing a crime scene could be as simple as closing and locking a door. It is also possible to use existing structures such as walls, gates, or trees. Although a variety of options exist, a clearly defined physical perimeter will ensure that there is no mistaking what is and what is not the crime scene, which will allow police personnel to best enforce access to the scene, and thus ensure its integrity. However, it should be kept in mind that this barrier is by no means fixed. The area of the defined crime scene is always subject to modification dependent upon investigative progress and process.

Remember, it is better to over-define an area rather than to underestimate and identify an area. A crime scene can always be shrunk, but once a perimeter is established it is nearly impossible to increase the perimeter later and be certain that any evidence found within the expanded area has maintained its integrity. First responders will find

Figure 4.1 A First Responder Secures the Perimeter of a Crime Scene.

that defining a perimeter (e.g., putting up barrier tape) is sometimes an invitation for onlookers to press forward to see what is going on. Interested parties will assemble as close to the barriers as possible to get their greatest vantage point. If the perimeter was incorrectly established, any evidence found outside the established perimeter most likely will be useless due to destruction and contamination (**Figure 4.2**).

In situations where it is feasible, the ideal procedure would be to establish a perimeter that involves **multilevel containment**. This method utilizes several perimeters, and is the most effective for ensuring integrity while also allowing a workable scene structure (**Figure 4.3**).

Three levels of multilevel containment are:

1. *The outer perimeter.* If there are media personnel present, a special area within this first level may be designated for their use. By providing the media with this area apart from the general public, a relationship of good will and cooperation is established between the reporters and law enforcement.
2. *Inside of the outer perimeter, adjacent to the actual crime scene itself.* This area is only to be accessed by police and emergency personnel. It is acceptable for emergency and crime scene vehicles to be within this area, as it will make investigative and processing efforts more streamlined. This area typically holds the **command post**,

Figure 4.2 Multilingual Barrier Tape.

which is established to coordinate on scene activities and efforts. Sometimes this is referred to as "incident command" under the incident command and control structure. Supervisory decisions are made and the scene is managed from this point. The incident commander is typically the first person on scene (first responder); however, this duty is passed along to higher ranking and specialized administrative personnel upon their arrival to the scene. The designated multilevel area allows higher ranking personnel access that is in closer proximity to the crime scene, without the chance of contaminating the actual scene. This area also is used by processing personnel to stage their equipment, as well as a place where they can take breaks.

3. *The perimeter that defines the specific crime scene target area.* This area should have the strictest level of access and control, and should only be entered by those actively involved in the processing and investigative efforts. Anyone not actively involved at the time, should withdraw to area 2.

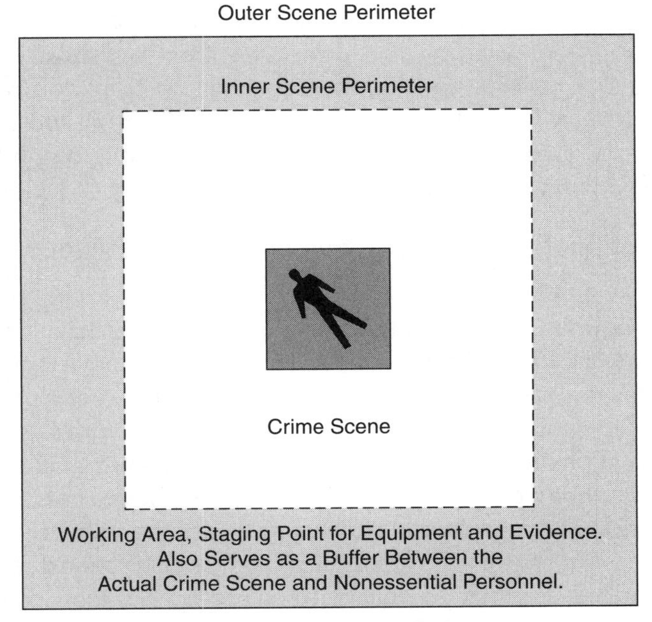

Figure 4.3 Multilevel Containment.

Source: Courtesy of Ellie Bruchez, University of Wisconsin–Platteville.

Exclusion of Official Visitors

One of the most difficult challenges is to prevent and/or prohibit so-called "official sightseers" such as other officers, superiors, or government officials who are not directly involved in the processing or investigatory efforts but yet demand to be admitted into the scene. Every attempt should be made to exclude these individuals, explaining the potential consequences of disturbing the scene, and stressing the importance of filing a narrative report noting observations and actions taken should they make entrance within the scene.

Sometimes the exclusion of unauthorized personnel proves to be a more difficult task than expected. Crimes of violence are especially susceptible to the attention of higher level officials and members of the press, in addition to the curiosity of neighbors and others. What must be remembered is that every individual who enters the crime scene is a potential destroyer of physical evidence, even if it is unintentional. One suggestion to aid with this difficult scenario is to establish multilevel containment quickly, as defined on the previous page. This would allow such official sightseers minimal access to the scene, but not to the target crime scene.

If proper control is to be exercised, the officer charged with the responsibility of securing and protecting the scene must have the authority to exclude anyone, including fellow police officers not directly involved with the investigatory process. Securing and isolating the scene is a critical step within an investigation, the accomplishment of which is a mark of a well-trained and professional CSI team.

Identification and Protection of Transient Evidence

As discussed in Chapter 1, Locard's exchange principle is important to remember when arriving at the crime scene and during the initial stages of crime scene efforts. There is trace evidence to be found at every crime scene and on every criminal. As time passes much of the trace matter and the ability to collect it is lost. It is crucial, therefore, to quickly identify, collect, and preserve trace and perishable evidence, both on the suspect and at the crime scene. Sometimes this evidence is referred to as transient. **Transient evidence** is physical evidence present at the crime scene that is either fragile or at great risk for loss, alteration, or destruction if not properly identified, documented, collected, and preserved as soon as possible.

Initial Scene Documentation

To ensure the integrity of a crime scene, once a scene is safe and is secured, a log should be established by the first responder to log all personnel who enter and leave the scene, including those present upon arrival. Most departments will have forms established for this task. This document is the initial documentation effort at the scene. It is important that the first responder record an accurate arrival time at the scene, and when the scene was secured. With modern technology and the advent of Computer Aided Dispatch (CAD), often these times are noted by dispatch after an officer makes radio transmissions regarding arrival and scene situation updates. Because a first responder may be called away to respond to another call before the crime scene is properly processed, this initial documentation is important as it will set the foundation for later documentation efforts by other crime scene personnel.

It is also important that the first responder note any additional departments and personnel who were involved, the agency case number(s), and the contact person from each department who was involved in the efforts (**Figure 4.4**).

A first responding officer should be cognizant of not becoming involved in lengthy interviews or interrogations with victims, suspects, or witnesses at a crime scene, as the officer cannot effectively protect the scene integrity when involved with such matters.

Figure 4.4 A First Responder Vehicle.

Processing Cases Involving Multiple Scenes

Sometimes a case will involve an expansive scene or multiple locations associated with the same criminal case. Typically, the **primary scene** is the first encountered location where evidence was located. This usually is the area where the officer was initially dispatched but can be the location of a witness who called in a complaint.

Other scenes may be identified later as being associated with the primary scene, however. These scenes are referred to as **secondary scenes**. An example would be a homicide where a victim is believed to have been killed within a residence, and then the body transported via an automobile and buried in a remote location, where it was discovered by a person walking a dog. In this case, the primary scene would be the location identified by the witness (i.e., the burial location), and the automobile and residence (when located) would be secondary scenes. In all three cases, each scene must be processed utilizing the same approach and guidelines discussed for dealing with responding to and properly processing a crime scene.

Sometimes, multiple scenes may mean the involvement of multiple jurisdictions in an investigation. This is not typically an issue, as most agencies have multijurisdictional or mutual aid agreements that predetermine areas of responsibilities in the event of a criminal event that encompasses more than one jurisdiction.

Scenes Involving Death

Chapter 17 covers the topic of death investigation scenes more in depth. With regards to the initial response, however, if the first responder to the scene is able to establish through ocular (visible) signs of death (rigor mortis, odor, lividity, bug activity, decomposition, etc.) or by other means that a person is dead, then the body should not be touched or moved until a detailed examination has been conducted. While in most states a physician, coroner, or other medical personnel is the only person able to declare a person legally dead, an officer need not be a doctor to note if a body present at a scene exhibits characteristics inconsistent with sustaining life (i.e., missing major portions, insects protruding from orifices, positioning, catastrophic damage, etc.).

Once it has been established that a person is dead and a brief survey of the scene has been conducted, a first responder should notify his or her supervisor about the situation. It is suggested that such notifications be made via telephone, as media staff do monitor police radio frequencies.

Next, the medical examiner (ME) or coroner should be contacted. Whether it is the ME or coroner who is contacted is a matter of local custom. Many agencies wait to make such notifications until departmental investigative personnel arrive and begin their investigation, which can save coroner's staff time by ensuring that they do not have to wait at a scene for the police to conclude their investigative efforts before they can enter and process the body. It is suggested that departments have policies in place to address such situations and order of notification.

The ME or coroner's office typically has legal jurisdiction over the body at the scene, and many times, all associated evidence within the proximity of the body. In jurisdictions where this is true, the body may not be moved or searched without the notification and consent of ME or coroner personnel.

In situations involving strangulation or hanging, where there is the presence of unmistakable signs of death, the officer should do nothing to the body. Again, no efforts should be made to cut down the body! If there appears the possibility that the hanging mechanism might break, the officer may attempt to support the corpse, but it should not be cut down. If, however, signs of life are present, the officer should make all efforts necessary to ensure that the victim lives. Knots and areas of articulation should be left intact, if possible. Knots can sometimes give investigators knowledge of certain occupation or skill level. If a rope must eventually be cut, the loose ends should be appropriately labeled. Another option to labeling the ends is to tie the ends to one another with string. These are not actions typically conducted by first responding personnel, however.

Scenes Involving Firearms and Ammunition

Chapter 13 is specifically dedicated to firearms- and ballistic-related evidence but as a general rule, any firearms and ammunition located at a crime scene should not be touched by first responders. Such items should remain **in situ** (as they are) until the arrival of investigative personnel who will make processing determinations. However, sometimes it becomes necessary, for safety and security reasons, to recover and secure weapons and ammunition. If the presence of the weapon and ammunition create an unsafe situation, if they may be inadvertently moved or compromised during the emergency care of an injured person, or if conditions exist such that the first responder cannot ensure the integrity and safety of the scene, which might lead to bystanders disturbing the weapon evidence, then the evidence may be collected and secured by the first responder.

If a weapon is to be collected, its collection should be made in such a way as to minimize the potential destruction of evidence, if at all possible. Nothing should be inserted within the bore of a firearm in order to lift the weapon. Doing so could dislodge trace evidence (e.g., gunshot residue, blood, particles of tissue, hair, etc.) that might be associated with the weapon's discharge. It is suggested that firearms be picked up by the gnarled pistol grip or other area less likely to destroy fingerprint evidence, which would allow for print preservation and subsequent recovery. If there is the time to do so, the weapon and ammunition evidence should be photographed and their positioning sketched prior to removal. If there is not time, the officer should do their best to make a mental or written note. Information such as positioning often leads to a determination of whether the event was a suicide or it could explain the positioning of the assailant. The positions of the hammer and safety should be noted also (Fisher, 2004).

When weapons are collected the officer should keep in mind the evidentiary possibilities for the item of evidence. There may be reason to believe that fingerprints could be present. If so, keep in mind that excessive heat could destroy such useful evidence and, therefore, the firearm should not be removed to an excessively hot area.

Dealing with Media

Sometimes media personnel may arrive at a crime scene before the first responder. In these instances, the first responders should not—in any circumstance—provide the media with information concerning the case.

Crime Scene "Don'ts"

At no time during the process should a first responder or any individual present within the crime scene target perimeter do any of the following:

- Eat
- Drink
- Smoke
- Use land lines (telephone, computer) present within the crime scene
- Use a radio or cell phone (if it's a bomb scene, this could serve as a detonator)
- Use the bathroom, if present within a scene

It may be permissible for personnel to do the aforementioned (except use electronic or radio equipment if involved with the investigation of a bomb scene) while the personnel are present within the outermost perimeter, or the middle perimeter of a multilevel containment situation, but such actions must never occur within the target crime scene area. Doing so will place the evidence in serious danger of contamination or destruction.

This duty falls to the police chief, sheriff, or public/press information officer (PIO). While an officer must not provide information, it is important that she or he behaves professionally with media personnel. The officer must remember that media personnel often provide invaluable assistance during the investigation of a crime. Reporters and staff are helpful in ensuring that a victim is identified or a suspect apprehended through distribution of photographs, videos, and other such evidence to the community. However, this information is never to be distributed by the first responder in the initial stages of the investigative efforts.

First Response Overview

The following suggested process is to be followed by first responders, as written by the National Institute of Justice (2004) Special Report entitled, "Crime Scene Investigation: A Reference for Law Enforcement Training."

The First Response Process

1. Initial Response/Receipt of Information

   Principle: One of the most important aspects of securing the crime scene is to preserve the scene with minimal contamination and disturbance of physical evidence. The initial response to an incident shall be expeditious and methodical. Upon arrival, the officer(s) shall assess the scene and treat the incident as a crime scene.

   Policy: The initial responding officer(s) shall promptly, yet cautiously, approach and enter crime scenes, remaining observant of any persons, vehicles, events, potential evidence, and environmental conditions.

   Procedure: The initial responding officer(s) should:

   a. Note or log dispatch information (e.g., address/location, time, date, type of call, parties involved).
   b. Be aware of any persons or vehicles leaving the crime scene.
   c. Approach the scene cautiously, scan the entire area to thoroughly assess the scene, and note any possible secondary crime scenes. Be aware of any persons and vehicles in the vicinity that may be related to the crime.
   d. Make initial observations (look, listen, smell) to assess the scene and ensure officer safety before proceeding.
   e. Remain alert and attentive. Assume the crime is ongoing until determined to be otherwise.
   f. Treat the location as a crime scene until assessed and determined to be otherwise.

(continues)

The First Response Process (Continued)

Arriving at the Scene: Initial Response/Prioritization of Efforts

1. Initial Response/Receipt of Information

   Summary: It is important for the initial responding officer(s) to be observant when approaching, entering, and exiting a crime scene.

2. Safety Procedures

   Principle: The safety and physical well-being of officers and other individuals, in and around the crime scene, are the initial responding officer(s') first priority.

   Policy: The initial responding officer(s) arriving at the scene shall identify and control any dangerous situations or persons.

   Procedure: The initial responding officer(s) should:

   a. Ensure that there is no immediate threat to other responders — scan area for sights, sounds, and smells that may present danger to personnel (e.g., hazardous materials such as gasoline, natural gas). If the situation involves a clandestine drug laboratory, biological weapons, or radiological or chemical threats, the appropriate personnel/agency should be contacted prior to entering the scene.
   b. Approach the scene in a manner designed to reduce risk of harm to officer(s) while maximizing the safety of victims, witnesses, and others in the area.
   c. Survey the scene for dangerous persons and control the situation.
   d. Notify supervisory personnel and call for assistance/backup.

   Summary: The control of physical threats will ensure the safety of officers and others present.

3. Emergency Care

   Principle: After controlling any dangerous situations or persons, the initial responding officer(s') next responsibility is to ensure that medical attention is provided to injured persons while minimizing contamination of the scene.

   Policy: The initial responding officer(s) shall ensure that medical attention is provided with minimal contamination of the scene.

   Procedure: The initial responding officer(s) should:

   a. Assess the victim(s) for signs of life and medical needs and provide immediate medical attention.
   b. Call for medical personnel.
   c. Guide medical personnel to the victim to minimize contamination/ alteration of the crime scene.
   d. Point out potential physical evidence to medical personnel, instruct them to minimize contact with such evidence (e.g., ensure that medical personnel preserve all clothing and personal effects without cutting through bullet holes, knife tears), and document movement of persons or items by medical personnel.

The First Response Process

e. Instruct medical personnel not to "clean up" the scene and to avoid removal or alteration of items originating from the scene.

f. If medical personnel arrived first, obtain the name, unit, and telephone number of attending personnel, and the name and location of the medical facility where the victim is to be taken.

g. If there is a chance the victim may die, attempt to obtain "dying declaration."

h. Document any statements/comments made by victims, suspects, or witnesses at the scene.

i. If the victim or suspect is transported to a medical facility, send a law enforcement official with the victim or suspect to document any comments made and preserve evidence. (If no officers are available to accompany the victim/suspect, stay at the scene and request medical personnel to preserve evidence and document any comments made by the victim or suspect.)

Summary: Assisting, guiding, and instructing medical personnel during the care and removal of injured persons will diminish the risk of contamination and loss of evidence.

4. Secure and Control Persons at the Scene

Principle: Controlling, identifying, and removing persons at the crime scene and limiting the number of persons who enter the crime scene and the movement of such persons is an important function of the initial responding officer(s) in protecting the crime scene.

Policy: The initial responding officer(s) shall identify persons at the crime scene and control their movement.

Procedure: The initial responding officer(s) should:

a. Control all individuals at the scene—prevent individuals from altering/destroying physical evidence by restricting movement, location, and activity while ensuring and maintaining safety at the scene.

b. Identify all individuals at the scene, such as:
   - Suspects: Secure and separate.
   - Witnesses: Secure and separate.
   - Bystanders: Determine whether a witness, if so treat as above, if not, remove from the scene.
   - Victims/family/friends: Control while showing compassion.
   - Medical and other assisting personnel.

c. Exclude unauthorized and nonessential personnel from the scene (e.g., law enforcement officials not working the case, politicians, media).

(continues)

The First Response Process (Continued)

Summary: Controlling the movement of persons at the crime scene and limiting the number of persons who enter the crime scene is essential to maintaining scene integrity, safeguarding evidence, and minimizing contamination.

5. Boundaries: Identify, Establish, Protect, and Secure

Principle: Defining and controlling boundaries provide a means for protecting and securing the crime scene(s). The number of crime scenes and their boundaries are determined by their location(s) and the type of crime. Boundaries shall be established beyond the initial scope of the crime scene(s) with the understanding that the boundaries can be reduced in size if necessary but cannot be as easily expanded.

Policy: The initial responding officer(s) at the scene shall conduct an initial assessment to establish and control the crime scene(s) and its boundaries.

Procedure: The initial responding officer(s) should:

a. Establish boundaries of the scene(s), starting at the focal point and extending outward to include:
   - Where the crime occurred.
   - Potential points and paths of exit and entry of suspects and witnesses.
   - Places where the victim/evidence may have been moved (be aware of trace and impression evidence while assessing the scene).

b. Set up physical barriers (e.g., ropes, cones, crime scene barrier tape, available vehicles, personnel, other equipment) or use existing boundaries (e.g., doors, walls, gates).

c. Document the entry/exit of all people entering and leaving the scene, once boundaries have been established.

d. Control the flow of personnel and animals entering and leaving the scene to maintain integrity of the scene.

e. Effect measures to preserve/protect evidence that may be lost or compromised (e.g., protect from the elements, such as rain, snow, wind, and from footsteps, tire tracks, sprinklers).

f. Document the original location of the victim or objects that you observe being moved.

g. Consider search and seizure issues to determine the necessity of obtaining consent to search and/or obtaining a search warrant.

Note: Persons should not smoke, chew tobacco, use the telephone or bathroom, eat or drink, move any items including weapons (unless necessary for the safety and well-being of persons at the

The First Response Process

scene), adjust the thermostat or open windows or doors (maintain scene as found), touch anything unnecessarily (note and document any items moved), reposition moved items, litter, or spit within the established boundaries of the scene.

Summary: Establishing boundaries is a critical aspect in controlling the integrity of evidentiary material.

6. Turn over the Scene and Brief Investigator(s) in Charge

Principle: Briefing the investigator(s) taking charge assists in controlling the crime scene and helps establish further investigative responsibilities.

Policy: The initial responding officer(s) at the scene shall provide a detailed crime scene briefing to the investigator(s) in charge of the scene.

Procedure: The initial responding officer(s) should:

a. Brief the investigator(s) taking charge.
b. Assist in controlling the scene.
c. Turn over responsibility for the documentation of entry/exit.
d. Remain at the scene until relieved of duty.

Summary: The scene briefing is the only opportunity for the next in command to obtain initial aspects of the crime scene prior to subsequent investigation.

7. Document Actions and Observations

Principle: All activities conducted and observations made at the crime scene must be documented as soon as possible after the event to preserve information.

Policy: Documentation must be maintained as a permanent record.

Procedure: The initial responding officer(s) should document:

a. Observations of the crime scene, including the location of persons and items within the crime scene and the appearance and condition of the scene upon arrival.
b. Conditions upon arrival (e.g., lights on/off; shades up/down, open/closed; doors, windows, open/closed; smells; ice, liquids; movable furniture; weather; temperature; and personal items.)
c. Personal information from witnesses, victims, suspects, and any statements or comments made.
d. Own actions and actions of others.

Summary: The initial responding officer(s) at the crime scene must produce clear, concise, documented information encompassing his or her observations and actions. This documentation is vital in providing information to substantiate investigative considerations.

Chapter Summary

The basic overview of duties for the first responder is as follows:

- Assist any victims.
- Search for, apprehend, and arrest suspects still on the scene or in the area of the crime scene.
- Detain and separate witnesses.
- Protect the integrity of the crime scene.
- Communicate with supervisors and appropriate agencies.
- Document all actions taken and observations made.

The first responder is responsible for ensuring that the initial response is accomplished quickly and competently. Once the scene is secure, and the integrity of the scene is ensured, the investigatory process begins.

Review Questions

1. What is meant by the term first responder?
2. Discuss instances when the loss or destruction of evidence at a crime scene is acceptable.
3. Explain what is meant by the term perimeter as it relates to crime scene investigation.
4. Utilizing multiple perimeter areas in order to segregate the crime scene, media, and crime scene processing personnel is referred to as ______________.
5. What is transient evidence and how does it differ from other types of evidence in relation to how a first responder must respond to it?
6. Explain the difference between a primary and a secondary scene.
7. What are official visitors and how should they be treated in relation to other crime scene personnel at a crime scene?
8. ______________ is the area from which the critical incident is managed and overseen.

References

Doyle, A. C. (1892). The Adventure of the Copper Beeches. In The Adventures of Sherlock Holmes. *Strand Magazine.*

Fish, J. T., Miller, L. S., & Braswell M. C. (2007). *Crime scene investigation: An introduction.* New York: LexisNexis Anderson Publishing.

Fisher, B. A. J. (2004). *Techniques of crime scene investigation*, 7th ed. Boca Raton, FL: CRC Press.

National Institute of Justice. (2004). *Crime scene investigation: A reference for law enforcement training*, June 2004 Special Report, U.S.DOJ, Office of Justice Programs. NCJ200160. Retrieved July 20, 2009, from http://nij.ncjrs.gov/publications/search_form.asp

Wisconsin Department of Justice. (2001). *Physical evidence handbook*, 6th ed. Madison, WI: Wisconsin Department of Administration.

Specialized Personnel and Safety Considerations

You see, but you do not observe.

Sir Arthur Conan Doyle (1859–1930)
Sherlock Holmes, in *A Scandal in Bohemia*

LEARNING OBJECTIVES

- Explain the purpose of a walk-thru.
- Identify the types of personnel involved in crime scene investigative efforts and list their job descriptions.
- Define and discuss the function of PPE
- Understand the various levels of PPE.
- Name the pathogens and other most common health risks to CSI personnel.

KEY TERMS

Barrier Protection
Bloodborne Pathogen
Cross Contamination
IDHL Environment
Personal Protective Equipment (PPE)
Universal Precautions
Walk-Thru

Investigative Personnel

Once the first responder has secured the crime scene and all associated evidence, she or he will then contact their agency and request investigative personnel to respond to the incident. In some smaller agencies and in situations where the crime scene is of a minor nature, this investigative role may be handled by the first responder. If this is the case, efforts still are made to ensure the integrity of the crime scene while the officer is conducting their investigation.

If other investigative personnel are requested, this typically will be a detective or investigator, depending on the label given to such personnel within the department in question. Sometimes the first responder will speak with the investigator via phone and articulate to them the situation at hand. Often this will result in the investigator acquiring enough information to form an opinion about the manner in which the investigation should proceed, which can sometimes mean that the investigator will not respond to the scene, but rather will instruct the officer on what to do before reporting back on the findings.

If investigative personnel respond to the scene, this can be in a variety of degrees. As discussed in Chapter 1, detectives/investigators are responsible for conducting the overall investigative efforts associated with the scene. These are the individuals who lay out the plan for the processing and investigative efforts. If the department in question has personnel who serve in a crime scene technician, evidence technician, or crime scene investigator capacity, such processing issues may be left up to them while the investigator concentrates on interviews, interrogations, and interagency cooperation efforts.

Once the scene is secure, investigative personnel should conduct a preliminary scene survey, which is also known as a **walk-thru**. The preliminary scene survey will have the greatest informational possibilities if the first responder is available to accompany the investigative personnel. This is because it is the first responder who has the most direct knowledge of what the scene originally looked like when law enforcement responded to the event. He or she also should know of any changes made to the scene since that initial response. It is very important that investigators are well briefed by first responders regarding the case before conducting their examination of the scene. This ensures that the preliminary scene survey will result in maximum information gathering, while minimizing scene contamination and evidence destruction. The primary purpose of the preliminary scene survey is to assess the scene for logistical and safety considerations. The following is the suggested process to be followed by investigative personnel in conducting a preliminary scene survey, as written by the National Institute of Justice (2004) Special Report entitled, "Crime Scene Investigation: A Reference for Law Enforcement Training.

Until this point nothing should have been touched or moved by personnel unless there was cause to move an injured person, or conditions were present (e.g., foul weather) that necessitated the movement or collection of some evidence in order to best preserve it. With an unaltered (or minimally altered) scene as a continued goal, the survey is begun. The path should attempt to follow the indirect path that was cleared or identified by first responders to minimize evidence destruction and disturbance of the scene.

The following list contains ten suggested matters to consider while conducting a walk-thru:

1. As with first responder efforts, make note of transient evidence present within the scene, and efforts needed to properly document,

collect, and preserve such evidence. If steps not already taken to do so, it may be necessary at this point to document, collect, package, and preserve such evidence.

2. Make note of weather and climate conditions (both indoors and outdoors).
3. Note whether lights are turned on or off.
4. Document whether doors and windows are locked, unlocked, open, closed, or if there appears to be evidence of forced entry.
5. Note the presence of any particular odors that may be connected to an individual (perfume, cologne) or an event (gas, smoke, chemicals, etc.).
6. Look for signs of activity (meal preparation, house tidy or disheveled, etc.) or struggle.
7. If timing is of great concern, look for date and time indicators such as on food, newspapers, mail, etc.

Preliminary Scene Survey Process

Principle:
The scene "walk-thru" provides an overview of the entire scene, identifies any threats to scene integrity, and ensures protection of physical evidence. Written and photographic documentation provides a permanent record.

Policy:
The investigator(s) in charge shall conduct a walk-thru of the scene. The walk-thru shall be conducted with individuals responsible for processing the scene.

Procedure:
During the scene walk-thru, the investigator(s) in charge should:

a. Avoid contaminating the scene by using the established path of entry.
b. Prepare preliminary documentation of the scene as observed.
c. Identify and protect fragile and/or perishable evidence (e.g., consider climatic conditions, crowds/hostile environment). Ensure that all evidence that may be compromised is immediately documented, photographed, and collected.

Summary:
Conducting a scene walk-thru provides the investigator(s) in charge with an overview of the entire scene. The walk-thru provides the first opportunity to identify valuable and/or fragile evidence and determine initial investigative procedures, providing for a systematic examination and documentation of the scene. Written and photographic documentation records the condition of the scene as first observed, providing a permanent record.

8. Attempt to locate the most probable point of entry, point of exit, paths between them, and any other areas of apparent action within the scene. These areas should be noted to ensure that processing personnel will reduce their movements in such areas to allow for the optimum opportunity to discover and collect physical evidence within the scene.
9. Attempt to answer the questions of: Who? What? When? Where? How? and Why? as they pertain to the scene and the crime in question.
10. Assess the scene for personnel (How many? Specialized?), equipment (How much? What kind?), and logistical concerns (How long? Food needs for personnel? Bathroom needs? Media considerations? Budgetary issues?).

This walk-thru should be conducted in a cautious and aware manner. As with the first responders, nothing should be touched or moved. At this point, the scene should be secure and the evidence safe from harm as well. Typically there is no need to touch or move evidence. The walk-thru is a minimally invasive information-gathering event, and not an evidence search or collection effort.

There are two schools of thought about whether investigative personnel should wear gloves while conducting this scene survey. One view is that if personnel wear protective gloves they will be more inclined to touch items and, therefore, they should not wear gloves and should adhere to the "hands in pockets" approach. The other view is that personnel should always wear gloves whenever they are inside of a crime scene. The author agrees with the latter line of thinking for several reasons. First, the purpose of gloves is to both protect the wearer from contamination, and to protect any item touched from contamination by the wearer. While it is true that in the walk-thru there should be no touching of items, this is not to say that transient evidence will not be discovered that necessitates movement or collection. Having gloves on will ensure that such evidence is minimally damaged if such contact is necessary. Also, a "hands in the pocket approach" is not realistic because the point of a walk-thru is to document conditions present throughout the scene. The investigator most certainly will have her or his hands outside of any pockets, and will be writing and pointing throughout the process. It is best to have personnel gloved up with the thought in the forefront of her or his mind that nothing is to be touched unless it is absolutely imperative.

After the initial walk-thru has been conducted, investigative personnel should have the information necessary that they need to apprise supervisors of the situation and to lay out the crime scene processing strategy. At this stage there may be a call for more specialized personnel. Some of these personnel may be from within the ranks of law enforcement. Other specialists such as entomologists or engineers may be necessary to provide technical assistance that is outside of the training and education of those in law enforcement. Agencies are encouraged to think broadly and utilize sources such as local universities and other private, local, state, and federal agencies to maximize the investigatory potential. If an individual has not been trained to collect or document certain evidence, then they should not; instead they should rely upon experts to do so.

A brief, and by no means all-inclusive, list of personnel who may be called upon to assist with the investigative efforts is in this section.

Crime Scene Investigator/Crime Scene Technician

These are police or civilian personnel who are specially trained to process a crime scene. Their purpose is twofold: to collect and preserve physical evidence.

Identification (ID) Officers

They are responsible for photographing the scene and searching for latent fingerprints, but they are not responsible for other types of physical evidence. Often these individuals are fingerprint experts who later will perform comparative analyses.

Forensic Surveyors

Often they are used to provide an accurate architectural rendition of the crime scene. They typically utilize Computer Aided Drafting (CAD) to assist them with their documentation efforts.

Forensic Photographers

Specialized photography (low light, aerial, infrared, underwater, etc.) demands specific skills. These photographers have advanced training in photographic concepts and specialized situations.

Forensic Scientist/Criminalist

This person has gained specialized training and education in chemistry and biology as applied to the recognition, identification (ID), collection, and preservation of physical evidence.

Medical Examiner/Coroner

Typically, a forensic pathologist is responsible for performing autopsies in criminal cases. This may include providing an ID of the deceased; determining cause, manner, and time of death; and taking custody of the remains.

Forensic Nurse
This licensed nurse has specialized training in proper evidence collection, and most often is utilized in sexual assault investigations. Such nurses usually are certified Sexual Assault Nurse Examiners (SANE).

District Attorney
When called upon, this person provides a search warrant or a court order to obtain known specimens from a defendant. A district attorney may operate in an advisory capacity when a case involves a police officer (e.g., a police-related shooting, in-custody death of suspect, etc.).

Hazardous Materials Specialists
These experts assist with recognition, collection, destruction, clean-up, disposal, and preservation of hazardous materials at the crime scene.

Forensic Engineers
This type of engineer analyzes the structural integrity of a building or other structures in accident investigations.

Firearms Examiners
This expert assists in crime scene ballistic recovery and can assess the trajectory of fired weapons. He or she also may assist in determining whether a shooting was accidental or intentional.

K-9 Officers
This sworn officer and trained dog may be called upon to assist with searches and tracking of individuals; if the individual is believed to be dead or buried, cadaver dogs may be utilized. Cadaver dogs are specially trained to recognize the scent of decaying remains.

Federal Authorities
Numerous federal agencies can be called in to assist or take over a crime scene involving mass disasters, terrorist acts, bombings, major fires, and bank robberies. Some examples of these agencies include: Federal Bureau of Investigation (FBI), Drug Enforcement Administration (DEA), Bureau of Alcohol Tobacco Firearms and Explosives (ATF), and the United States Secret Service (USSS; **Figure 5.1**).

■ Crime Scene Investigation Processing Models

With regards to the utilization of specialized personnel, there are several different models for assembling a response to investigating and processing a crime scene. Each has its positive and negative attributes. The factors determining which model will be implemented include: available resources, manpower and training, type of crime investigated, and the degree of scientific and technological advancement of involved agencies.

Figure 5.1 Example of Specialized Personnel (Federal).

Local or Traditional Model

Many agencies choose to employ a more traditional approach to crime scene processing efforts. In this case, patrol officers and sometimes investigators are utilized as the primary crime scene personnel. This is seen as being beneficial to those agencies with limited budgets, minimal crime rates, and where the need for crime scene processing efforts is very low. This is also the default model in the absence of any other model. Negative attributes of this model include the fact that typically personnel are not highly trained in specific areas and are more generalists than specialists. Also, crime scene processing efforts may interfere with designated patrol duties.

Crime Scene Investigator/Crime Scene Technician Model

Some departments employ dedicated civilian or sworn personnel who are dedicated to crime scene documentation and physical evidence collection and preservation efforts. This allows such individuals to become specialized and have a better understanding for the procedures involved with such processes. Most times these individuals work the crime scene from a technical standpoint, but are not actively involved in the investigative process. Such matters are left up to the investigator or detective, who typically has considerably more investigatory experience.

Crime Lab Model

Some crime labs, specifically state and federal labs, operate a crime scene response team that will respond to serious cases. These teams are comprised of forensic scientists/criminalists with specific job skills who are responsible for individual areas of crime scene documentation and processing efforts. These individuals are typically highly skilled but lack experience due to their limited use. Also, it must be remembered that these individuals are performing a scientific or technical function and usually lack investigative experience; therefore, the information derived from their efforts must be passed along to the appropriate investigatory personnel for use within the case.

The Team Model

This is the most effective but also the most specialized model. It incorporates highly trained personnel at all levels of the investigation, and has its greatest effect when all levels are able to utilize personnel with experience and who are not mandatorily rotated out or lost due to promotion, retirement, or other reason. In this model, the individual team members are assigned specific crime scene duties, dependent upon their specific training and experience. This results in the most efficient and successful crime scene investigation. Utilization of the team model requires cooperation and collaboration between agency departments as well as between agencies. Cooperation, coordination, and communication are the trident by which this model operates.

VIEW FROM AN EXPERT

The Globalization of Forensic Science Research

There have been a considerable number of advances in the various forensic disciplines over the years. Unfortunately, not all of these discoveries are introduced into the peer-reviewed journal process. Even for the articles that are accepted into the pages of a peer-reviewed journal, there are simply too many journals worldwide with which to keep apprised. In some cases, these discoveries may be presented as a lecture or workshop at a particular conference. The difficulty that then arises is the sheer number of such conferences, symposia, and meetings that occur around the world. Who has the time, or more importantly the resources, to attend all of the relevant conferences each year? Compounding this problem is the fact that resources in the forensic science field are very scarce, especially in the area of research and development.

The scientific community settled this quandary back in 1931. It was in that year that Professor Neil E. Gordon, a member of the chemistry faculty at Johns Hopkins University, organized the first meeting of what would ultimately be called the Gordon Research Conferences (GRC)[1]. His goal was to assemble the most active researchers in a particular field of study and bring them together in a relaxed environment suitable for rigorous formal and informal discussions. He strove to make each meeting in a particular field the best overall conference for that discipline. One of the GRC's goals is to encourage young scientists in their field of expertise as well as to provide contacts and career guidance. To encourage free exchange of information, all presentations given at these conferences are considered private communications and cannot be disseminated without the presenter's approval or until it has been published in a peer-reviewed journal.

With regard to the forensic sciences, at least one discipline has established a similar solution. There are few research groups around the world that have active, robust programs in the area of visualization of latent print residue using chemical, optical, and physical means. Since the early 1970s, attempts have been made to bring together these researchers at various irregularly scheduled meetings. Eventually in 1999 this group, officially known as the International Fingerprint Research Group (IFRG), began to assemble the most active researchers in this field on a bi-annual basis. Some of the earliest meetings were held in the United Kingdom and sponsored by the Home Office. Subsequent meetings have been held in the United States, Israel, Canada, Germany, The Netherlands, Australia and Switzerland.

As with the Gordon Research Conferences, the IFRG tries to restrict attendance to only those who are the most active in the field. It is the responsibility of the meeting chair (from the host country) to issue invitations for the meeting. These meetings typically occur during the course of a week (Monday through Friday), when as many as 50 or more lectures are generally presented. Social events are scheduled throughout the week to allow participants to bond and to encourage working partnerships and collaborations.

These efforts relate primarily to the scientific aspects of a particular field rather than to policy-making. In the United States, a number of policy setting groups for several different forensic disciplines have undertaken the task of setting policies and guidelines. A number of scientific working groups (SWGs; formerly known as technical working groups [TWGs]), have been established, including: Friction Ridge Analysis, Study, and Technology (SWGFAST); Imaging Technology (SWGIT); Digital Evidence (SWGDE); Education and Training in Forensic Science (SWGED); Shoeprint and Tire Tread Evidence (SWGTREAD); DNA Analysis Methods (SWGDAM); Questioned Documents (SWGDOC); Analysis of Seized Drugs (SWGDRUG); Firearms and Toolmarks (SWGGUN); Illicit Business Records (SWGIBRA); Materials (SWGMAT); Bloodstain Pattern Analysis (SWGSTAIN); Dog and Orthogonal Detector Guidelines (SWGDOG); Forensic Analysis of Chemical Terrorism (SWGFACT); and Microbial Genetics and Forensics (SWGMGF)[2]. These groups have a diverse membership, drawing from across the country and, in some cases, from throughout the international community. Each group has its own Web site where policies are posted for review by the community. Issues like competency testing, training guidelines, examination methodologies, proficiency testing, minimum qualifications, professional conduct, quality assurance, and validation of research and technology, are addressed by such policies.

Another similar set of working groups was established in Europe in the mid-1990s. This organization, known collectively as the European Network of Forensic Science Institutes (ENFSI), performs a similar role throughout the continent. The first organizational

(continues)

meeting occurred in 1993, when representatives from eleven countries met in the Netherland[3]. ENFSI was formally established on October 20, 1995 and now consists of 54 representatives from different laboratories in 31 countries. Its official mission is to share knowledge, exchange experiences, and establish mutual agreements across the various disciplines within forensic science. Among its primary goals is to encourage ENFSI laboratories to comply with best practice and international standards for quality and competency. Standing committees within ENFSI include: digital imaging, DNA, documents, explosives, fibers, fingerprints, firearms, fire and explosive investigation, forensic information technology, forensic speech and audio analysis, handwriting, marks, paint and glass, road accident analysis, and scenes of crime.

The establishment of a set of internationally recognized standards is another critical element in the global forensic science research effort. The first organization to contribute substantially in this area was the American Society for Testing Materials (ASTM)[4]. Interestingly, the birth of this organization came out of a dispute over quality standards for steel rails used by the Pennsylvania Railroad, which was the largest company in the world during the 19th century. The efforts of Dr. Charles Dudley, head of the Pennsylvania Railroad's chemistry department, ultimately led to the formation of the ASTM in Philadelphia in 1898. Quite a number of forensic disciplines have ASTM guidelines. One field in particular—questioned document analysis—has generated many standards for the discipline. Examples include the Standard Guide for Test Methods for Forensic Writing Ink Comparison (E1422) and the Standard Guide for Examination of Handwritten Items (E2290). Overall, ASTM guidelines impact nearly every aspect of forensic work performed either in the laboratory or at a crime scene. Additional examples include: the Standard Guide for Microscopic Examination of Textile Fibers (E2228); the Standard Test Method for Fiber Analysis of Paper and Paperboard (D1030); the Standard Guide for Forensic Paint Analysis and Comparison (E1610-94); and Practice for Receiving, Documenting, Storing, and Retrieving Evidence in a Forensic Laboratory (E1492). Today, ASTM has a truly global outreach, with over 30,000 members in more than 120 countries.

Another standards organization that is specifically designed for the forensic community is the American Society of Crime Laboratory Directors/Laboratory Accreditation Board International (ASCLD/LAB—International)[5]. This organization combines its own objectives with the general requirements for the competence of testing and calibration laboratories (17025) of the International Organization for Standardization/International Electrotechnical Commission (ISO/IEC). The objective of this organization is to improve the quality of laboratory services; maintain standards; provide an independent and impartial operational review of laboratories; and to publicly recognize those laboratories that have demonstrated compliance with the high level of requirements. ASCLD/LAB—International currently offers accreditation in the following disciplines: controlled substances, toxicology, trace evidence, biology, firearms/toolmarks, questioned documents, latent prints, crime scene, and digital and multimedia evidence. The accreditation process is quite rigorous and demanding. However, for those laboratories that do achieve accreditation, there is a greater appreciation of the laboratory's quality among the public as well as the judicial system. There is a growing worldwide interest in having forensic laboratories become accredited by some recognized, external quality assurance organization.

Another source for forensic science standards is the National Institutes of Science and Technology's (NIST) Office of Law Enforcement Standards (OLES)[6]. NIST OLES conducts numerous studies and has published many articles, guides, and standards. Some of its more recent projects include a National Software Reference Library, which is comprised of a collection of commercially available software products and a validated database of known software, file profiles, and file signatures in different formats. NIST OLES also provides the Standard Casing Reference Material, which is a collection of traceable standards that can be used for equipment calibration and inter-laboratory evaluations. NIST OLES also can supply traceable glass standards for refractive index measurements. This organization additionally has provided a significant number of standards for DNA analysis and quantification, as well.

Large-scale criminal and terrorist groups have demonstrated that their reach is truly global in scope. Efforts on the part of police and intelligence services around the world now can be coordinated as never before. Forensic research and development as well as crime scene work will be a critical component of the law enforcement's response to these activities. To further that endeavor, more cooperative efforts will have to be initiated to maximize the meager resources available. International standards will be a critical part of this effort, because all laboratories must adhere to a high level of quality to ensure that evidence collected and analyzed will remain admissible in court.

Robert Ramotowski, M.S.
Chemist
U.S. Secret Service
Forensic Services Division

1. Gordon Research Conferences. (updated 2008). Retrieved July 21, 2009, from http://www.grc.org/history.aspx
2. Scientific Working Group on Dog & Orthogonal Detector Guidelines. (updated 2009). Retrieved July 21, 2009, from http://www.ifri.fiu.edu/SWGDOG.htm
3. European Network of Forensic Science Institutes. (updated 2009). Retrieved July 21, 2009, from http://www.enfsi.eu/
4. ASTM International. Standards Worldwide.(updated 2009). Retrieved July 21, 2009, from http://www.astm.org/
5. ASCLD-LAB International. Accreditation program. Retrieved July 21, 2009, from http://www.ascld-lab.org/international/pdf/alpd3013.pdf
6. Electronics and Electrical Engineering Laboratory, Office of Law Enforcement Standards. (updated 2008). Retrieved July 21, 2009, from http://www.eeel.nist.gov/oles/forensics.html

Crime Scene Safety Concerns

Risk of Infection and Disease

Certainly crime scenes always have been hazardous environments. While such a phenomenon is by no means new, the last decade has brought significant awareness and has seen crime scene personnel be at far greater risk of infection by a pathogen while engaged in their job-related duties. HIV, AIDS, hepatitis B and C, tuberculosis, and tetanus, as well as other bacterial- and virus-caused diseases are quite common in today's society. Such blood and airborne pathogens can be transmitted via minimal contact, such as by simply touching or breathing in confined spaces (Lee, Palmbach, & Miller, 2001). Although this might appear disconcerting to crime scene personnel, the chances of acquiring such diseases is dramatically reduced by adherence to some basic **universal precautions**. These are typical good hygiene habits: hand washing and use of barriers and aseptic techniques, and appropriate handling of cutting or puncturing objects like needles, syringes, and razor blades (called *sharps*).

At the scene of a crime, it is up to every individual to be aware of the safety issues and hazardous conditions that may be present and potentially create dangerous situations for emergency responders and investigative personnel. This section is included within Specialized Personnel and Safety Considerations so that individuals will learn the importance of developing routine personal security measures that will reduce their potential exposure to situations that might be hazardous to health. Each department or agency has its own guidelines pertaining to employee safety considerations. Information is provided here as part of this text so that the safety of crime scene personnel will be enhanced, and the possibility of **cross contamination** of physical evidence is minimized. Cross contamination refers to the unintended movement or transfer of material between two objects, and should be avoided and/or minimized during processing efforts. While the primary objective is the protection of crime scene personnel, with the continued modernization and sensitivity of forensic equipment and analysis capabilities, the potential for contamination of physical evidence by investigators has dramatically increased. Through adhering to the use of barrier protection, the possibility of crime scene personnel depositing their own DNA or causing a secondary transfer of unrelated material between two or more objects is significantly reduced (Lee, Palmbach, & Miller, 2001).

Personal Protective Equipment

Crime scene personnel, who include the first responders, never should enter the crime scene without having the proper **personal protective**

equipment (PPE). Most likely first responders will not know the appropriate level of protection when they arrive on the scene, and safety and security issues often do not allow for them to assemble the proper equipment prior to entry. However, once they have secured the scene and have an initial assessment of the situation, those observations should be passed along to subsequent personnel so that those individuals enter the scene properly protected.

PPE equipment typically consists of gloves, Tyvek suits, shoe covers, eye protection, and respiratory equipment. A crime scene is home to many hazards, both obvious and hidden, which demand that crime scene personnel remain vigilant and approach the scene from a contamination point of view. This means utilizing **barrier protection** when processing a crime scene. Barrier protection involves creating a barrier between the personnel and their surroundings to ensure that they are not contaminated by the scene and that they themselves do not contaminate the scene and evidence within. It is imperative that universal precautions be followed to minimize the risk to personnel of coming into contact with potentially infectious materials.

The Occupational Safety and Health Administration (OSHA) mandates that all employers must train their employees about job-related safety strategies as well as supply them with appropriate levels of PPE for the specific job functions performed by their employees. Even with this measure, personnel must remember that the use of PPE does not guarantee total protection; rather, it is used to minimize the exposure contact possibility. Keeping up with immunizations is important medical prophylaxis. After an exposure, a medical checkup or monitoring might be necessary.

PPE Levels

The Environmental Protection Agency (EPA, 2009) has defined a tiered level of protection according to the severity of hazardous material response. These PPE levels are general guidelines as some situations will necessitate other combinations of equipment. The four PPE levels and examples are presented in this section.

Level A (Figure 5.2)

This event, termed an *immediately dangerous to life or health* **(IDLH) environment**, contains the greatest risk for exposure to potentially life-threatening biological hazards. It requires the maximum respiratory, eye, and skin protection. Equipment could include: a totally encapsulated chemical- and vapor-protective suit; inner and outer chemical-resistant gloves; positive-pressure, full face-piece self-contained breathing apparatus (SCBA) or positive pressure supplied air respirator (SAR) with escape SCBA; and disposable protective suit, gloves, and boots.

Level B (Figure 5.3)

This level of protection is required in circumstances where the responder needs the highest level of respiratory protection but a moderate level of skin protection. This level is the minimum required PPE for entering an unknown hazardous environment. The full-face SCBA or a positive-pressure SAR with escape SCBA should be used. All clothing should be

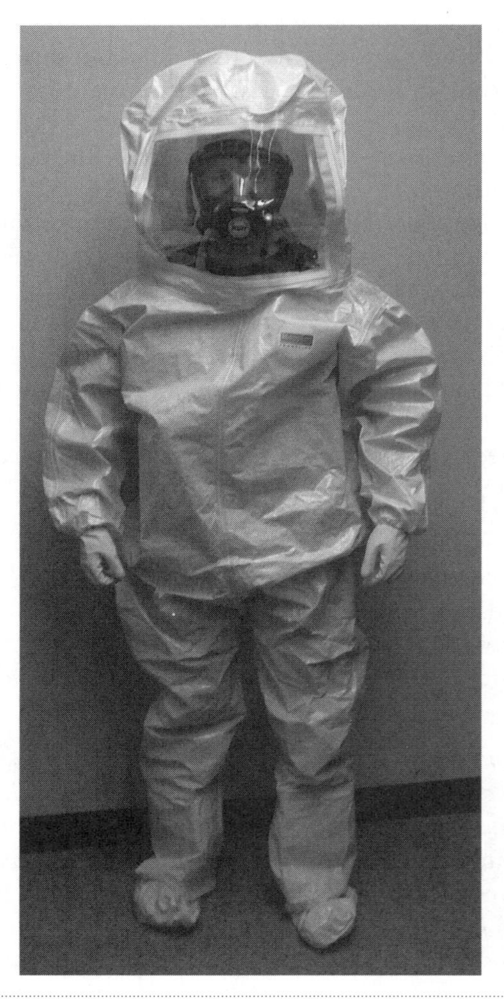

Figure 5.2 Level A PPE Example.

chemical-resistant and include hooded coveralls, inner and outer gloves, boots, and a face shield.

Level C (Figure 5.4)

When the concentration and type of airborne substance(s) are known and the criteria for an air-purifying respirator (APR) are met, responders should wear Level C protection. Besides the APR, the same types of chemical-resistant clothing should be worn as those for Level B hazards. This is the highest level for using an APR.

Figure 5.3 Level B PPE Example.

Level D (Figure 5.5)

For situations where no contaminants are present or may involve splashes, immersion, or a potential for unexpected contact/inhalation of chemicals, only the minimum protection is required. Level D equipment might include gloves, coveralls, safety glasses, face shield, and chemical-resistant steel-toed boots.

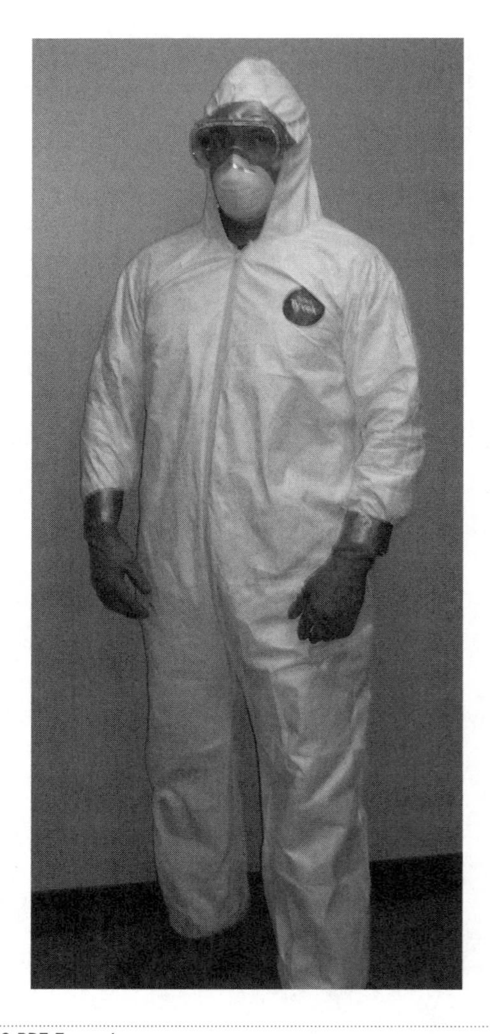

Figure 5.4 Level C PPE Example.

Types of Respiratory Protection (Figures 5.6 and 5.7)
SCBA, SAR, and APR apparatus must be approved by the National Institute of Occupational Safety and Health (NIOSH) and certified according to PPE level. If it is reusable, it must be decontaminated according to standard practices for the particular biological agent before taking it off. This may be as simple as a thorough scrubbing

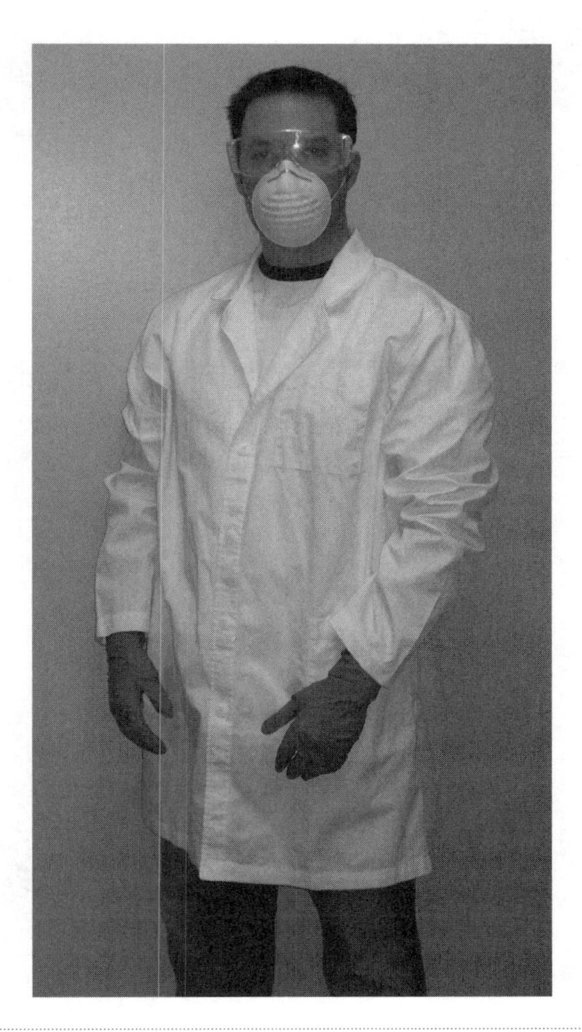

Figure 5.5 Level D PPE Example.

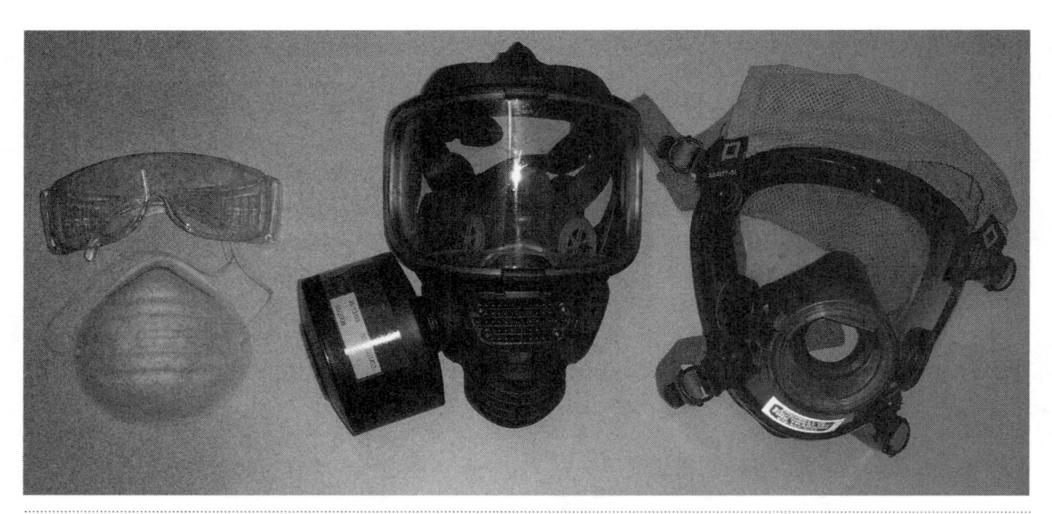

Figure 5.6 Examples of Various Respiratory Equipment.

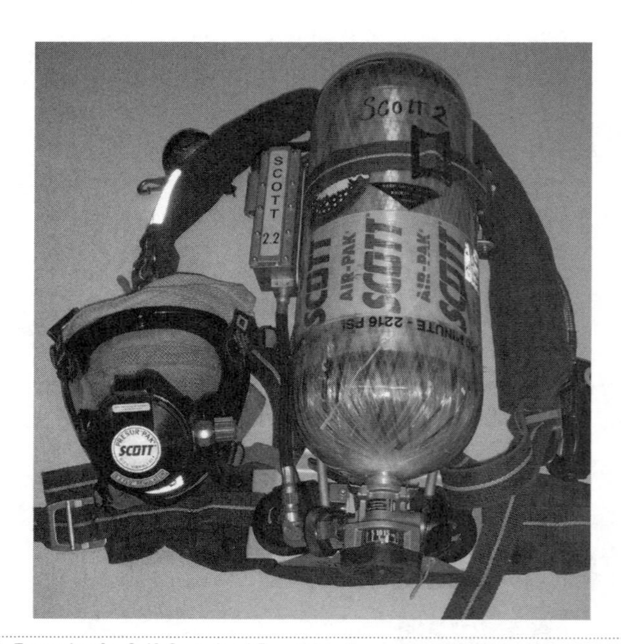

Figure 5.7 Example of a Self-Contained Breathing Apparatus (SCBA).

with soap and water and/or bleach, but depends upon the material, contact time with the hazard, and the specific contaminant. Disposable gear is worn only once and should be disposed of according to departmental regulations to preserve safety and security.

Common Crime Scene–Related Pathogens

Bloodborne pathogens are microorganisms found in the blood that can cause infection and disease, and may be transported in biological fluids. However, it bears mentioning that fluids other than blood may contain pathogens because they contain microscopic traces of blood within them. These include, but are not limited to: feces, urine, tears, sweat, saliva, semen, vaginal secretions, nasal secretions, breast milk, and vomitus. It also should be noted that feces, nasal secretions, sweat, tears, vomit, and urine (when not contaminated with blood) are not addressed under universal precautionary materials. However, they can transmit other infectious diseases so caution should be used with anything produced by the human body. The most commonly encountered pathogens are human immunodeficiency virus (HIV)/acquired immunodeficiency deficiency virus (AIDS), hepatitis, and tuberculosis (TB).

Human Immunodeficiency Virus and Acquired Immunodeficiency Deficiency Virus

This disease starts off as HIV, which is a retrovirus. Most people will not know that they are infected and highly contagious until seroconversion. This involves four clinical stages of illness beginning with a fever, joint pain, rash, and enlarged lymph nodes. Symptoms slowly (5–10 years) become more severe, and by stage 4 signals that the HIV has progressed to AIDS. By this time, the retrovirus has invaded the individual's DNA and thereafter depresses the immune system, causing an irreversible reduction in the body's ability to fight off infection and making the individual more susceptible to such opportunistic infections as pneumonia or HIV-related cancer. Exposure to any AIDS-contaminated blood or body fluids carries a risk of infection. There is no known cure for AIDS to date. Vaccines and medications can delay disease progression but the outcome is death.

Hepatitis

Transmitted by way of blood and body fluids, this is an infectious disease of the liver. The three different forms of this virus are known as types A, B, and C. Of these, hepatitis B and C are the most deadly. There are vaccinations for both hepatitis A and B but none for type C.

Hepatitis B is sometimes known as *serum hepatitis* and can result in jaundice, cirrhosis, and in some cases cancer of the liver.

Tuberculosis (TB)

Unlike the aforementioned pathogens, TB is a bacteria (*Mycobacterium tuberculosis*) that it airborne. Infection occurs when sputnum is aerosolized by an infected person's cough, shouting, or spitting, and the individual breathes in the minute particles. This disease affects the lungs and most cases can be treated successfully via antibiotics. A small group of organisms known as atypical TB can live alongside the body's normal bacteria and not cause any damage, meaning that the infected individual is a carrier of TB but does not get sick. Because this strain is more resistant to antibiotics, it can take up to a year and a half of medication before being cured.

Crime Scene Precautionary Steps to Reduce Risk

There are several simple habits that will help allay or prevent infection at a crime scene.

- Consider all blood and body fluids, wet or dry, to be infectious. Current research indicates that the infectious activity of some organisms persists more than several days after drying.
- Protect eyes, nose, and mouth by using goggles and a mask. These provide a barrier to spraying, splashing, or aerosol transmission of infectious materials.
- Be extremely aware and cautious when handling sharp items. Never shear, break, or bend a contaminated sharp.
- Place all syringes, razor blades, needles, and other sharp instruments in a puncture-resistant sharps container.
- Latex or nitrile gloves should be worn when handling potentially contaminated items. Remove jewelry before putting on gloves. Most latex gloves have microscopic holes and openings, so double gloving (wearing two pairs of gloves at the same time) and changing gloves frequently aids in protection.
- Avoid hand-to-face contact (eating, smoking, drinking, rubbing eyes, rubbing nose, etc.) when processing a crime scene.
- Immediately wash hands, arms, face, and any exposed skin after removing and disposing of barrier equipment when crime scene processing efforts are completed. Frequent hand washing is a good safety practice.

- Any contaminated equipment should be decontaminated with a solution of one part household bleach to nine parts water. Afterwards, the item can be wiped down with soap and water or isopropyl alcohol to remove the smell of bleach.

Steps to Take for Exposure to Blood/Body Fluids

Immediately following an exposure:

- Wash needle sticks and cuts with soap and water.
- Flush splashes/ splatters to the nose, mouth, or skin with water.
- Irrigate eyes with clean water, saline, or sterile irrigates.
- There is no scientific evidence to show that using antiseptics or squeezing the wound will reduce the risk of transmission of a bloodborne pathogen.
- Using a caustic agent such as bleach is not recommended.
- Report the exposure to the department (e.g., occupational health, infection control) responsible for managing exposures. Prompt reporting is essential because, in some cases, post-exposure treatment may be recommended and it should be started as soon as possible.

Chapter Summary

Each crime scene encountered by investigative personnel is unique to itself. Although it may appear similar to others, there is none exactly like it. Investigators must decide the most appropriate CSI model to utilize in the investigatory and processing efforts. This often involves the utilization of many different specialized personnel. Because many of these specialized personnel may not necessarily come from law enforcement backgrounds, it is imperative that all involved make use of strong interpersonal skills. Working together with the understanding that all concerned are working towards the same goal, and working in a spirit of mutual cooperation and respect will ensure a professional and successful investigation.

However, it is not simply enough for such individuals to work together. They must work in a manner that ensures each of their safety and well-being. For this reason, all personnel must be cognizant of health and safety risks and put into practice universal precautions within their processing efforts. Doing so will ensure that all personnel will literally live to fight another day.

Review Questions

1. A preliminary scene survey is often referred to as a ______________.
2. Explain the fundamental differences between the four CSI processing models.
3. PPE, as it relates to CSI, is an acronym for ______________.
4. ______________ refers to the unintended movement or transfer of material between two items.
5. Name three bloodborne pathogen risks to crime scene personnel that can be encountered during their processing efforts.
6. Explain what is meant by barrier protection as it relates to processing hazardous material scenes.
7. List things for the investigator to consider and do during the walk-thru.
8. Explain why the team method is the most effective way to process a crime scene.
9. What is OSHA, and what is the organization's purpose with regards to CSI?

References

Doyle, A. C. (1892). A Scandal in Bohemia. In The Adventures of Sherlock Holmes. *Strand Magazine.*

Environmental Protection Agency (EPA). (updated 2009). Personal protective equipment. Retrieved July 21, 2009, from http://www.epa.gov/emergencies/content/hazsubs/equip.htm

Fish, J. T., Miller, L. S., & Braswell M. C. (2007). *Crime scene investigation: An introduction.* New York: LexisNexis Anderson Publishing.

Lee, H. C., Palmbach, T., Miller M. T. (2001). *Henry Lee's crime scene handbook.* San Diego: Academic Press.

National Institute of Justice. (2004). *Crime scene investigation: A reference for law enforcement training*, June 2004 Special Report, U.S. DOJ, Office of Justice Programs. NCJ200160. Retrieved July 21, 2009, from http://nij.ncjrs.gov/publications/search_form.asp

Occupational Safety and Health Administration (OSHA). Retrieved July 21, 2009, from http://www.osha.gov/

Methodical Approach to Processing the Crime Scene

It is a capital mistake to theorize before one has data. Insensibly one begins to twist facts to suit theories, instead of theories to suit facts.

Sir Arthur Conan Doyle (1859–1930)
Sherlock Holmes, in *A Scandal in Bohemia*

LEARNING OBJECTIVES

- Understand the basic components of crime scene documentation.
- Identify the three photographic ranges in crime scene photography.
- Understand the concept of "fair and accurate" standard of evidence photography.
- Identify the various types and perspectives of crime scene sketches.
- Differentiate between sketching and mapping.
- Identify the necessary components of a final sketch.
- Identify the various methods of crime scene mapping and know which is best for different crime scenes.
- List the crime scene search patterns and when/where they are most effective.

KEY TERMS

Close-Up Photographs
Crime Scene Sketch
Final Sketch
Legend
Mapping
Mid-range photographs
Overall photographs
Photo Log
Photo Placard
Rough Sketch
Swath

Methodical Approach to Crime Scene Processing

Crime scenes are complex and confusing creatures. The first step in crime scene processing is to establish a plan. In Shakespeare's 1600 play, *Hamlet*, Polonius says, "Though this be madness, yet there is method in it" (Act 2, Scene 2). All steps of crime scene response should be calculated and methodical to ensure the most positive result. It is for this reason that investigative personnel should take the information garnered from their walk-thru and develop a systematic plan for proceeding with the processing efforts. A systematic plan will ensure that nothing is overlooked and no pertinent evidence is lost in the course of the subsequent investigation.

As stated in Chapter 1, all crime scenes are different but there are guidelines that exist in all cases that serve as a framework for the processing efforts. The general crime scene processing structure is as follows:

- Initial Scene Assessment
- Search for and Recognition of Physical Evidence
- Documentation of Physical Evidence
- Collection of Physical Evidence
- Packaging and Preservation of Physical Evidence
- Crime Scene Reconstruction

These are guidelines for the overview of efforts involved with the processing of a crime scene. However, often these tasks are not separate from one another but may overlap. This will be addressed as the chapter unfolds. In any case, investigative and processing efforts should start in the least intrusive and destructive manners and progress to the most intrusive and destructive. Processing the scene this way will ensure evidence integrity for as long as possible. Chapters 4 and 5 covered the first responder's duties at the crime scene and the initial assessment stages. This chapter concentrates on the processing activities following that initial assessment. The first phase is documentation.

Documenting the Crime Scene

Documentation efforts at the crime scene begin the moment that an officer gets a call and continue until the case is closed. This is often the most time-consuming but also the most important step in crime scene investigation. It is the purpose of crime scene documentation to record and preserve the location and relationship of discovered evidence as well as the condition of the crime scene as it was when the documenter was observing it. For the purposes of this text, there are four primary methods of documentation that are involved in CSI. These are:

1. Reports and note-taking (sometimes audio)
2. Photographs
3. Videography
4. Crime scene sketching and mapping

The end purpose of documentation should be the successful notation of all observations made within the scene of the crime, which will ensure the individual engaged in the documentation efforts will best be able to recall the events in the future. Importantly, this information may be presented in court. As Sherlock Holmes explains in *The Five*

Orange Pips, "The observer who has thoroughly understood one link in a series of incidents, should be able accurately to state all the other ones, both before and after" (Doyle, 1892, p. 81).

Each of these methods is an integral part of crime scene documentation. None is a substitute for another. While some of the methods might appear to be redundant, this serves to corroborate the other methods, ensures that nothing is overlooked, and all areas are accounted for. Notes and reports are not sufficient by themselves because they do not accurately portray the scene in detail the way photographs can. However, photographs are not sufficient by themselves, as they often need more explanation, which is the purpose of reports and notes. Sometimes notes are dictated into a tape or digital recording device, yet at some point are transcribed into a written format for court purposes. Here, therefore, notes and reports are defined as being both audio and written. While photographs are a good tool for documenting the visual aspect of a scene, nothing brings the scene to life as much as videotaping. However, video cannot be used in the same manner as photographs from a forensic analysis standpoint when documenting physical evidence.

Because each type of record has its place in documentation efforts, all must be considered and utilized when available and appropriate.

Crime Scene Photography

Entire texts have been written solely on this topic. Photographers are urged to seek out books and courses that will help them to continually refine their skills. This introduction comprises a succinct but thorough overview of the purpose and skills involved in crime scene photography.

The purpose of crime scene photography is to capture adequate images for the best possible documentation and reproduction of the reality present at the moment in time when the scene was photographed. Whether a person is using digital or film imaging equipment, accuracy is the key (Weiss, 2009). When attempting to shoot precisely, one must remember that photography is a mechanical means of retaining vision. When properly taken, a photograph is one of the only ways to capture an instant of time. However, the camera was never intended to replace vision, because it certainly cannot (Weiss, 2009). Crime scene photography is visual storytelling, and as such, the photographs should be a fair and accurate representation of the scene about which the story is being told.

Photographs are almost universally accepted by the courts and allowed into evidence irrespective of their image quality so long as the images contained within them are not inflammatory or prejudicial in nature (Weiss, 2009). Although it used to be necessary for a person to also be able to testify as to how a photo was developed or processed, now this is rarely the case, as the images themselves are not the evidence but rather what they represent.

Photographers often may attempt to create photographs of objects or scenes "as seen" by someone else. Undoubtedly this is an impossible undertaking, as no one can accurately document an item or moment as someone else saw it. Instead, it is an appropriate step to document the image or scene from the perspective of the viewer in approximately the same position, although not at the same moment in time (Weiss, 2009).

VIEW FROM AN EXPERT

Fair and Accurate

Photographic images of evidence have been presented in court for over a century. The admissibility of the images is decided by the judge. The judge may use the credibility and competency of the witness presenting the images, plus other important factors, as parameters for the decision. Opposition counsel has the right to challenge the images and to try to demonstrate that they are not, in fact, accurate representations of the evidence.

Photographs are usually allowed into court regardless of their quality. The actual photographer is rarely asked to testify about image accuracy, because the photographs are not considered as evidence but simply representations of the physical evidence.

Photographic images, whether captured on film or digital sensors, may be questioned regarding their degree of accuracy as representations of the imaged subject. In the courtroom, photographs must provide the best possible illustration of a very specific reality. In order to promote an industry-acceptable degree of quality and credibility in evidentiary photographs, professional organizations, including the Evidence Photographers International Council (EPIC), have published standards for evidence photographs.

A *standard* is an established norm or requirement. It is a formal, peer-accepted document that establishes uniform engineering or technical criteria, methods, processes and practices. All standards for evidence photographs define *admissibility* essentially as a matter of a "fair and accurate representation of the subject portrayed."

How are photographers and the court to interpret *fair and accurate*? *Fair* is a relative term. The judge is tasked with making the determination of what is fair, and often this call will be based on the credibility of the witness. Even poor photographic quality will not necessarily cause an image to be inadmissible if the judge believes the image is fair and relevant to the proceedings. It should be noted that there is not a standard definition or set of parameters for the term "*accurate*" Judge John Panos, a state-court judge in DeKalb County, Georgia, stated, "I would like to see a standard definition of 'accurate' made and published. This can then be referred to as the standard of the industry." Why is this definition necessary? The broad definition of accurate includes some of the subtler and more technical aspects of photography, including, perspective, angle of view, and dimensionality. It is conceivable that any photographic image could be questioned.

Terms such as color management, dynamic range, resolution, perspective, angle of view, or dimensionality may not be fully understood by the professional photographer or the attorney, let alone the juror. How many people can properly explain the difference between vision and perception, and articulate how this correlates to the accuracy of a photographic representation? It cannot be taken for granted that anyone in the courtroom understands photography on that level.

Photographers may attempt to create photographs of objects or scenes "as seen" by someone at the moment in question. Of all the purposes or goals that apply to photography, probably the most impossible is to create an image of anything exactly as another person would have seen it. It is, however, possible and much easier to make images and then use those images to help explain how it looked to you.

So in the real world, the definition of fair and accurate might be what the image is intended to show. In most cases, in order for an image to be a fair and accurate representation, it should show the questioned area or object in its most natural state. For example, if an attorney wishes to show the approximate physical or general area of involvement, then the judge may not be too strict in interpreting the term fair and accurate. In this case, a photograph of the scene would suffice for the purpose of identifying a location. On the other hand, if the primary purpose of the image is to illustrate exact details of a scene or object, such as the measured distance between two objects or the details of a latent print, then determining whether the image is fair and accurate will require much closer scrutiny.

If an attorney wants to show the exact distance between a crosswalk and a traffic signal, photographic experience and expertise becomes very important. If that photo, submitted as evidence was created by someone lacking a high level of experience and expertise, a judge may not allow a witness to testify that, based on the photograph, the distance between the crosswalk and the traffic signal is 25 feet.

Fair and accurate may amount to different things at different trials. It will always be a measure of the competence and credibility of the person presenting the photographs, rather than the sophistication of the camera and equipment used to create the images.

Sanford Weiss, EPIC Crime Scene Photographer
Author, *Forensic Photography: The Importance of Accuracy*

Photographic Ranges and Perspectives

In keeping with the storytelling theme, the first photos taken at a scene should not be of gore or an item of physical evidence. Instead, they should be of the overall crime scene. It should set the stage for the beginning of the story. As such, there are three important ranges of photographs that are taken at the scene of a crime: overall photographs, midrange/evidence-establishing photographs, and close-up/comparison/examination photographs.

Also, it is important to remember to take a photograph of a photo placard as the first photo taken at the crime scene. A **photo placard** is a handwritten or agency-developed sheet (**Figure 6.1**) that lists pertinent case information for the photographs to follow. Taking a photo of this as the first photo on a roll of film or as the first digital photo of a case will ensure that personnel are familiar with which photographs pertain to which case, and the name of the photographer. Only one case should be photographed on a roll of film; however, with today's digital media, often several (if not many) cases are photographed on a single digital media card prior to downloading onto a computer. Photographing a photo placard will serve as a separator between the cases, so that case photos will not become commingled.

Overall Photographs (Figure 6.2)

Overall photographs are exposed with a wide angle lens or in such a fashion that allows the viewer to see a large area in the scene at eye-level. Their function is to document the condition and layout of the scene as found. They help eliminate issues of subsequent contamination (e.g.,

Cityville Police Department

Case #: ____________________

Date: ____________________

Location: ____________________

Photographer Name: ____________________

Photographer ID: ____________________

Roll # (if applicable) ____________________

Figure 6.1 Example of a Photo Placard.

Figure 6.2 Example of an Overall Photograph.

tracked blood, movement of items). Typically these are shot from the four corners of the crime scene. If indoors, usually they are taken from the corners of the room, shooting towards the center. If outdoors, they are often shot from the direction of a cardinal heading (North, South, East, and West). These four photographs most likely will capture the entire scene. If not, then additional photographs from an appropriate vantage point can be taken. These overall photographs set the scene and should include street signs and addresses if possible. Also, it may be necessary to not only take overall photos facing the building or scene in question, but also overall photos facing away from the scene to show the surrounding area.

Midrange/Evidence-Establishing Photographs (Figure 6.3)

The function of **midrange photographs** is to frame the item of evidence with an easily recognized landmark. This visually establishes the position of the evidence in the scene, with its relationship to the item's surroundings. These types of photographs are the most overlooked in crime scene work. They are taken of the evidence prior to movement or manipulation and should never include a scale of reference in the photo. The evidence-establishing photograph is not intended to show details, but simply to frame the item with a known landmark in the scene. The close-up and the evidence-establishing photograph go hand-in-hand.

Figure 6.3 Example of a Midrange Photograph.

Close-Up/Comparison/Examination Photographs (Figure 6.4)

The function of **close-up photographs** (also called *comparison*, *examination*, or *macro* photographs) is to allow the viewer to see all evident detail on the item of evidence. This photo should be close and fill the frame with the evidence itself. They are taken with and without a scale. It is extremely important that photographs of this type are first taken without a scale of reference, and then with a scale of reference. The first photo shows the scene prior to contamination or manipulation by the photographer or crime scene personnel. The second includes a scale of reference with which the viewer is able to gauge size of the item presented within the photograph. This scale will allow for a 1:1 ratio reproduction of the photograph (i.e., 1 inch equals 1 inch). Failure to photograph the close-up without a scale prior to incorporating a scale in the photo could result in the photo being inadmissible because of the allegation of scene tampering.

The preceding photographic ranges are used anytime there is an item of evidence that is important and will have a bearing on the investigation. While there might be a variety of perspectives photographed, any photograph taken at a crime scene will fall under one of the preceding ranges. For instance, photographs taken from

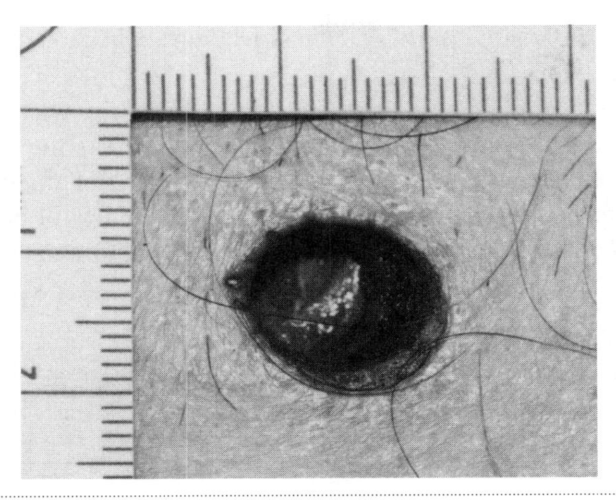

Figure 6.4 Example of a Close-Up/Comparison/Examination Photograph.

the reported position of a witness would fall into the overall range category. Those taken to show the address of a residence would fall into the midrange category if they showed more than simply the numbers/letters and included the façade of the house or entry to the home. However, if it was only of the letters/numbers this photograph would fall into the close-up range.

Photo Log

Regardless of the perspective or range taken, each photograph taken at a crime scene should be documented on a **photo log**. A photo log is a permanent record of all information pertaining to documentation by photographs. Department policy often dictates what is found within a photo log; however, if no policy exists, the following suggestions are offered (**Figure 6.5**). Information that should be included in a photo log includes:

- Title and information block consisting of: Date/Time/Case Number/Agency Name
- Photo equipment used
- Numerical ordering of each photo taken
- Brief description of each photo taken
- Direction facing for each photo taken

- Approximate distance from subject matter in each photo taken
- Shutter speed, aperture setting, and ISO for each photo. If photographed with conventional photography, then pertinent photographic information should be included for each. If photographs are taken in a digital format, documenting such information is not as imperative, because it will be digitally recorded when each photo is taken as part of the digital file for each photo.

Photograph List Case #

Code Section and Description	Month	Day	Year	Page of
Location of Incident	City			Time
Victim's Name	DOB		CDL	
Photographer/ID #	Scribe/ID #			
Camera, Lenses, and Flash Used				
Total Number of Rolls	Processing Log ID #	Film Type(s)		

#	Description of Photo	Polar Filter (Y or N)	Tripod (Y or N)	Lens Used (if zoom, length set on)	Flash (yes/no & normal, bounce, or off camera)	Direction Facing	F-Stop	SS	Distance from subject
1									
2									
3									
4									
5									
6									
7									
8									
9									
10									
11									
12									
13									
14									
15									
16									
17									
18									
19									
20									

Figure 6.5 Example of a Photo Log.

Order of Taking Photographs

While this manner of documentation is listed near the beginning, obviously taking overall photos is much less intrusive to a crime scene than taking close-up photos (due to movement of items and the addition of scales of reference). It therefore is important that you realize that although these are listed together, not all ranges of photographs are taken together or at the same time during a scene investigation. After the initial scene survey has been conducted, but before a detailed search or examination is undertaken, the crime scene should be photographed. However, usually this only includes the overall photographs, but if items of evidence have been located, then mid-ranges can be taken from a safe position. Close-ups are not typically taken until a thorough search of the scene has been conducted, unless the item is of a transient nature.

As the reader can see, crime scene processing must be very methodical and follow specific guidelines.

Guidelines for Crime Scene Photography

The following strategies have proven useful in crime scene investigations.

- Always use a photo placard on the first shot of each roll to demonstrate administrative data (see Figure 6.1).
- Always use a crime scene photo log (see Figure 6.5).
- Document the entire scene in situ as soon as possible using overall photographs.
- Photograph all fragile evidence as soon as possible.
- In the documentation stage, photograph all known evidence using evidence establishing and evidence close-up photos.
- As items are discovered in later stages, return and document them fully, including additional overall photographs if needed.
- Create photographs that fully demonstrate the results of additional examinations (e.g., latent prints, bloodstain pattern analysis, trajectory analysis).
- Try not to include the photographer or other people in the photographs, if possible.
- Shoot all close-up photographs with the use of a tripod.
- Close-up photos should be taken with and without a scale of reference.
- Be sure that the scale is on the same plane as the item of evidence being photographed.

- The subject matter should be parallel to the film plane/camera to eliminate distortion caused by skewed angle photographs.
- If in doubt, photograph it!

Videography

As a result of digital media gaining widespread acceptance within U.S. Courts, in the last few years videography has become a routine method of documenting major crime scenes. While this is an obvious and useful method of providing visual documentation of the conditions and items encountered at the crime scene, it must be remembered that doing so is not a substitute for still photography. Each has its merit.

Video is taken to record the scene in as close to its original condition as possible, as this is an easy method to employ and is relatively quick in its application. Oftentimes, video is shot while conducting the initial walk-thru as a way of recording the layout and conditions of the scene. This documentation is useful to supervisors and investigative personnel in determining logistic and equipment needs as well as reducing official visitors by giving them the opportunity to look at the crime scene without actually entering into it themselves. It also enables investigative personnel to later "enter" the scene as often as necessary through viewing the video without the need for a search warrant. This is especially useful if the crime scene is no longer available to personnel.

Videography is a useful method for documenting a crime scene. It can provide a perspective that is more easily understood and perceived by the viewer than those offered by notes, sketches, or still photographs. However, it must be remembered that this is a supplemental method and not a replacement for still photography or other documentation methods.

Guidelines for Videotaping a Crime Scene

While some of these points are similar to those for photography, a few key points are important to remember when shooting a moving data stream.

- Begin with introductory placard that states case number, date, time, location, and other pertinent case and chain of custody information.
- This video should be a storytelling event. Start with a general view of the area surrounding the crime scene. Following this should be an overview of the crime scene itself. It is suggested to take overalls from the cardinal compass directions (North, South, East, West) for orientation purposes.

- Turn off the audio on the video recorder unless you intend to narrate.
- Do not move the camera too quickly by panning (moving side-to-side), or zooming (moving in for a close-up view) as this results in abrupt motion and bad focus.
- Unless in sunlight, always use a video strobe. Never use a flashlight to illuminate the scene.
- Do not use the zoom unless it is necessary because of an inability to get physically closer to the subject matter, or if it is unsafe to do so. The human eye cannot zoom. If the video is to be a fair and accurate representation of how the videographer observed the scene, no zoom should be used.
- Video never should be edited or altered in any manner following the initial taping. The original copy should be kept as evidence, and duplicate copies should be made for viewing purposes.

Documentation/Reports

There is an adage in police work that "if it's not written down, it didn't happen." To a large extent, this is true. It is important that each step of the process and every action taken be documented extensively by using notes, photographs, sketches, and reports. The written notes begin with the first responder and continue throughout the investigative process. At each step, those individuals involved in the process are responsible for documenting all observations that they made, and all actions they performed. This includes documentation of efforts that resulted in negative findings as well. An example of a negative finding is a search for latent fingerprints that yielded nothing.

Each department typically has its own format and requirements for various levels of documentation within the investigative process. At the very basic level, written documentation consists of:

- Notification information
- Arrival information
- Scene description
- Victim description
- Crime scene team

Essentially there are two types of written documentation. The first are notes. *Notes* are brief, often in a bulletpoint format, documentation of efforts, observations, and actions. Notes are taken at the time of the incident and are informal. The second type of written documentation is a report.

Reports can be either fill-in-the-blank forms that are utilized to record pertinent information relating to a case or they can be of narratives. These are formal and are typically unique to a particular department and specific to a certain type of scene or case. Narrative reports are formally written, usually in the first person, active voice, and past tense. They document all actions taken by the report's author, and all observations he or she made.

Sketching and Mapping the Scene

Sketching

A **crime scene sketch** is a permanent record of the size and distance relationship of the crime scene and the physical evidence within it. The sketch serves to clarify the special information present within the photographs and video documentation, because the other methods do not allow the viewer to easily gauge distances and dimensions. A sketch is the most simplistic manner in which to present crime scene layout and measurements. Often photographer/camera positions may be noted within a sketch also.

Why is a sketch important to crime scene documentation?

- It accurately portrays the physical facts.
- It relates to the sequence of events at the scene.
- It establishes the precise location and relationship of objects and evidence at the scene.
- It helps to create a mental picture of the scene for those not present.
- It is a permanent record of the scene.
- It usually is admissible in court.
- It assists in interviewing and interrogating.
- It assists in preparing the written investigative report.
- It assists in presenting the case in court. Well-prepared sketches and drawings help judges, juries, witnesses, and others to visualize the crime scene.

When should sketches be made?

- Sketch all serious crimes and accident scenes after photographs have been taken and before anything is moved.
- Sketch the entire scene, the objects, and the evidence.

Two types of sketches are produced with regards to crime scene documentation: rough sketches, and final/finished sketches. **Rough sketches** (**Figure 6.6**) are developed while on-scene, typically during the crime scene assessment/preliminary scene evaluation phase to assist with

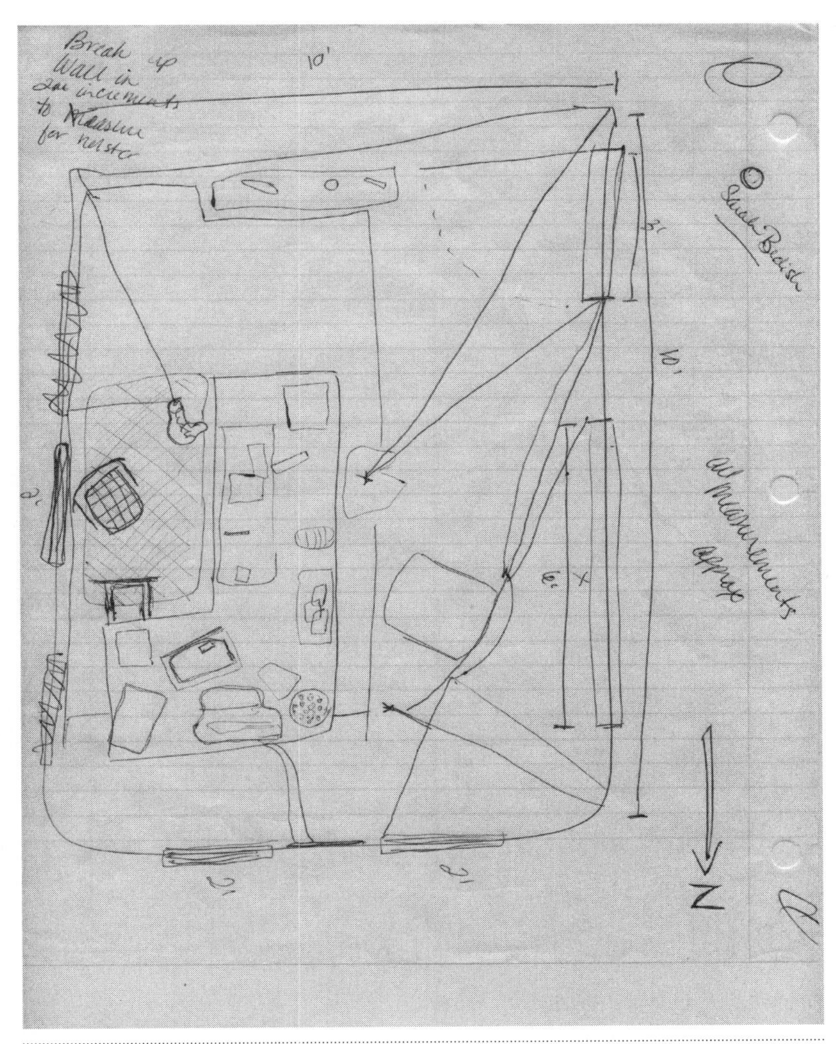

Figure 6.6 Example of a Rough Sketch.

Source: Courtesy of Sarah Bedish, University of Wisconsin–Platteville.

development of a strategic plan for processing. The sketch is not done to scale, can be drawn with any implement (crayon, chalk, pencil, pen, etc.), and is very rough artistically. As work progresses at the crime scene, the sketch will include not only the crude crime scene layout, but also will be used to record measurements of items and structures, and distances between items.

A **final sketch** (**Figures 6.7** and **6.8**) is a finished rendition of the rough sketch. They are usually prepared for courtroom presentation

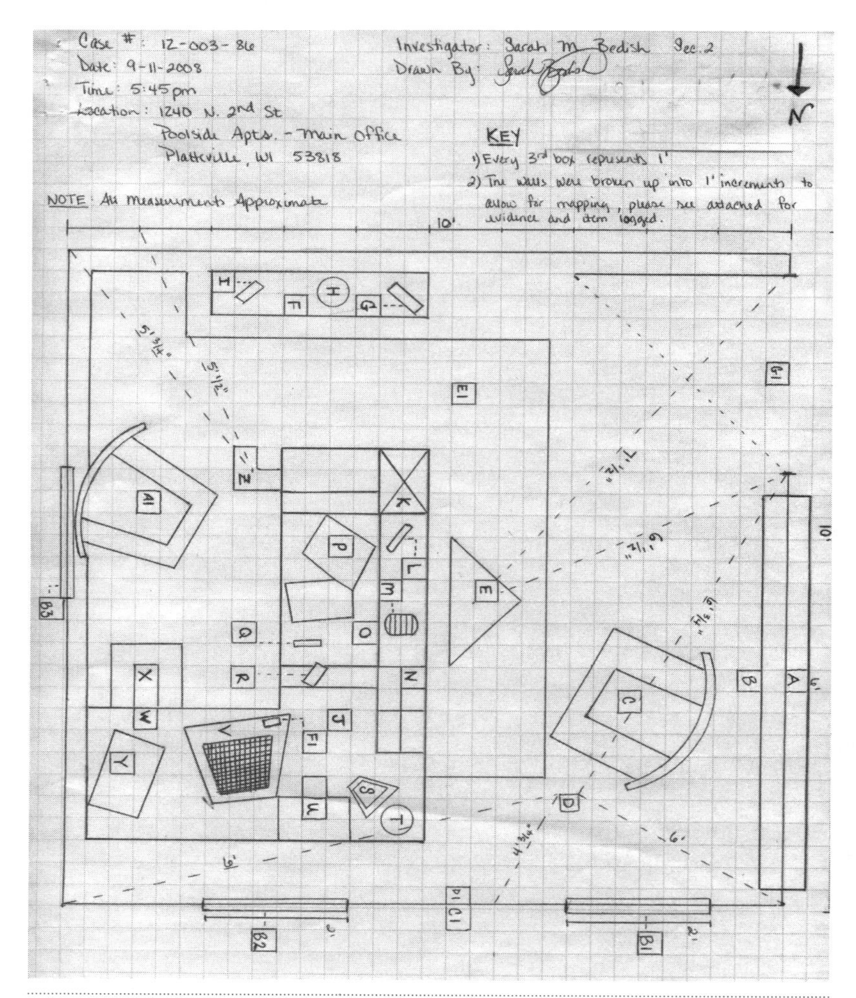

Figure 6.7 Example of a Final Hand-Drawn Sketch.

Source: Courtesy of Sarah Bedish, University of Wisconsin–Platteville.

and often will not show all measurements and distances originally recorded on the rough sketch. Only significant items and structures are typically present within a final sketch. A final sketch is produced in either ink or on a computer, in a manner that is not able to be modified (i.e., not in pencil!). The sketch should be clutter-free and should accurately depict all pertinent items of evidence, typically through the use of an accompanying legend. A **legend** is a note of explanation, outside of the sketch area, which relates to a specific item, symbol, or

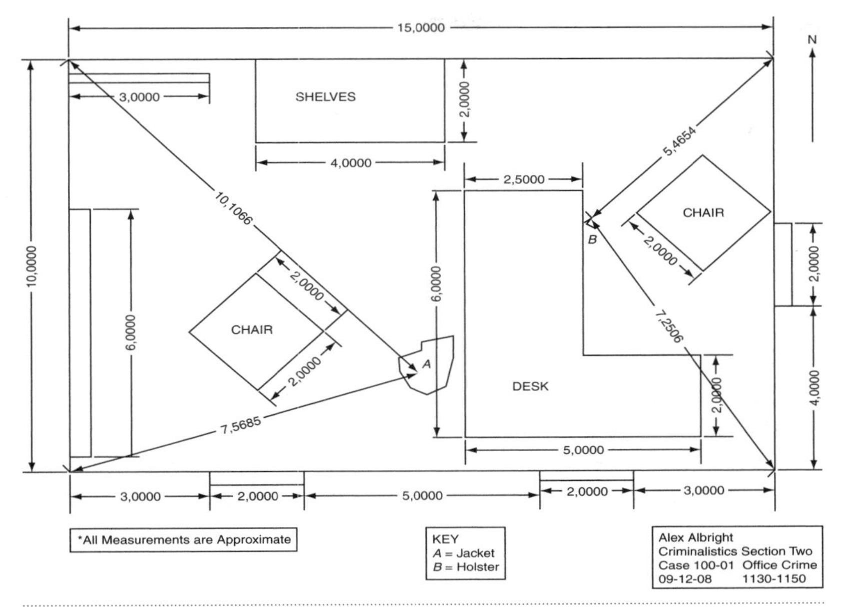

Figure 6.8 Example of a Final Computer-Generated Sketch.

Source: Courtesy of Alex Albright, University of Wisconsin–Platteville.

information contained within the graphical representation of a sketch. A final sketch should include:

- Title (What does the sketch represent? For example, Sketch of Bank ABC Robbery).
- Legend (What do symbols within the sketch mean?).
- Case Information. (i.e., date, time, place, case number).
- Initials/Name (person who drew the sketch).
- Indication of direction (e.g., North).
- Scale (e.g.: 1″ = 1′).
- Measurement table (If measurements are not represented within the confines of the sketch, an accompanying measurement table should be included to explain the distances and measurements associated with it.).
- There should be a notation following the scale or measurement table stating: "All Measurements are Approximate." This will ensure that the sketch's author does not get into a credibility argument in court that a measurement is documented as the listed measurement, but could in fact be greater or lesser due to rounding errors or other factors.

Four different crime scene perspectives can be represented within a sketch: (a) the bird's eye or overhead view (**Figure 6.9**), (b) the elevation or side view (**Figure 6.10**), and (c) the three-dimensional (3D) view (**Figure 6.11**). Sometimes personnel choose to incorporate several perspectives within a sketch (e.g., using both elevation and overhead sketches to draw an exploded or cross-sectional view of a scene; **Figure 6.12**).

An overhead or bird's eye view is the most common form of crime scene sketching. It is prepared with the perspective being as though the author was looking down upon the scene from above. This type shows the floor layout but cannot represent heights of items or show associated evidence on walls. In order to show such information, a person must sketch an elevation or side view sketch to show evidence located on a building façade, interior wall, or any item of which height is an important aspect (e.g., death involving a hanging). A 3D crime scene perspective is created with the aid of computers, and has its primary function as being

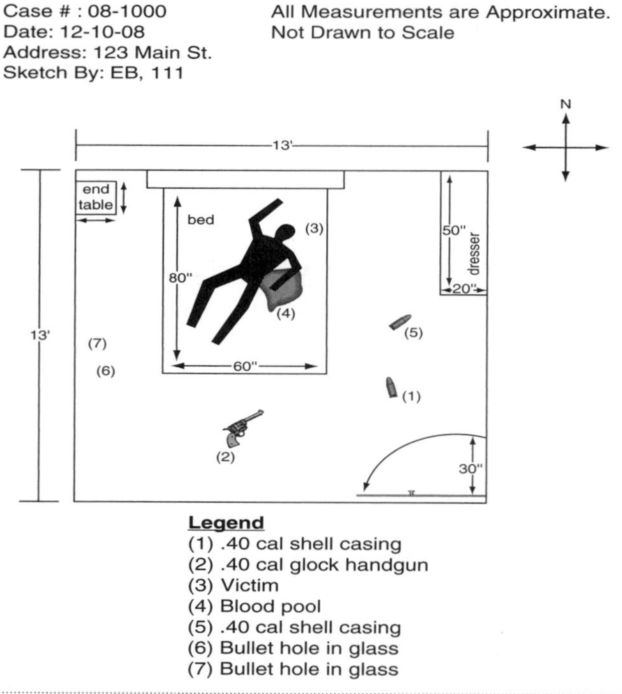

Figure 6.9 Example of an Overhead/Bird's Eye View Sketch.

Source: Courtesy of Ellie Bruchez, University of Wisconsin–Platteville.

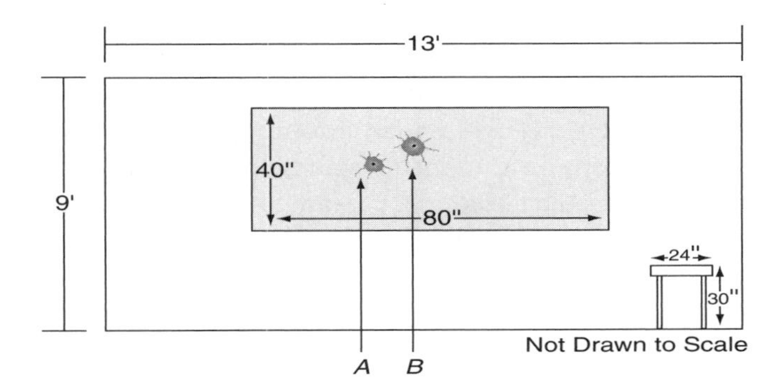

Figure 6.10 Example of an Elevation/Side View Sketch.

Source: Courtesy of Ellie Bruchez, University of Wisconsin–Platteville.

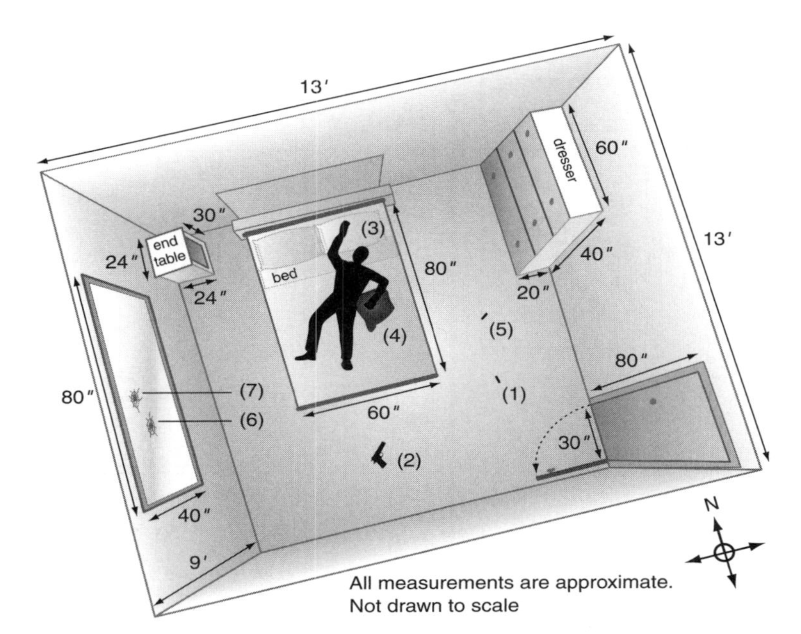

Figure 6.11 Example of a Three-Dimensional Crime Scene Sketch.

crime scene activity reconstruction, to help explain what happened and in what order.

Crime Scene Mapping

Mapping is the term associated with crime scene measurements. Sometimes a person may sketch but not map, meaning that he or she draws a sketch of an area but does not apply measurements to the sketch produced and items represented. Rarely, however, will one map without sketching (i.e., record measurements with no graphical representation for what the measurements represent). Sometimes this step is referred to as *measuring*. There are a variety of methods for mapping a crime scene, depending upon whether the crime scene is an interior or exterior scene. As this is an introductory text, only the most basic and most often used methods are covered here. The basic types of mapping methods utilized

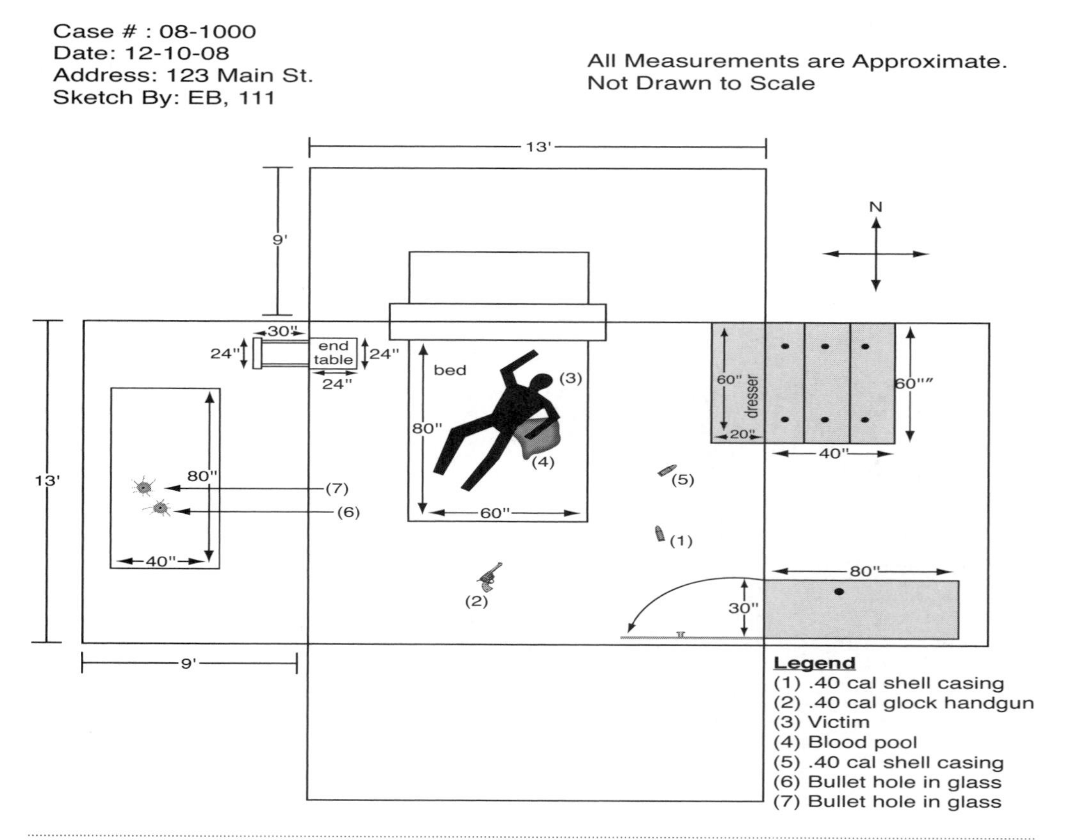

Figure 6.12 Example of an Exploded/Cross-Sectional Sketch.

Source: Courtesy of Ellie Bruchez, University of Wisconsin–Platteville.

for crime scene sketching and mapping are: (a) baseline, (b) rectangular coordinates, (c) triangulation, and (d) polar/grid coordinates.

Baseline Mapping

This is the most basic—and least accurate—form of crime scene mapping. For this method, a baseline is developed or identified from which to conduct measurements. This can be an existing area, such as the edge of a roadway, a wall, fence, etc., or it can be developed by personnel, such as by placing a string or tape measure through the scene and conducting measurements from there. In the case of the latter, the line should be run between two known fixed points, such as trees or other identifiable points, so that the points could be found in the future and the scene reconstructed if necessary (**Figure 6.13**). Once the baseline is established, measurements are taken from the baseline at an approximate 90 degree angle from the baseline to a point on the identified item or area of the crime scene. Typically, most measurements are made either to center mass of the item or to the nearest point of the item to the baseline. Because it is impossible to ensure that the measurement was taken at 90 degrees, the possibility exists that the measurement will be longer if the measurement was over 90 degrees from the baseline, or if it was less than 90 degrees from the baseline. For this reason, this method is not as accurate as some of the other methods; however, it is quick and extremely easy to use.

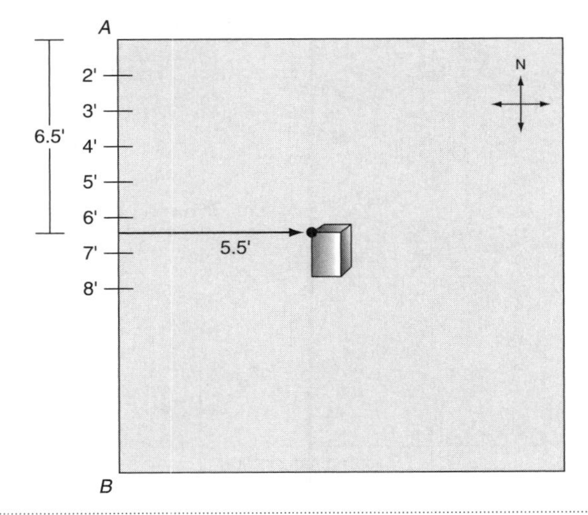

Figure 6.13 Example of a Baseline Map.

Source: Courtesy of Ellie Bruchez, University of Wisconsin–Platteville.

Rectangular Coordinate Mapping

The rectangular coordinate mapping method is a slightly more accurate variation of the baseline method because it utilizes two such baselines instead of one. Two measurements are taken to a point on an item or location at the scene. One from each identified baseline. Some personnel choose to measure to two or more points on an item, using multiple rectangular measurements as a way of increasing accuracy, while others simply choose to measure to an arbitrarily-identified center mass of the object in question or point to which the measurements are being taken. As with the baseline method, it cannot be determined that such measurements are taken precisely at 90 degree angles from the baseline, so there exists a greater possibility of errors than with some of the other methods. However, due to this method having two measurements, it has much greater accuracy than with the single line baseline method. This method is especially useful in confined spaces and smaller interior scenes (**Figure 6.14**).

Triangulation Mapping

This is the most accurate method that does not make use of advanced technology. While it is quite a bit more laborious and time-consuming, is sufficiently more accurate than the aforementioned methods of mapping to be worth the effort. The accuracy for this method comes in its foundation: two fixed points. From these two fixed points, measurements are taken to specified points on an item or within the crime

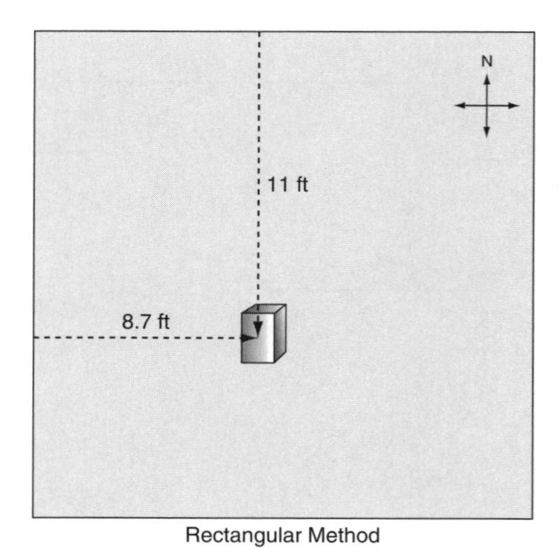

Figure 6.14 Example of a Rectangular Coordinate Map.

Source: Courtesy of Ellie Bruchez, University of Wisconsin–Platteville.

scene. There is no need to worry about whether or not measurements have been made at a right angle because the points derive from a known fixed point, such as the corner of a room, or edge of a door frame. From these fixed points, a minimum of two measurements are made to each identified point. If the object is of a fixed or constant shape (e.g., a firearm or item of furniture), then the object is measured to two points, from the two fixed points, for a total of four measurements. If the object is of a variable shape or size (e.g., a puddle of water, pool of blood, or pile of clothes), then the object is measured to an approximate center of mass (**Figure 6.15**).

Polar/Grid Coordinate Mapping

Utilizing polar coordinates is the third method of crime scene mapping used to document evidence location at a crime scene. Like those previously mentioned, this is a two-dimensional system that indicates the location of an object by providing the angle and distance from the fixed or known point. Obviously, in order to conduct measurements by this method a transit or compass is necessary to measure the angles and polar directions. This method is best utilized in large outdoor scenes with very few landmarks (e.g., a plane crash in forest or large field; (**Figure 6.16**).

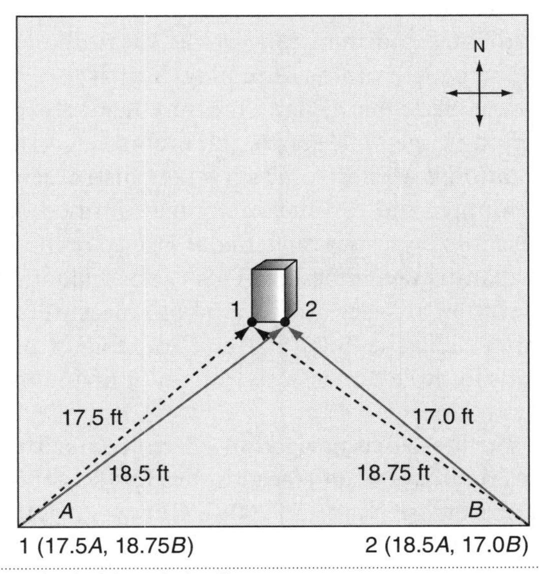

Figure 6.15 Example of a Triangulation Map.

Source: Courtesy of Ellie Bruchez, University of Wisconsin–Platteville.

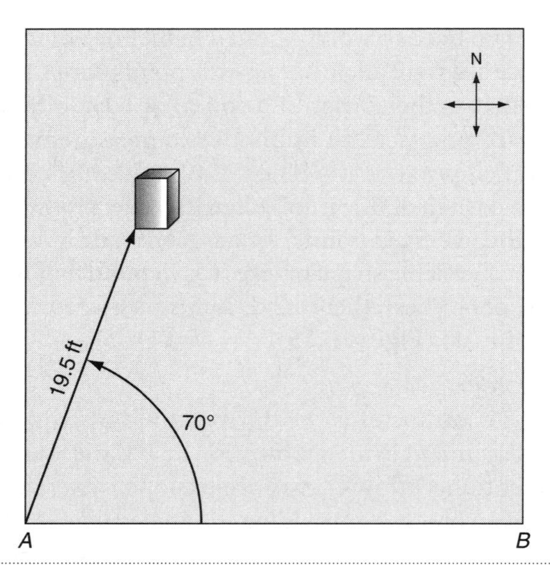

Figure 6.16 Example of a Polar/Grid Coordinate Map.

Source: Courtesy of Ellie Bruchez, University of Wisconsin–Platteville.

Advanced Mapping Techniques

Some departments may have the ability to make better utilization of modern technology, such as global positioning systems (GPS) and Total Stations, which are mapping systems that can take measurements in polar coordinates and then convert the measurements into grid coordinates. The benefit of this technology is that they both are able to provide precise electronic distance measurements and are extremely useful in mapping large-scale scenes and events.

A Total Station is an electronic surveying instrument that has an integrated computer and can measure angles in the horizontal and vertical planes, utilizing a laser rangefinder instead of the more archaic method of a manual tape measure. This is especially useful because changes in elevation are very difficult to both measure and depict on a crime scene sketch. The Total Station is capable of recording evidence positions in three dimensions, thus simplifying this otherwise complicated situation.

GPS is a satellite-based navigation system comprising a network of 24 satellites that have been placed in the Earth's orbit by the U.S. Department of Defense (Garmin, 2009). GPS was originally used by and intended for the military; however, in the 1980s the government made the technology available for civilian use. The benefit of GPS is that it works in any weather condition, anywhere in the world, 24 hours a day. There are no subscription fees or setup charges to utilize GPS. These satellites complete two very precise orbits of the Earth a day, during which they transmit signal information. It is these signals that GPS receivers gather and then use triangulation to calculate the user's location. A GPS receiver must be locked on to the position signal of at least three satellites in order to calculate a two-dimensional position (latitude and longitude) as well as track movements of an object. If the GPS receiver is able to lock onto four or more satellites, the receiver can determine the user's three-dimensional location (latitude, longitude, and altitude), along with object movement. The more satellites that the GPS is locked onto, the greater the accuracy of the position. Once the user's position has been determined, an additional service is that calculation of movement can provide GPS users the ability to record information such as speed, bearing, track, trip distance, distance to destination, sunrise, sunset, time, and many more possibilities (Garmin, 2009).

How accurate is GPS? In most cases, commercially-available GPS receivers are accurate to approximately 12 meters, with higher end units capable of accuracy in the 3- to 5-meter range. This is sufficiently accurate for large scenes that have no known/fixed landmarks. A GPS reading is typically used to "mark" a known point and then measurements are made from that location, thereby ensuring that any measurements taken will all be "off" by the same amount because they all originate from the same location.

As with all other crime scene measurements, all measurements are approximate, and are never documented as or testified to as being 100% accurate. Crime scene mapping is about doing the best possible documentation with the resources available, realizing that rounding and other factors inhibit the ability to be completely accurate.

Searching the Crime Scene

A variety of factors can affect a search method and these will determine the best, most accurate way to approach the scene.

Environment

Environmental conditions such as wind, rain, snow, heat, cold, etc., will have an impact on the method chosen due to how it affects the scene and the personnel involved.

Object Being Searched For

Obviously, a larger item will not entail the same level of searching detail as would a smaller item (e.g., a handgun versus a bullet).

Number of Available Personnel

Some search methods are designed to incorporate a greater number of searchers in order to be most effective. If such personnel are not available, a method that utilizes fewer personnel needs to be considered.

Terrain

Obstructions (trees, buildings), ground cover (asphalt, grass), and grade (steep, flat) will all impact the type of method employed, as they will have a bearing on the ability of searchers to perform the task, and the ability to properly locate the necessary items of evidence.

Exigency

In cases of lost children, search for a loaded handgun (public safety issue), and other events, often there is the need for exigency that trumps the more detailed search patterns that would be preferable. Therefore, a quick and efficient method should be chosen, making use of the maximum number of resources available in the quickest manner possible.

Swath Size

A **swath** is the effective area that a searcher can cover while conducting a search. Swath is affected by all of the aforementioned matters and is itself a consideration in the determination of a proper search method to employ. If looking for a firearm, a larger swath would be possible in a parking lot than in high grass for instance. Also, a search conducted at night or in low light would have an impacted swath due to the ability of a flashlight to illuminate the area of responsibility.

Types of Crime Scene Search Patterns

Depending on the aforementioned factors, a variety of crime scene search patterns exist that can be employed at a crime scene. Regardless of the search pattern chosen, the crime scene investigator must be sure that the search is conducted in a systematic and thorough manner. This will ensure that all evidence is properly located, documented, and collected.

Lane/Strip Search

This type of search pattern breaks the scene up into manageable lanes in which the searcher(s) proceed back and forth, in a slightly overlapping fashion. This is similar to mowing one's lawn. This method is typically conducted by only one person (**Figure 6.17**).

Line Search

This method is incorporated when there is a large number of personnel available, often volunteers. In this method, searchers assemble in a line

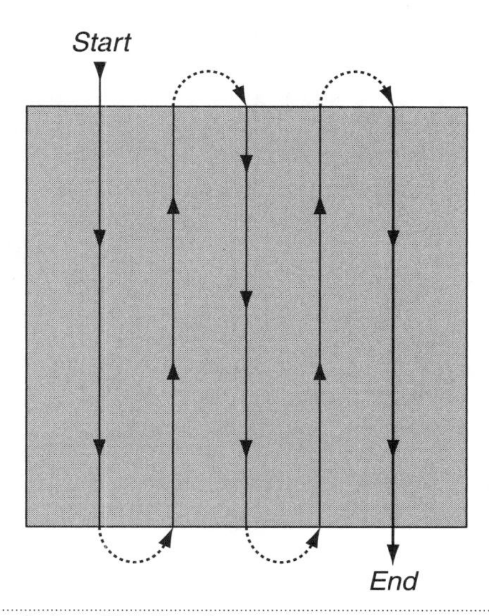

Figure 6.17 Example of a Lane/Strip Search.

Source: Courtesy of Ellie Bruchez, University of Wisconsin–Platteville.

that runs along a chosen edge of the crime scene. Searchers stand side by side and spread apart, incorporating a manageable swath distance between each person. A search coordinator should place her or himself in the middle of this line to make certain that everyone walks forward in as straight as possible a line. If one end begins to lag, then the other end is requested to slow up. At no point should anyone be encouraged to search faster! Keeping all searchers in a straight line reduces the possibility of missing an area and thus not discovering potential evidence. This method is the most commonly employed type during an exigent search for an item or person, especially when a large number of people are available (**Figure 6.18**).

Grid Search

This sometimes is referred to as a *double strip* or *double lane* method. In this method, a lane is searched in one direction, similar to the line search method. However, at the lane's terminate, a 90-degree direction change is made and another lane is searched. This can either occur through the use of two searchers (one responsible for one direction, and the other for the perpendicular direction), or else it can utilize a large number of searchers incorporating the line method as described earlier, and then turning 90-degrees and performing a second line search perpendicular to the original lane. While quite time-consuming, this method allows the same area to be searched two separate times,

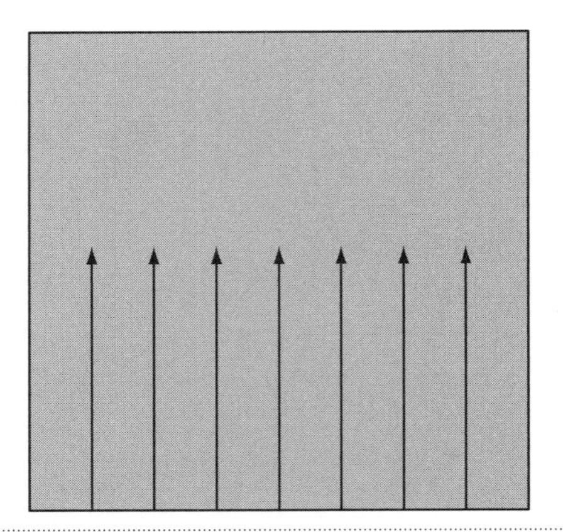

Figure 6.18 Example of a Line Search.

Source: Courtesy of Ellie Bruchez, University of Wisconsin–Platteville.

and at different angles. This redundancy will reduce searcher boredom, and will change the lighting and obstruction conditions present, thus increasing the ability of the searcher to locate evidence (**Figure 6.19**).

Zone Search

This method is typically utilized in an area that is already broken up into defined or manageable zones (e.g., a house or car). It is typically used indoors, but may be used outdoors if the areas are broken down into defined zones. Zones can be searched independently and later re-searched by different search personnel to ensure that no evidence has been overlooked. This method also can be used as a way to break up a larger crime scene, so the search coordinator then can choose from any of the search methods to cover a zone area (**Figure 6.20**).

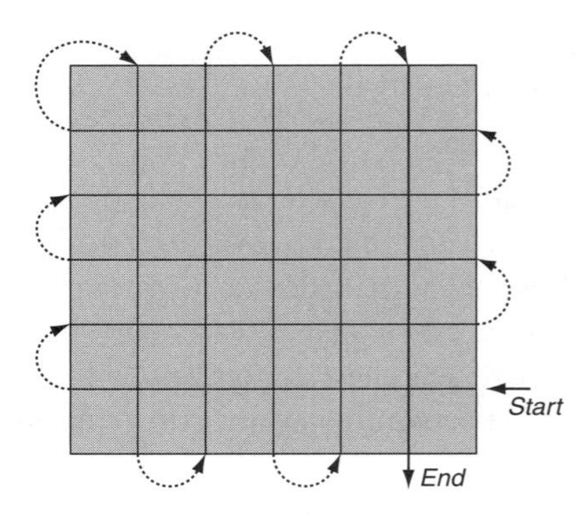

Figure 6.19 Example of a Grid Search.

Source: Courtesy of Ellie Bruchez, University of Wisconsin–Platteville.

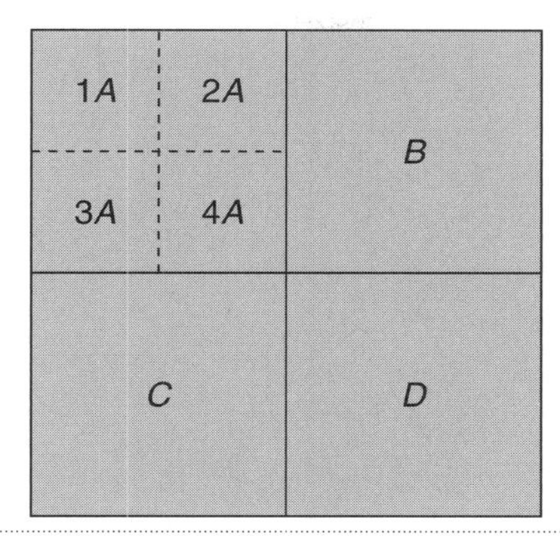

Figure 6.20 Example of a Zone/Quadrant Search.

Source: Courtesy of Ellie Bruchez, University of Wisconsin–Platteville.

Circle/Spiral Search

This is a very specialized search pattern method that is seldom utilized; however, it does have its usefulness and merit. In this method, searchers can either start at a defined outer boundary and circle or spiral in towards the defined critical point, or else they can begin at the critical point and circle or spiral outward towards the crime scene perimeter. Physical obstructions and barriers within the scene will present problems with this method. This method is typically employed in bomb or explosive scenes with a defined seat of explosion. It may be used in underwater or open water searches where there was a last known location for an item, vessel, or victim. If using a circling rather than a spiraling pattern, to ensure thoroughness, it is suggested that a central point and an effective swath width be determined. Once this is done, searchers should move out in concentric circles, often through the use of a lanyard affixed to a point at the center of the scene (especially true for underwater searches). The searcher proceeds to search in a 360-degree manner, around the central point, and once they reach the end of their circuit, they let out the lanyard a pre-determined amount, incorporating manageable swath width, and then proceed to conduct another 360-degree circuit of the scene. It is suggested that this new circuit be in the opposite direction of the previous circuit to both reduce the possibility of entanglement, and also to reduce the searcher's vertigo issues from walking in a continuous circle (**Figure 6.21**).

Important things to remember when conducting a search are:

- Do not touch, handle, or move evidence.
- Mark or designate found items without altering them.
- Found evidence must be documented before any evidence can be moved or collected.

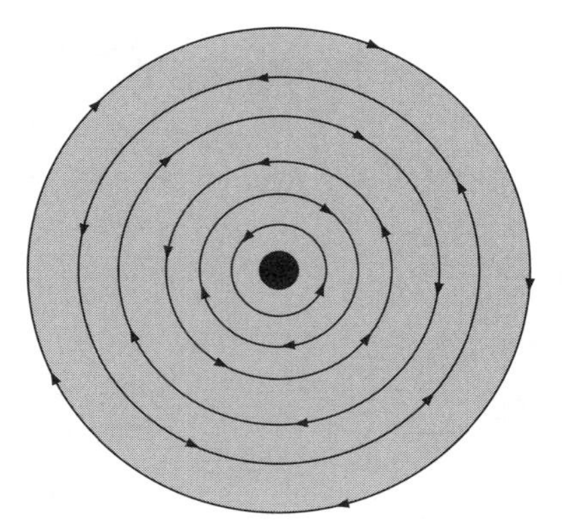

Figure 6.21 Example of a Circle/Spiral Search.

Source: Courtesy of Ellie Bruchez, University of Wisconsin–Platteville.

Collecting, Packaging, and Preserving Physical Evidence

After intensive crime scene search and documentation, collection and preservation of evidence should begin. The remainder of this text will be devoted to the different manners of collection and preservation specific to the type of evidence presented. At this point in the process, the following strategies should be done to ensure the most thorough and accurate investigation:

- One person should be designated as the evidence collector/custodian (this ensures that nothing is missed).
- Document, collect, package, mark, seal, and preserve.
- Transient, fragile, or easily lost evidence should be collected first.
- Paper is the preferred packaging.
- Package items separately.
- Containers should be properly marked.
- Containers should be properly sealed.
- Seals should be marked with initials and date/time.

Chapter Summary

Scientific crime scene investigation is the best methodology to ensure that an investigation is properly conducted and that justice is served. Use of this methodology will prevent the abrupt end of an incomplete investigation and allow for the best use of the physical evidence found at crime scenes. The general rule relating to crime scene documentation is "if it isn't written down, it didn't happen." This is important to remember when conducting the various steps of crime scene documentation. It reminds the individual to be as thorough and precise as possible to correctly retain and be able to recall the events, items, and locations involved with a crime scene.

Review Questions

1. ______________ is a note of explanation, outside of the sketch area, that helps to relate or give information on a specific item or area within a sketch.
2. ______________ is a drawing or graphical representation that forms a permanent record of the size and distance relationship of the crime scene.
3. ______________ is the term associated with crime scene measurements.
4. In order to ensure all photographs, information contained within them, and equipment that was used to capture them are properly documented, a CSI should make use of a ______________.
5. What factors can affect the choice of search method at a crime scene?
6. What is the difference between a strip/lane search pattern and a line search pattern?
7. When conducting a search of a crime scene, ______________ refers to the effective area that a searcher can cover.
8. Regardless of the search pattern chosen, the crime scene investigator must be sure that the search is conducted in a ______________ and ______________ manner.

References

Bellis, M. (n.d.) About.com: Inventors. History of the global positioning system – GPS. Retrieved July 23, 2009, from http://inventors.about.com/od/gstartinventions/a/gps.htm

Doyle, A. C. (1892). The Five Orange Pips. In The Adventures of Sherlock Holmes. *Strand Magazine.*

Doyle, A. C. (1892). A Scandal in Bohemia. In The Adventures of Sherlock Holmes. *Strand Magazine.*

Garmin. (2009) *What is GPS?* Retrieved July 23, 2009, from http://www8.garmin.com/aboutGPS/

Weiss, S. L. (2009). *Forensic photography: The importance of accuracy.* Upper Saddle River, NJ: Pearson Education Inc.

The Forensic Laboratory

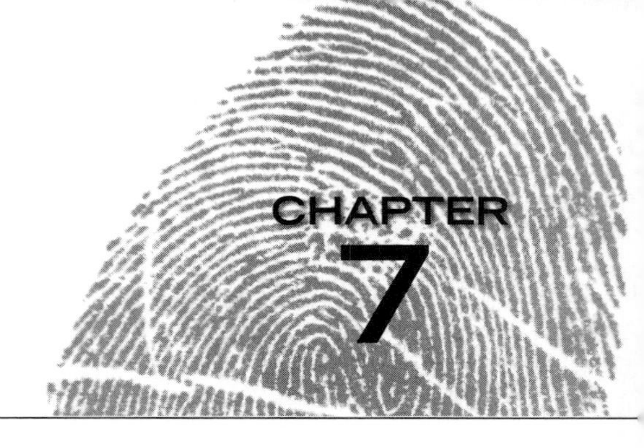

I am glad of all details, whether they seem to be relevant or not.

Sir Arthur Conan Doyle (1859–1930)
Sherlock Holmes, *The Adventure of the Copper Beeches*

LEARNING OBJECTIVES

- Identify the factors responsible for the growth of crime labs within the United States.
- Understand the functions of the modern forensic laboratory.
- Describe the various types of forensic laboratories.
- List and explain the roles of a criminalist.
- Explain the different levels of accreditation, why accreditation is necessary for a crime laboratory, and name the accreditation organizations involved with CSI.

KEY TERMS

Accreditation
Backlogged
Certification
Forensic Crime Laboratory
Known Evidence
Miranda Ruling
Proficiency Testing
Protocols
Quality Assurance (QA)
Quality Control
Questioned Evidence

History of the Forensic Laboratory

Before crime scene investigators or other police personnel can know what to collect and determine the forensic potential of the evidence, they must understand the abilities and inabilities of their local forensic laboratory. As discussed in Chapter 1, Edmond Locard was responsible for the development of what is considered to be the first forensic laboratory in the world. Locard's educational background was in medicine and law, the foundation of what would emerge as the science of criminal investigation. By 1910, he had persuaded the police department in Lyon, France to provide work space to analyze trace materials associated with crime scenes and criminal investigations.

It should come as no surprise that a society as dependent on science as ours would turn to that discipline to solve crimes. Forensic laboratories were first introduced as reform efforts to provide a reliable alternative to the vagaries of eyewitness testimony and the "third degree" abuses used in eliciting confessions. Nevertheless, their history is checkered with periods of rapid growth followed by years of neglect. Moreover, their place in the police organizational structure remains questionable at times, and their function in the adversarial court system is often at issue.

Experts disagree about which group initiated the first crime laboratory in the United States. Some credit August Vollmer for helping to establish a crime laboratory within the Los Angeles Police Department (LAPD) in 1923 (Saferstein, 2007) and others identify the Chicago Police Department as initiating the oldest crime laboratory in 1929 (FBI, 2008). Most of this controversy depends upon the definition of a "crime laboratory" at that time.

Vollmer, a police chief within the Los Angeles area, was assisted by Dr. Paul Kirk (who later established the Forensics program at University of California Berkeley) in developing the LAPD Crime Lab. However, during approximately the same time period, the Chicago police established their own laboratory. The Chicago Lab owes its roots to the infamous 1929 St. Valentine's Day Massacre, when five gangsters and two of their acquaintances were gunned down by assailants dressed in police uniforms and who fled the scene in a police squad car. The Massacre attracted national attention. As part of the investigative efforts, Colonel Calvin Goddard, who maintained an independent firearms laboratory in New York, was asked to analyze the crime scene bullets and cartridge cases. Goddard's findings subsequently excluded all police-issued guns as the murder weapons and months later matched the bullets to two machine guns seized from the home of Fred Burke, a suspect in the killings. As a result of these efforts, a member of the coroner's jury was so impressed with Goddard's work that he funded what became The Scientific Crime Laboratory of Chicago. This Laboratory, with Col. Goddard as its first director, was housed at the Law School of Northwestern University during the winter of 1929–1930; the city of Chicago took over its operations in 1938 (Muehlberger, 1955).

The oldest federal laboratory owes its beginnings to Director of the Federal Bureau of Investigation (FBI), J. Edgar Hoover. Intrigued with the idea of creating a place to incorporate the latest scientific methods into criminal investigations, Hoover sent an FBI agent, Charles Appel, to attend a training course offered at Goddard's lab in 1931. The Criminological Laboratory opened in 1932 and by the next year it was known as the Technical Laboratory. Ten years later it was renamed the FBI Laboratory and this name remains today. At its inception, the FBI Lab could provide firearms identification ("ballistics") and fingerprint examiners. "During its first month of service, the FBI Laboratory examiners handled

20 cases. In its first full year of operation, the volume increased to a total of 963 examinations. By the next year that figure had more than doubled" (FBI, 2008). Handwriting comparisons, the examination of various types of trace evidence (e.g., hairs, fibers, soils), and serological testing of blood and semen were added later (FBI, 2008).

Growth of the Modern Crime Laboratory

Currently there are over 390 publicly-funded crime laboratories within the United States (USDOJ, 2008). According to the U.S. Department of Justice (USDOJ), a **forensic crime laboratory** is defined as a scientific laboratory (with at least one full-time natural scientist) that examines physical evidence in criminal matters, and provides reports and opinion testimony with respect to such physical evidence in courts of law (USDOJ, 2008). These laboratories provide services for various levels of government: federal, state, county, and municipal. This number has dramatically increased since the mid 1960s, when there were approximately 100 operational U.S. labs (Girard, 2008) (**Figure 7.1**).

This tremendous growth can be attributed to three primary historical factors. One reason is a direct result of Supreme Court decisions, specifically the **Miranda ruling**. This ruling, coming out of the 1966 *Miranda v. Arizona* case, placed greater emphasis on the collection,

Figure 7.1 Example of a Modern Forensic Laboratory.

preservation, and analysis of physical evidence. According to this decision, individuals now have two Constitutional rights: police must advise individuals that they may have an attorney present during questioning, and they have the right to remain silent. This all but eliminated suspects' confessions and placed the emphasis on physical evidence relating to the crime in question.

A second factor can be traced to the explosive increase in crime rates across the United States since the 1960s. The introduction and proliferation of illicit drugs have been directly linked to the onslaught of crimes against property and persons. Because the crime of possession of a drug or controlled substance is the physical possession of the item—not an action that is done—every item or substance must be tested before the possessor can be charged with the offense. The ever-increasing numbers of drug samples to be tested are a burden for crime labs to handle efficiently and accurately. Only confirmatory testing of an illicit possession by the lab can be used as evidence in court.

The final reason for the dramatic growth in the number of crime labs is due to scientific advances and the CSI effect (see Chapter 2). As a result of public awareness and scientific as well as technological advances, there is renewed emphasis on the analysis of physical evidence. This has led to monies being spent to train individuals associated with such matters, and to design instrumentation and equipment to aid in such analysis.

Today, laboratories all over the United States struggle with the backlog of evidence as a result of the aforementioned reasons. A case is classified as **backlogged** if the analysis is not completed within 30 days of the item being submitted to the lab. As of the most recent USDOJ census, an estimated 359,000 cases were backlogged in the nation's publicly-funded crime labs. This is a 24% increase from the estimated 287,000 cases recorded as backlogged during the census taken at the end of 2002 (USDOJ, 2008). The areas encountering the highest levels of backlogging were DNA analysis (40%), and firearms and toolmark analysis (30%) (USDOJ, 2008).

Each forensic laboratory has its own organizational structure and each is responsible for its own jurisdictional region. Regardless of its structure or location, each lab has the underlying purpose of providing a fair and accurate analysis of the submitted item(s).

Types of Forensic Laboratories

Federal

While the FBI Laboratory in Quantico, Virginia is what most would consider the national forensic laboratory, in truth, the U.S. federal government does not have a single laboratory with unlimited jurisdiction. Instead, as evidence of continued advancement and emphasis on forensic matters, crime laboratories are located across the United States that are designed to process evidence from suspected federal violations. These include the following:

- *Department of the Treasury.* National laboratory located in Chicago, IL.
- *Bureau of Alcohol, Tobacco, Firearms, and Explosives (ATF).* National laboratory located in Ammendale, MD.
  - Regional Forensic Laboratories are located in Atlanta, GA, and San Francisco, CA.
- *United States Secret Service.* In 2003, the U.S. Secret Service was moved from the Treasury Department to the Department of Homeland Security (DHS), which has its national laboratory located in Washington, D.C.
- *Federal Bureau of Investigation.* National laboratory (Forensic Science Research and Training Center is also located here) in Quantico, VA.
  - Sixteen Regional Digital Forensic Laboratories throughout the United States.
- *Drug Enforcement Administration.* National laboratory located in Washington, D.C.
  - Regional laboratories are located in New York City; Largo, MD; Miami, FL; Chicago, IL; Dallas, TX; San Francisco, CA; and San Diego, CA. The specialized laboratories are in Sterling and Lorton, VA, both part of the metropolitan Washington, D.C. area. Subregional laboratories are in Kansas City, MO and San Juan, PR, with the mobile laboratory currently stationed at El Paso, TX.
- *U.S. Postal Service.* National laboratory located in Dulles, VA.
- *Department of the Interior.* National laboratory in Ashland, OR that provides forensic services in support of investigators who patrol the national parks.

State

Every state has developed its own crime laboratory system that serves both state and those local law enforcement agencies that lack such abilities within their locale. Some states have gone so far as to develop a specialized laboratory structure that minimizes duplication of services. These states include, California, New York, Illinois, Michigan, Texas, Florida, and Virginia) (Girard, 2008).

Local/Regional

Local laboratories are operated by municipal or county governments to serve primarily their locality. However, many have developed specialized sectors that can make such services available to outside agencies through inter-laboratory and inter-agency cooperation agreements. While many may think of municipal laboratories as being small, some are significantly larger than their state lab counterparts (the NYPD lab for instance).

As costs of associated services have risen, some jurisdictions have seen fit to combine resources and create regional laboratories. In these instances, several counties or municipalities have pooled their personnel, budgets, and equipment to service the most common forensic needs. This reduces overall expenses for each while continuing to maintain regional forensic capabilities without relying upon a state system that typically has a considerable wait time for forensic analyses.

Private

At this time, the majority of forensic labs within the United States are public; they are funded by taxes and operated by government agencies. However, the recent explosion in demand for DNA technology and analyses of everything from criminal investigations to paternity testing, combined with the subsequent backlog at law enforcement forensic labs has led to the creation of numerous private forensic labs throughout the United States. These private laboratories charge a fee for their services and relatively anyone with the financial ability can request the use of their services.

In fact, some corporations have developed their own forensic labs in an attempt to reduce inventory shrinkage and increase convictions of theft, fraud, and other retail crimes. Target is but one example of a corporation with forensic labs; locations are in Las Vegas and Minneapolis. There has been much debate about the jurisdiction, accreditation, and legal admissibility of evidence analyzed by private labs, as well as the training and certification of their employees. However, it should be pointed out that these are the same issues found in public labs.

Publicly funded forensic labs are also making use of the private sector in order to meet their demands for forensic services. In fact, according to the U.S. Department of Justice 2008 Bulletin from their Bureau of Justice Statistics section, a recently completed census of publicly funded crime labs reported that 54% utilized the services of private laboratories in order to aid with their backlog. Of these, almost 30% outsourced DNA casework. This was a 10% increase from an earlier conducted census taken in 2002 (USDOJ, 2008).

RIPPED FROM THE HEADLINES

CSI: Target

The $63 billion (revenue) retailer, Target Stores has entered the world of forensics. In 2003, in an effort to investigate safeness, theft, fraud, and other crimes involving its more than 1,600 stores, Target developed its own forensic lab in Minneapolis, Minnesota. In 2005, another followed in Las Vegas, Nevada, to help assist with the growing workload. Although the primary purpose of Target's forensic labs is to investigate occurrences within its stores, the investigators and analysts assigned to the labs typically spend only 70% of their time doing such. The remainder of their time is frequently spent assisting local, state, and federal law enforcement in investigating crimes unrelated to Target. In fact, since their inception, Target's labs have been instrumental in solving a number of felony criminal cases both in the United States and internationally.

Much of the equipment is similar to what would be found within a publicly funded crime lab. But, says Craig Thrane, a video analyst for the retailer, "Target investigators can work more quickly."

As with many other retail giants, Target installed cameras in most of its stores during the 1980s; however, they found that was not sufficient to significantly impact in-store thefts. "We had a volume of evidence from our cameras but no expertise," says Fredrick Lautenbach, the retailer's crime lab manager. Target did not want to burden already overloaded law enforcement with case after case of petty theft (which quickly adds up for the retailer but provides for numerous cases for law enforcement). In order to both combat the shrinkage of its inventory and not unduly burden law enforcement, Target created its first lab in Minnesota, hiring experts with prior law enforcement investigative and forensic experience to staff the lab. The labs are best known for their ability and technology associated with digital video analysis. Due to several successful analyses relating to law enforcement cases, Target has been overwhelmed with requests to restore or analyze tapes, track cell phones, and also to process and analyze fingerprints. Due to the tremendous amount of requests from outside agencies, Target decided to limit its volunteer work to cases that involve murder, sexual assault, or armed robbery. The work is considered voluntary, as part of Target Stores partnership with law enforcement. It does not charge for its services but instead asks law enforcement agencies to donate one of their agency patches when the work performed by Target assists them with an investigation. As of spring, 2008 Target had 136 patches on display in its corporate office, located in Brooklyn Park, Minnesota.

Source: Information retrieved from Egan, M. S. (April 21, 2008). CSI: Target. *Forbes Magazine*. Retrieved August 1, 2009, from http://www.forbes.com/forbes/2008/0421/102.html.

Private laboratories have another vital function within the legal system. They are able to provide independent analyses of forensic evidence for those accused of crimes. Typically public laboratories will not provide services for individuals, but rather only for government agencies or departments. However, most private laboratories are niche agencies, meaning that they specialize in one or two specific areas, such as DNA, digital technology, toxicology, drugs, or questioned documents, among other investigative technologies.

Services Offered

The experienced staff of a forensic laboratory typically examines and identifies materials such as suspected accelerants, drugs, explosive residues, fingerprints, glass, paints, fibers, metals, biological stains, toxins, and soils to extract as much information as possible (Wisconsin Department of Justice [WIDOJ], 2001) . The laboratory examines **questioned evidence** and compares them with **known evidence** to determine whether or not the source is the same, or whether or not a relationship exists between them (WIDOJ, 2001). A *questioned* item refers to any type of evidence or material of an unknown, unacknowledged, or unaccounted for source. (i.e., glass particles recovered from the body of a victim of a hit-and-run motor vehicle accident). *Known* items refer to any type of evidence that originate from a known, acknowledged, or accounted for source that is to be compared to an unknown or questioned material (i.e., glass pieces from a motor vehicle to be used to compare against glass particles found on a victim of a hit-and-run motor vehicle accident). This is done for many reasons; however, often it is to provide an investigatory focus, which may lead investigators to follow an entirely different direction from that originally thought to be the most significant, prior to the analysis.

Organization of Forensic Laboratories

Regardless of the jurisdiction serviced or the agency that operates the laboratory, each offers different levels of service. For instance, in a statewide system a minimum of one laboratory will operate as a full-service forensic lab, whereas established regional labs typically only offer a limited number of forensic services. Wisconsin, for example, operates two full service laboratories in Milwaukee and Madison. However, the North Central Laboratory in Wausau supports only sections for drug analysis, fingerprint/footwear analysis, imaging, audio tape enhancement, and field response. This is significantly less than those services offered by the other two full-service laboratories. However, regardless of size or location, crime labs are typically responsible for several types of analytical services. As of 2005, USDOJ statistics reported that on average, crime labs typically provided a median number of six functions of analysis. Identification of controlled substances was the type of analysis performed by the largest percentage of labs (89%), while computer crime-related analysis was the function reported by the fewest number of laboratories (12%) (USDOJ, 2008).

The following section addresses the services offered by a full-service forensic laboratory. However, it is worth mentioning that *full service* means different things to different agencies and states. Because there are an inordinate number of forensic specialties, it is unrealistic to have every state lab provide all services. However, there are a few recognized primary sections that are almost universally included within most forensic full service labs.

Firearms/Ballistics

This unit is responsible for the examination of firearms and ballistic components, which can include: shells, cartridge casings, discharged bullets, and ammunition. Sometimes other objects are examined to determine the presence of discharge residues and to approximate muzzle to target distance at the time of the weapon being discharged. Often laboratories also will make use of the principles used within this section to analyze striation and compression marks made by tools. Approximately 60% of the publicly funded labs in the United States examine firearms-related evidence (USDOJ, 2008).

Toxicology/Chemistry/Drug Analysis

This unit is responsible for the examination of body fluids and organs in order to determine whether or not drugs or poisons are present or, in some cases, absent. This section also typically makes use of the intoxilyzer, a machine used to measure the intoxication level of individuals who have consumed alcohol. In some laboratories this section incorporates all other analyses that require chemical analysis. This can include drug identification, explosives analysis, and accelerant analysis from suspected arson scenes.

Trace

The trace unit has as its operational foundation Locard's Exchange Principle (see Chapter 1). This principle drives all efforts within the trace unit. *Trace evidence* is microscopic but there may be infinite amounts of the material or substance. These may include: paint, glass, fibers, hair, soil, and cosmetics, to name a few.

Biology/Serology/DNA

This section may be known by a number of different titles but its basic mission remains the same: the identification of biological components, which are typically associated with blood. Conventional serology (i.e., blood typing) has largely been replaced by DNA analysis. The types of substances typically examined by this section include: blood, saliva, semen, vaginal secretions, and hair (if not by the trace section). DNA testing is performed by over half of all publicly funded forensic laboratories (USDOJ, 2008).

Document

The document section analyzes handwriting, digital documents, and typewriting on questioned documents to determine authenticity, originality, and source. Some of the related responsibilities of this unit include ink and paper analysis and analysis of indented writings, erasures, obliterations, and burned documents.

Photography

The photography unit may be involved in both examination and documentation of evidence. While providing services to examine submitted photographic evidence, the unit often is responsible for photographically documenting other types of physical evidence in an effort to aid in its subsequent analysis. This unit makes use of highly specialized photographic techniques that may include: digital imaging, ultraviolet photography, x-ray photography, infrared, aerial photography, and underwater photography.

Fingerprints

This unit is responsible for processing and examining submitted evidence (typically in conjunction with other lab examinations) for the presence of latent fingerprints. In the case of visible prints, and processed latent prints it is this unit responsible for the identification of the prints. More than half (55%) of crime laboratories perform fingerprint analyses (USDOJ, 2008.)

Tire Track/Footwear/Toolmark Impressions

If not incorporated within the firearms unit, the impression section provides analysis and examination services to provide information about and aid in identifying the source of an impression.

Polygraph/Voice Stress Analysis Unit

While this area of forensic technology has increasingly found its way into the investigatory sector, it was originally housed, and in some cases continues to be housed, within the confines of the forensic laboratory.

It is this section that is responsible for utilizing the polygraph and voice stress analyzer (VSA) to determine the truthfulness of testimonial evidence. This is the only section within a forensic lab whose primary function does not relate to the analysis of physical evidence, but rather to the examination of testimonial evidence.

Crime Scene Field Response

Many laboratories have begun to incorporate crime scene response units within their structure as a way of ensuring that crime scenes are properly processed, and the evidence properly collected and preserved to ensure the ability to subsequently analyze it. This unit typically will only respond in the cases of felonies and only when requested by an agency to do so, if the crime scene does not already fall within the jurisdiction of the lab in question.

Functions of a Criminalist

Today, the nation's crime labs employ over 12,000 full-time personnel whose responsibility is to analyze and examine the various forms of physical evidence that are submitted (USDOJ, 2008). The role of the criminalist in CSI efforts is to provide investigative leads through the scientific analysis and examination of physical evidence and reconstruction efforts. This role is supported by accompanying reports of actions taken, observations made, and results and conclusions noted. In addition to preparing and returning a written report relating to the scientific findings, laboratory staff may be required to appear in court as expert witnesses to give testimony explaining and supporting their analyses.

Investigator

While criminalists are typically involved in investigatory activities at the crime scene, they must make use of their own ability to be inquisitive and do their best to investigate the information available to them. Unlike the scenarios shown on television, criminalists are not responsible for the investigation efforts; however, if they are to know what to analyze and what types of analyses to perform on the submitted evidence, they do require some background information relating to the case. For this reason, communication between the investigating officer and the criminalist is imperative. Paul Kirk addressed this point in his 1974 book entitled, *Crime Investigation*:

> The efficiency of the laboratory usually bears a direct relationship to the willingness of the police officer to keep the laboratory workers informed of all pertinent facts. Complete frankness and confidence between the two types of investigators is most desirable and profitable and unfortunately

not so common as it might be. Only good understanding, by both, of their reciprocal functions can completely eliminate this barrier to the realization of the full benefits of a well managed crime laboratory.

Educator

An often unrecognized and overlooked area of importance related to the criminalist is the role of educator. During his or her tenure as a criminalist, an individual will have many opportunities to educate law enforcement officers, attorneys, crime scene personnel, judges, and medical personnel. The training performed will vary depending upon the recipient but typically centers around the correct way to collect, package, and preserve evidence in order to ensure the most successful analysis. The criminalist also is responsible for educating with regards to the interpretation and meaning behind the scientific conclusions to ensure the correct probative value. As always, the results or conclusions of these analyses will be limited by the quality, quantity, and type of the submitted physical evidence (James & Nordby, 2005).

Student

The dynamic world of forensic science requires that the criminalist be a life-long student with a thirst for continuing education. Improved technology and new crime trends demand that the criminalist stay abreast of current literature, continue to perform experiments, conduct research, and attend professional seminars, conferences, and workshops. Of particular value is the *Journal of Forensic Science*, which is published by the American Academy of Forensic Science. It is also wise for a criminalist to be aware of historical forensic materials such as those produced by innovators like Edmund Locard, Paul Kirk, Hans Gross, Charles O'Hara, Calvin Goddard, James Osterburg, and Matthew Orfilia. In perusing the works by these contributors to forensic science, the criminalist will note that while technology may change and increase the efficiency of forensic analysis, the fundamental principles and approach to analysis remains the same.

For the most part, gone are the days of the criminalist being a generalist, where the individual knows a little bit about a lot of areas of forensic science. Today's criminalist is typically a specialist within a specific area and must remain adept at the latest instrumental and analytical techniques, technologies utilized, and the areas where further research must be conducted to close gaps within his or her area of specialty. If criminalists do not adhere to the concept of continuing education, the techniques and concepts they learned and used initially very quickly will become obsolete, as will their ability to provide effective analyses.

VIEW FROM AN EXPERT

A Day in the Life of a Criminalist

The field of forensic science intrigued me before I began college and I knew I wanted to work in a crime laboratory. Growing up I learned that participating in teamwork and collaborating with others suited me well. Of course, it was not the gray concrete walls and fluorescent lighting that beckoned me to work a crime lab; it was the aspiration to be a scientist and desire to help the community. A crime lab consists of a group of scientists functioning in unison to put together the pieces from a crime by analyzing evidence, reporting on the findings, and ultimately, testifying in a court of law. While both victims and suspects may lie to cover up a crime; the silent witness, the evidence, never lies and it is a forensic scientist's task to analyze the evidence to unveil the truth.

My path to the crime lab started with my education. I received my Associates in Science from the College of Lake County in Illinois and then continued to the University of Wisconsin-Platteville (UW-P) and achieved a Bachelor of Science in Chemistry-Criminalistics. A part of the chemistry curriculum at UW-P was to complete a chemistry-related internship. Fortunately, I had the opportunity to intern at the Northeastern Illinois Regional Crime Laboratory (NIRCL) for two summers. During my first internship, I had the opportunity to work in the Toxicology section where I assisted in a validation study for the detection of drugs in blood using a gas chromatograph/mass spectrometer (GC/MS). My second internship was in Biology/DNA where I validated an instrument called the CentriVap™ DNA Concentrator, which concentrates a liquid DNA sample into a solid, making it stable at room temperature. These two internships were very much eye-opening experiences. I gained valuable knowledge about the sections I interned in and observed scientists analyzing evidence from various crimes; some of which happened in my very own community. I was very fortunate to have worked in two sections and acquire all of the irreplaceable experiences. From then on, I realized my aspirations were to work in a crime lab and I eagerly wanted to return.

The laboratory participates in a stringent audit held by the American Society of Crime Laboratory Directors-Lab Accreditation Board (ASCLD-LAB). This is a voluntary audit where the lab is thoroughly inspected to ensure employees are following the standard operating procedures and other implemented manuals, participating in proficiency testing and adhering to the quality assurance program. The testing methods utilized in the laboratory are reviewed to see if they are accepted in the scientific community. It is of great accomplishment that our lab successfully passed these inspections every time.

Fortunately, luck came knocking at my door. I was offered a job in the drug chemistry section at NIRCL, to which I enthusiastically accepted. My first day of work started off on the fast track, diving right into lectures lead by the Director of the lab, who has been a chemist for over fifteen years. My training consisted of lectures,

(continues)

readings, written and oral tests, and a series of moot courts. A typical workday in drug chemistry consists of analyzing powder substances, tablets, liquids, residues, and plant material. The types of tests and instrumentation I use are a balance, GC/MS, Fourier Transform Infrared Spectrometry (FTIR), and chemical-color tests. On average, I analyze thousands of cases every year.

I quickly realized there were more opportunities within this profession than the areas I worked in, such as the other forensic sections, a safety program, quality assurance program, and accreditations programs that crime laboratories participate in nationally. Working at a small lab has the benefit of allowing scientists to not only work in their section, but also assist in other sections or areas of the laboratory. Recently, I took on the role of the safety manager, where I strive to maintain a safe working environment by monitoring the storage and proper use of chemicals, gas cylinders, first aid kits, personal protective equipment, conduct a tornado and fire drill and an annual safety test, and many more. Additionally, not long after my chemistry training was complete, I branched into the biology/DNA section where I work part-time and learned to screen evidence for the presence of bodily fluids and preserve samples for DNA analysis.

I am also a part of a professional organization called the Midwestern Association of Forensic Scientists. This organization furthers the branching and networking of scientists across the USA and the world. This allows for various new techniques and findings to be spread and shared, in hopes to constantly improve the detection and identification of evidence. Without this networking between examiners, laboratories, and countries, I feel the sections that I work in today would not be as advanced as they are today.

My workday is as multifaceted as my personality. I can be working on chemistry cases, screening biology cases, perform a safety inspection and assist with an internal lab audit all in one day. I love my profession because each day brings something new to the table and I am constantly learning. I feel better leaving at the end of the day knowing I made a positive difference that will not only benefit the community, but hopefully someone's life.

Sarah E. Owen
Forensic Scientist
Chemistry & Biology Sections
Safety Manager
Northeastern Illinois Regional Crime Laboratory
A Nationally Accredited Laboratory (ASCLD/LAB)

■ Requesting Aid

When a forensic laboratory's assistance is desired during the course of an investigation, it is suggested that the district attorney of the appropriate county be advised that an investigation is being undertaken and that the laboratory's services are necessary and are requested. It bears mentioning that most forensic labs reserve the right (as authorized by state statutes) to decline to provide services in any matter not involving a potential felony charge.

Regardless of whether or not a lab accepts a case, they are typically available for consultation relating to forensic matters. If questions arise during the course of investigative efforts that relate to the collection, preservation, or potential analysis of physical evidence, law enforcement officials should communicate with laboratory personnel for advice, counsel, and/or recommendations relevant to the challenges confronting them within their investigation (WIDOJ, 2001).

Submitting Evidence to the Laboratory

After evidence is collected at the crime scene, it is transported to and stored at an evidence facility, typically found at a police department or other governmental agency.

Evidence is submitted to the crime laboratory either by personal delivery or via mail. The U.S. Postal Service (USPS) and other parcel carriers have specific policies and procedures that apply to the shipment of evidence from a police department or crime scene to a forensic laboratory. Investigators and evidence technicians should check with the carrier prior to shipping items of concern. For instance, postal regulations prohibit shipping live ammunition via the USPS, but both FedEx and UPS do permit live ammunition shipments.

Every item shipped to the laboratory for analysis should be packaged separately. Every shipment should include an evidence submission form (either an agency-specific form, or one prepared by the laboratory for that purpose). The submission form typically will list specific details about the item to be analyzed, the submitter's name and agency, an agency case number, a list of requested analyses to be performed, and a brief case synopsis. It should be noted, however, that often the submitting agency or individual is not always completely aware of the forensic capabilities of the crime laboratory. Therefore, although there is a list of requested analyses to be performed, the criminalist is not limited by those requests and is encouraged to use his or her training and experience to perform the most scientifically correct examinations. Sometimes this in fact may mean not performing an examination at all, if the criminalist determines that either the sample size is too small, the sample is too degraded, or there simply is no way to garner any useful information from the submitted sample.

Accreditation and Certification

In 2008, the USDOJ reported that 80% of publicly funded crime laboratories were accredited (USDOJ, 2008). The American Society of Crime Lab Directors (ASCLD) is responsible for the accreditation of forensic laboratories within the United States. The term **accreditation** refers to the endorsement of a forensic laboratory's policies and procedures. Accreditation is a sign of industry recognition and acceptance for the manner in which the laboratory performs its forensic analyses. As mentioned in Chapter 3, this is a necessary component of establishing credibility within the court system. In order to qualify for accreditation, a crime laboratory must meet minimum requirements set forth by the certifying authority. A number of requirements must be met, but the following policies and programs are typical:

- Develop a quality control manual. **Quality control** measures within a laboratory ensure that analysis results meet a specified standard of quality.
- Develop a lab testing protocol. **Protocols** are the steps and processes undertaken by the laboratory to ensure that the correct tests are performed and that they are performed with accuracy.
- Develop a **quality assurance (QA)** manual. QA is a way of ensuring and verifying quality control, as well as providing for standardized methods of measurement. A laboratory's QA assessment measures are necessary to oversee, verify, and document the performance of the laboratory.
- Develop a program for proficiency examination. **Proficiency testing** is a measure for determining whether or not the individuals working in the lab and the laboratory as a whole are operating at an industry-established standard. For this testing process, a criminalist is given a sample for which the analytical results are already known, and if the results of the subsequent analysis do not match up, then there is an identified problem within the lab's analytical operation. Tests are given in one of two fashions, either blind (the criminalist is not aware that an examination is being issued) or known (the criminalist is cognizant of the examination and is free to consult any resources they deem necessary) (Girard, 2008).

One of the steps that will assist with the accreditation process and lend professional credibility to a laboratory is the employment of certified criminalists. **Certification** is a voluntary process of peer review by which a practitioner is recognized as having attained the professional qualifications necessary to practice in one or more disciplines of criminalistics (American Board of Criminalistics [ABC], 2008). There are a number of organizations that are involved in certification processes and to which employees within a lab may be members of in order to receive continuing education within their respective fields.

American Board of Criminalistics

The ABC offers a certificate in comprehensive criminalistics as well as in the specialty disciplines of molecular biology, drug chemistry, fire debris analysis, trace evidence (i.e. hairs, fibers, paints, and polymers) (ABC, 2008). The ABC lists the following four "Areas of Testing" for certification of criminalists:

1. The philosophical, conceptual, and scientific basis of criminalistics.
2. Basic technical subjects of criminalistics, including:
   - Drug analysis
   - Crime scene reconstruction
   - Firearms and toolmarks
   - Molecular biology and DNA
   - Fire debris and explosives
   - Photography
   - Trace evidence
   - Safety
3. Ethics.
4. Appropriate areas of civil and criminal law.

International Association for Identification

The International Association for Identification (IAI) is one of the oldest professional organizations to certify crime scene–related personnel. For individuals employed in the field of crime scene processing, belonging to this outstanding organization will help them to maintain competency and professionalism. The IAI is responsible for publication of the monthly peer-reviewed *Journal of Forensic Identification*. Many states have IAI divisions and one may consider becoming a member to stay abreast of local forensic and crime scene-related advances and rulings. The IAI is responsible for numerous seminars and training opportunities around the globe each year where members are able to receive the latest information and training relating to their respective field of employment. With over 6000 members from the United States and many other countries, the advancement of forensic discipline–related education and certification continues to be one of the IAI's top priorities (IAI, 2008). The IAI offers certification in the following fields:

- Bloodstain Pattern Examination
- Crime Scene
- Footwear Examination
- Forensic Art
- Forensic Photography
- Latent Print Examination
- Tenprint Fingerprint Examination

American Academy of Forensic Science

The American Academy of Forensic Science (AAFS) is a multidisciplinary professional organization that provides leadership with the goal of advancing science and its application within the legal system. The objectives of the organization are to promote education, foster research, improve practice, and to encourage collaboration within the forensic sciences (AAFS, 2009). One way that the AAFS meets its goals is through publication of the *Journal of Forensic Science*, produced six times a year. The organization's approximately 6000 members are divided into eleven different sections of forensic specialty. The AAFS also conducts various conferences and meetings that bring together professionals from around the globe to exchange research findings and initiate new research.

Chapter Summary

The future of the forensic crime laboratory within the United States is secure. Whether privately or publicly funded, there will continue to be a need for crime labs and for an increasing number of criminalists to be employed within them in order to keep up with forensic analysis-related requests. Based on its most recent census of publicly-funded crime labs, the USDOJ estimated that laboratories performing DNA analysis would need an estimated 73% more staff to complete requests for DNA analysis. Biological screening (typically a task performed in preparation of DNA analysis) represented the next highest need for an increase in full-time criminalists (57%), followed by firearm and toolmark analysis (46%) and examination of trace evidence such as hair and fibers (43%) (2008). There is increasing need for specialists within each of these fields in both the private and public sectors.

The forensic laboratory relies upon the knowledge and proficiency of those within the field to properly identify, document, collect, and preserve physical evidence. Without that quality of evidence, the laboratory is a useless commodity. As a result of courtroom decisions, social changes, and historical events, the crime lab industry and the number of submitted items of physical evidence continues to grow at a rapid pace. Such growth requires that crime laboratories maintain or improve upon their levels of proficiency and reliability. This is accomplished through accreditation of the laboratories and certification of its employees. Continuing to strive for technological efficiency and the proper forensic knowledge requires a dedication to employee continuing education and constant re-examination of a laboratory's analysis techniques and processes.

Review Questions

1. ______________ was responsible for the development of what is considered the first forensic laboratory, which was located in ______________.
2. The first crime laboratories in the United States began in which decade?
3. Who was Colonel Calvin Goddard, and what role did he play in the early forensic laboratory?
4. List and explain three reasons for the tremendous growth in the number of crime laboratories in the United States.
5. Which landmark case is partially responsible for the growth of the number of crime laboratories in the United States?
6. A case is classified as ______________ if the analysis is not completed within 30 days of the item being submitted to the lab.
7. A ______________ item refers to any type of evidence or material of an unknown or unacknowledged or unaccounted for source.
8. What is a VSA and how is it used in the crime laboratory?
9. Why have many crime laboratories incorporated crime scene field response teams within their structure?
10. What is the acronym ASCLD and what is its role in crime laboratories?

References

American Board of Criminalistics (ABC). (2009). What is the American Board of Criminalistics? Retrieved August 1, 2009, from http://www.criminalistics.com/

American Academy of Forensic Sciences (AAFS). (2009). Homepage. Retrieved August 1, 2009, from http://www.aafs.org/

Doyle, A. C. (1892). The Adventure of the Copper Beeches. In The Adventures of Sherlock Holmes. *Strand Magazine.*

Egan, M. S. (April 21, 2008). CSI: Target. *Forbes Magazine.* Retrieved August 1, 2009, from http://www.forbes.com/forbes/2008/0421/102.html

Girard, J. E. (2008). *Principles of environmental chemistry.* Sudbury, MA: Jones and Bartlett Publishers.

Kirk, P. L. (1974). *Crime investigation*, 2nd ed. New York: John Wiley & Sons.

International Association of Identification (IAI). (2008). Welcome to the IAI! Retrieved August 1, 2009, from http://www.theiai.org

Muehlberger, C. W. (1955). Col. Calvin Hooker Goddard 1891–1955. *The Journal of Criminal Law, Criminology, and Police Science, 46,* 103–104.

James, S. H., Nordby, J. J. (2005). *Forensic science: An introduction to scientific and investigative techniques* (2nd ed.). Boca Raton, FL: CRC Press.

Saferstein, R. (2007). *Criminalistics: An introduction to forensic science* (9th ed.). Upper Saddle River, NJ: Pearson Prentice Hall.

U.S. Department of Justice (USDOJ). (2008). *Census of publicly funded forensic crime laboratories, 2005.* Bureau of Justice Statistics Bulletin, NCJ 222181.

U.S. Department of Justice (USDOJ), Federal Bureau of Investigation (FBI). (2008). Birth of the FBI's technical laboratory 1924–1935. Retrieved August 1, 2009, from http://www.fbi.gov/libref/historic/history/birthtechlab.htm

Wisconsin Department of Justice (WIDOJ). (2001). *Physical evidence handbook* (6th ed.). Madison, WI: Wisconsin Department of Administration.

SECTION II

Physical Evidence

Fingerprint Evidence

Every human being carries with him from his cradle to his grave certain physical marks which do not change their character, and by which he can always be identified . . . This autograph consists of the delicate lines or corrugations with which Nature marks the insides of the hands and the soles of the feet.

Mark Twain (1835–1910)
Pudd'nhead Wilson, 1894

LEARNING OBJECTIVES

- List the contributions of historical individuals who developed and refined fingerprint evidence.
- Describe the categorical methods for fingerprint processing.
- Explain ACE-V.
- Distinguish between the two primary methods of U.S. fingerprint classification.
- Define AFIS.
- Explain SWGFAST and how it relates to fingerprints.

KEY TERMS

ACE-V
Automated Fingerprint Identification System (AFIS)
Alternate Light Sources (ALS)
Anthropometry
Cyanoacrylate Ester Fuming
Classification
FBI National Crime Information Center–Fingerprint Classification (NCIC-FPC)
Henry System
Latent prints
Minutia
Ninhydrin
Patent Prints
Plastic Prints
Points of Comparison
Small Particle Reagent (SPR)
Scientific Working Group on Friction Ridge Analysis, Study, and Technology (SWGFAST)

History of Fingerprints

Since the beginning of humankind, there has been a desire and attempt to identify individuals. Fingerprints are among the oldest and most probative types of forensic evidence. There is evidence throughout history, from Neolithic cave carvings to Chinese artifacts over 5,000 years old, that humans have had some inclination of the individuality inherent in fingerprints.

But while artifacts upon which this historical premise is based are thousands of years of age, modern study and understanding of fingerprints has its foundation in 1684.

Nehemiah Grew (1641–1712)

British horticulturalist Nehemiah Grew is the first documented person to study and accurately describe the ridge patterns present on the surfaces of the hands and feet. In addition to writing on the topic, he also published detailed drawings of finger and palm patterns and descriptions of pore detail (Lee & Gaensslen, 2001).

Although Grew published these findings, more than 200 years would pass before the permanency, classification, and individualized identification of fingerprints were studied in depth and presented to the world. Those influential in such matters are listed in the next sections.

Alphonse Bertillon (1853–1914)

Bertillon's system, entitled **anthropometry**, was a series of eleven body measurements of the bony parts of the body, and an in-depth description of marks (scars, moles, warts, tattoos, etc.) on the surface of the body. Anthropometric measurements could be taken from individuals who were over the age of twenty, as it was assumed that such individuals had completed their vertical growth so their measurements would remain constant from that point onward. Bertillon's system of identification was the accepted method for policing agencies until 1903, when an incident at the federal penitentiary in Leavenworth, Kansas identified the shortcomings of the system.

Upon being incarcerated, Will West was photographed and his Bertillon measurements taken. The resulting photograph and measurements were nearly identical to those of another prisoner, William West, already on file. Will West denied having ever been incarcerated at Leavenworth, and a subsequent investigation led authorities to realize that they had two separate prisoners—Will and William West—both incarcerated within the walls of the penitentiary, who bore an almost identical resemblance both visually and in their anthropometric measurements. However, when authorities decided to collect fingerprints from the two men, they realized that each had unique fingerprints from the other. This event led authorities to realize that Bertillon anthropologic measurements were an unreliable method for personal identification, and that two people who have nearly identical facial features and measurements nevertheless have unique fingerprints.

Sir William J. Herschel (1833–1917)

As an Englishman stationed in India during the mid 1800s, Herschel was flummoxed with designing a method for having Indian workers sign legal documents for the British government in a way that was not subject to ease of forgery, such as signing with an "X" or other non-unique mark. He began experimenting with having the workers "sign" legal documents by applying their inked palm and, later, thumb impressions to the documents. Through examination of hundreds such document signatures and examination of his own fingerprints for over 50 years, he noted that the prints did not change over time. He attempted—unsuccessfully—to convince others within the government to implement his practices (James & Norby, 2005).

Henry Faulds (1843–1930)

Faulds, a Scottish physician working in a hospital in Tokyo, Japan, became involved in local archeological digs. Noticing the fingerprints of artisans left in pottery shards, he began to study contemporary hand and fingerprints. In 1880 he is reported to have written a letter to the scientist Charles Darwin about his findings, suggesting that fingerprints could be classified and noting that their ridge details appeared to be unique between individuals. Faulds also mentioned that he had used these unique print details to apprehend criminals and to exonerate the innocent. Darwin forwarded Faulds' letter to his half-cousin, Sir Francis Galton, an English scientist.

Sir Francis Galton (1822–1911)

In 1892 Galton published the first recognized in-depth study of fingerprint science, entitled "Finger Prints." This text included the first classification system for fingerprints, wherein Galton identified the characteristics that enable fingerprints to be identified. These minute variances within fingerprints are termed **minutia**, but are often referred to as *Galton points* or *Galton details* in honor of his recognition of such characteristics. However, although Galton's work in fingerprints became the foundation for modern fingerprint science, his method of classification was much too awkward and involved to make it practical for use in criminal investigations.

Sir Edward Henry (1850–1931)

In 1897, a trainee of Galton's developed a more functional classification system independent of Galton, while working in India. In 1901, Scotland Yard appointed Edward Henry their Assistant Commissioner of Police. While working in this capacity, he introduced his fingerprint system, and within a decade, Henry's system had been adopted by police and prison forces in most English-speaking countries around the globe, and it remains in use to this day (Girard, 2007).

Another individual, however, is responsible for development of a classification system used in many non-English–speaking countries.

Juan Vucetich (1855–1925)

The Argentinean police official Vucetich came to understand the value of fingerprints as a method of criminal identification because of his correspondence with Galton. Through his growing interest in the matter, he developed his own system for classifying fingerprints and named it *Vucetichissimo*. By 1896, Argentine police had implemented Vucetich's system of identification involving fingerprints and had abandoned anthropometry. Vucetich has the distinction of being involved with the first recorded case in which fingerprints were used to solve a crime. This case took place in 1892 and involved the homicide of illegitimate children. As a result of his experience with this case and his continued study of fingerprints, Vucetich wrote *Dactiloscopia Comparada* in 1904 (Caplan, 2001). The system of classification developed by Vucetich is still in use in many Spanish-speaking countries, particularly in South America.

What Are Fingerprints?

Structure of Fingerprints

Skin, the outer covering of the body, is the largest and heaviest organ. Volar skin, found on the soles of the feet, palms of the hands, and on the underside of the fingers and toes, is furrowed and only contains sweat glands. Volar skin does not secrete sebaceous oils, but bodily oils are found on volar skin (especially the hands). Skin in volar areas lacks pigmentation. Smooth skin contains hair, sebaceous glands, and sweat glands. Friction ridges are a textured surface, continuously corrugated with narrow ridges and found on the surface of volar skin. The purpose of friction ridges is to increase friction between the volar surfaces and any other surface they contact.

Composition of Fingerprints

You will observe that these dainty curving lines...form various clearly defined patterns, such as arches, circles, long curves, whorls, etc., and that these patterns differ on the different fingers. The patterns on the right hand are not the same as those on the left. Taken finger for finger, your patterns differ from your neighbor's.

Mark Twain
Pudd'nhead Wilson, 1894

The term *fingerprints* actually refer to oil, perspiration, and other residue left behind by the friction ridge skin present on a person's friction ridge skin after they have touched something. Friction ridge skin is characterized by hills called *ridges* and valleys called *furrows.* These ridges and furrows form three basic fingerprint patterns—arches, loops, and whorls—with each pattern type having several subtypes: ulnar and radial loops, plain and tented arches, and plain whorls, central pocket loop whorls, double loop whorls, and accidental whorls (**Figure 8.1**).

In addition to the overall patterns, there are many tiny variations and irregularities within the ridges themselves, termed *minutia* or *ridge characteristics.* The ridges of the fingerprint form the minutiae by doing one of three things: ending abruptly (ending ridge); splitting into two ridges (bifurcation); or by forming ridge dots.

These ridge characteristics result in individuality not simply between individuals but between the fingers themselves. No two prints have been found to be the same. The FBI has over 50 million fingerprints on file and not one is the same as another. While it is impossible to physically observe each and every fingerprint in the world throughout history, the study of fingerprints within the past 200 years has failed to produce two prints with identical features. This has been known for over a century before the discovery of DNA. In fact, fingerprints are more unique than DNA, for while each person has his or her own DNA, monozygotic (identical) twins, triplets, and other multiples have the same DNA but each has individual fingerprints. In in his book *Pudd'nhead Wilson* (1894), Mark Twain wrote about the individuality of fingerprints between twins:

> The patterns of a twin's right hand are not the same as those on his left. One twin's patterns are never the same as his fellow-twin's patterns. You have often heard of twins who were so exactly alike that when dressed alike their own parents could not tell them apart. Yet there was never a twin born into this world that did not carry from birth to death, a sure identifier in this mysterious and marvelous natal autograph. That once known to you, his fellow-twin could never personate him and deceive you.

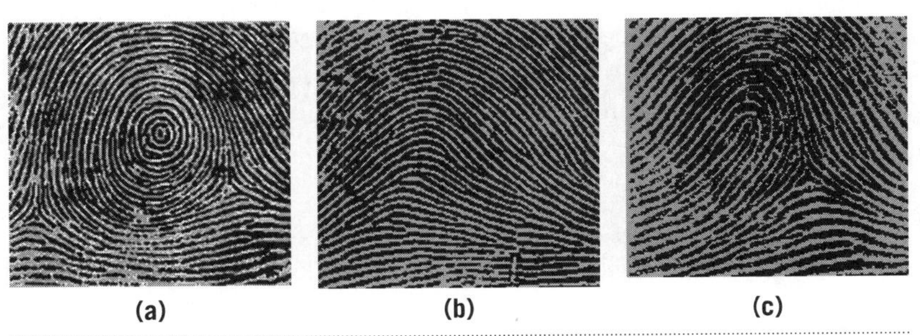

Figure 8.1 Basic Fingerprint Patterns. (a) Whorl Pattern, (b) Plain Arch, and (c) Loop Pattern.

Types of Fingerprints

In addition to three distinct categories of fingerprints (arches, loops, and whorls), the patterns can leave three different types of transfers. The three types of fingerprints are: **plastic**, **patent**, and **latent** impressions.

Plastic Fingerprints
These impressions are made in soft material like wax, paint, putty, tar, etc. They have a distinct three-dimensional appearance and often do not require further processing.

Patent Fingerprints
Made by blood, dirt, ink, or grease, these prints require no processing to be recognizable as a fingerprint and are even suitable for comparison.

Latent Fingerprints
These prints require additional processing to be rendered visible and suitable for comparison. Processing of latent prints is called development, enhancement, or visualization.

Searching for and Processing Latent Prints

Although one might think that due to the general knowledge that fingerprints are individualistic and that finding prints at a scene might implicate someone, many criminals continue to deposit their prints at crime scenes. Therefore, fingerprints should be sought at all types of crime scenes, especially at scenes of crimes committed by unknown perpetrators. Recognition of fingerprint evidence requires training and experience, however.

Persons processing a crime scene should keep in mind two seemingly paradoxical truths: (1) Latent prints can be found on almost any type of surface; and (2) latent prints will not be developed on every attempt. Past studies have shown that many departments locate usable latent prints at 30% to 50% of the scenes visited, and in some of those cases the prints belong to persons with legitimate access. Investigators should not be discouraged by recovery rates as low as one in three crime scenes. The likelihood of the recovery of usable latent prints is increased by the resourcefulness and diligence of the person conducting the search.

General Rule

Evidence should be collected intact and submitted to the lab for processing and examination. If impossible or impractical, apply latent development techniques at the scene, preferably by trained personnel.

Another important consideration is to initiate a search for latent prints as soon as possible after the discovery of the crime and to protect areas to be processed for prints from adverse weather conditions. Ordinarily, fingerprints are primarily composed of water and body fats and oils. These can evaporate if not processed in a timely manner or if exposed to sun, heat, or wind. Prints may be washed away by rain or dew if not protected.

Searches for latent prints should progress from the least invasive and destructive method to the most invasive and destructive method in hopes of minimizing potential evidence damage and maximizing the evidence potential. Suggested search and processing guidelines are in the Box.

Once a print has been located, it must be documented with photography and then a processing methodology must be determined. How a print is developed, enhanced, or visualized is dependent upon a number of factors including: substrate of the material on which the print is located, age of the print, color of the background on which

Search and Processing Guidelines for Fingerprints

Visual Examination

- Sometimes all that is needed to visualize a print is to use oblique lighting.
- After a visual exam, use a laser/alternative light/ultraviolet (UV) light search.
- Photograph all patent prints and other evidence in the impressions prior to removal or tape lifting.

Processing with Physical or Chemical Methods

- Photograph latent prints after development or visualization.
- Draw sketches of the location and orientation of the latent impression on the lift card or in the investigator's notes. The documentation of the *location* and *orientation* of the latent impressions detected will provide details for any reconstruction efforts for testing statements by a suspect with regard to innocent placement of the latent impressions.
- As a rule, wet items or surfaces should be allowed to air dry without the use of heat or forced air before processing.
- Items in freezing weather should be allowed to warm to room temperature before dusting.
- When drying is not feasible, the item may be processed for latent prints by applying small particle reagent (SPR; discussed later in this chapter).

Casting Plastic Impressions

- Impressions should be photographed with oblique lighting and a cast prepared from silicone casting material in the case of indented patent impressions.

Remember at all times: Crime scene fingerprints are perishable.

the print is found, environment (wet/dry/humid), and other factors. While there are over 80 ways to process a fingerprint, they can be grouped into four methods: physical, chemical, special illumination, and a combination approach.

Physical Methods

Methods that do not involve chemicals or reactions are physical methods. These utilize the application of fine particles to the fingerprint residue thereby creating a contrast between the ridges and background (**Figure 8.2**). The most common physical method is *powder dusting* with a brush and inorganic powders. Another variant is *magnetic powder dusting*, which has the advantage of being gentler and not as destructive as inorganic dusting because no bristles make contact with the print. A third common physical method is SPR, which typically is used on evidence that has been wet.

Powder and Brush

Fingerprints on smooth, nonporous surfaces such as glass, paint, glossy plastics, and other polished surfaces can usually be developed with inorganic (non-carbon based) powders. The fingerprint brush should be clean and free from oils or other materials that may affect the efficiency of dusting with the brush. The brush should be swirled vigorously to remove excess powder, and then dipped lightly into the powder with a swirling action, lifted, swirled again, and finally applied lightly to the surface in a circular manner (**Figure 8.3**). Once the latent

Figure 8.2 Close-up Photo of a Powder Dusted Latent Fingerprint.

becomes visible, the print should be dusted lightly in the direction of the ridges until clearly visible. It is important to avoid over-dusting, because the print may be wiped clean by too much dusting. The print is then lifted with fingerprint tape. The fingerprint tape should be applied by releasing an adequate length of tape from the roll, placing the leading edge to the side of and over the print, and then sliding the finger down the tape to cover the entire area of the print while holding the tape roll in the other hand so that it does not fall onto the surface of the print. The tape should be rubbed sufficiently to remove any bubbles present. If the tape cannot be removed without destroying the print, then the tape is left on the surface and the object collected. If the tape can be removed, the tape should be pulled up from the end away from the tape roll, and the tape should then be transferred to a latent card. The card should be labeled immediately and a sketch placed on the card illustrating the location and orientation of the print.

Magnetic Powder Dusting

Magnetic powder has been available since the early 1960s and adds a wide range of flexibility to fingerprint processing techniques. Typically, magnetic powder is used on nonmagnetic surfaces, and inorganic powder on iron-based surfaces. However, the crime scene investigator will find that inorganic powder is inappropriate for some surfaces, including many textured and plastic surfaces (i.e., vinyl imitation-leather, lightly textured automobile dashboards, automobile door panels), where magnetic powder performs quite well.

One of the primary advantages of magnetic powder over inorganic powder dusting is that with magnetic powder, there is no brush to touch and possibly damage the print. Nothing but the powder itself touches the print **(Figure 8.4)**.

Figure 8.3 Inorganic Powder and Brush.

Figure 8.4 Magnetic Powder and Applicator.

Small Particle Reagent

Small Particle Reagent (SPR) is a suspension of molybdenum sulfide grains in water and a detergent solution. The grains adhere to the fatty components of a latent print deposit, and assist with visualizing latent fingerprints. The reagent is first shaken to disperse the molybdenum sulfide grains in the liquid and then sprayed onto the surface suspected of bearing latent deposits. The surface is next sprayed with clean distilled water to remove excess reagent. Developed impressions are then photographed or lifted with tape after drying. The SPR method has the advantage that it can be used on wet and/or dirty/greasy surfaces; however, it also can be used on dry surfaces. In any case, SPR must be used with the understanding that the possible benefit of latent recovery must outweigh the possibility of water damage to the object.

Chemical Methods

Chemical methods of fingerprint processing are those that involve a chemical reaction taking place in order to enhance, develop, or visualize a latent fingerprint. The two most frequently utilized methods are ninhydrin and cyanoacrylate fuming.

Ninhydrin

The amino acid reagent **ninhydrin** has been available to crime scene use since 1910. This chemical is used to detect ammonia or amino acids within print residue, which reacts to form a bluish-purple color (**Figure 8.5**). It is most useful on porous surfaces (e.g., paper and raw wood) and is primarily used in document processing efforts. Ninhydrin may also be used as a preliminary treatment prior to the use of other chemicals or the use of laser or an alternate light source (ALS). Heat and humidity expedite the development process (**Figure 8.6**). Because

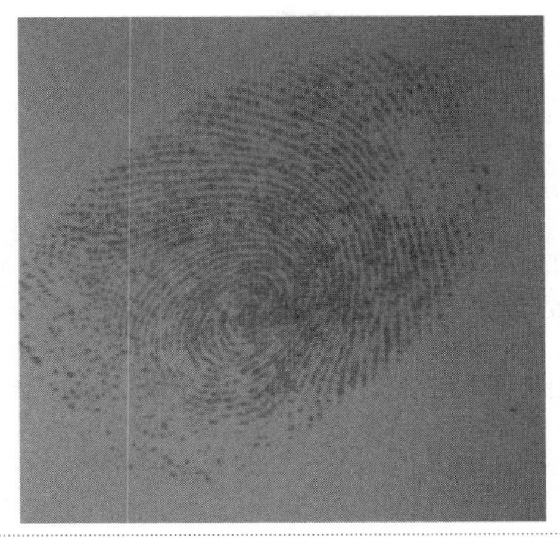

Figure 8.5 Close-up of a Latent Print Processed Using Ninhydrin.

ninhydrin reacts with amino acids, it should only be applied in well-ventilated areas to prevent serious health complications ensuing from its improper usage.

Cyanoacrylate Ester Fuming

Cyanaocrylate ester fuming (also called *super glue*) is a technique that stabilizes latent prints. Super glue fuming has been used since the early 1980s. In this method super glue is induced to fume and

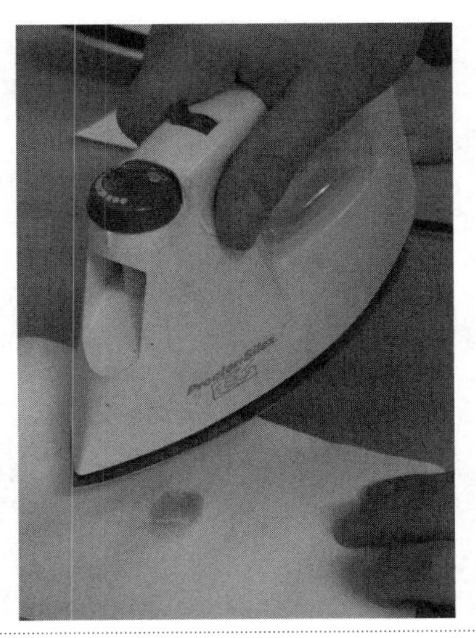

Figure 8.6 An Evidence Technician Using Heat and Humidity to Expedite Ninhydrin Processing.

the fumes interact with latent fingerprint residue by polymerizing them, yielding a stable friction ridge impression off-white in color (**Figure 8.7**). This process can be accomplished by placing the items to be processed in a fuming chamber and then fuming with any of a number of commercially available kits or with kits prepared by the analyst. Of particular interest is the development of the super glue fuming wand for cyanoacrylate fuming (**Figure 8.8**). This technique should allow for effortless fuming in the field by investigators. The cartridges can be ordered with dye added, which can be visualized with fluorescent lighting and thus eliminate the additional step of treating the fumed impressions with a fluorescent dye or powder. The process requires humidity (moisture source). It is primarily used on non-porous surfaces and is an initial step in fingerprint processing. Due to the fact that cyanoacrylate ester encapsulates the latent print in an off-white polymer shroud, it does not necessarily lend itself to effective visualization for identification purposes. Prints often will need to be dusted or further processed to result in the best visualization or enhancement. This method is also used as a way of "fixing" a print prior to the object being transported. This method safely encapsulates and protects the print from any rubbing damage during transport and will allow the lab to continue visualization and enhancement methods once at the lab.

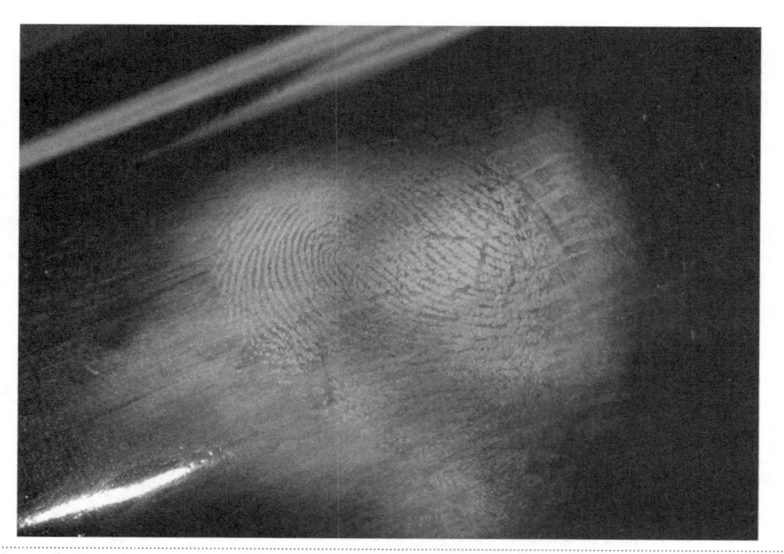

Figure 8.7 An Example of a Latent Print Processed Using Cyanoacrylate Fuming.

Figure 8.8 Cyanoacrylate Fuming Wand and Cartridges.

Special Illumination

Sometimes all that is required to visualize a latent fingerprint is oblique lighting. Light is a basic tool for crime scene searches. Clean white light is necessary for basic observation; however, often specialized lighting is necessary. **Alternate light sources (ALS)** are light-emitting devices supplied with colored filters that filter the source light so that the developed latent print can be viewed with light of a narrow wavelength range, rather than at the usual full spectrum ("white light") viewing range. (**Figure 8.9**). The ALS produces additional light energy to visualize different types of evidence and can be used for more than simply visualizing latent prints.

ALS will help visualize:

- Fluids and biological matter
- Fibers and some hairs
- Bruises or bite marks
- Enhances nearly invisible bloodstains
- Alterations to documents

Combined Approach

While many fingerprint development techniques can be divided into either physical or chemical types, often a combined approach will yield the best enhancement of a latent print. An example of a combined approach is utilizing the chemical method of cyanoacrylate ester fuming to locate and

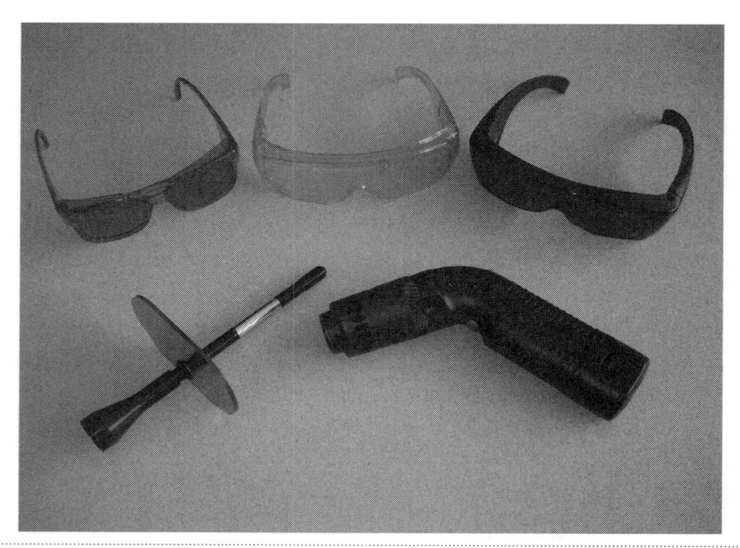

Figure 8.9 Alternate Light Sources (ALS) and Associated Barrier Filters.

protect the print, and then following up this method with powder dusting to create the best contrast for visualization.

Advanced Lifting Techniques

For the majority of scenes, a simplified approach to latent processing, such as powder dusting, will be sufficient to properly locate and preserve latent fingerprint evidence. Sometimes, however, the crime scene investigator will encounter a surface believed to contain latent prints but does not lend itself to processing in the typical manner (**Figure 8.10**). Irregular surfaces such as the dimpled surface of steering wheels and vehicle dashboards, or curved surfaces such as door handles or wood stairway spindles may prove challenging for lifting prints.

A suggestion for dealing with these difficult surfaces is to utilize a combined approach of powder dusting along with lifting utilizing forensic casting gel versus tape. The method is described below:

1. Used on rough or irregularly shaped objects.
   - Print is dusted with powder.
   - A casting gel or silicon (i.e., Mikrosil™ casting putty) is applied to lift the latent.
2. Certain surfaces are textured so that even with gelatin lifters, it is hard to lift powdered fingerprints in their entirety. Examples are styrofoam and rough leather.
   - In these cases, white or black silicone rubber casting compound can be used. The casting compound is mixed and placed over

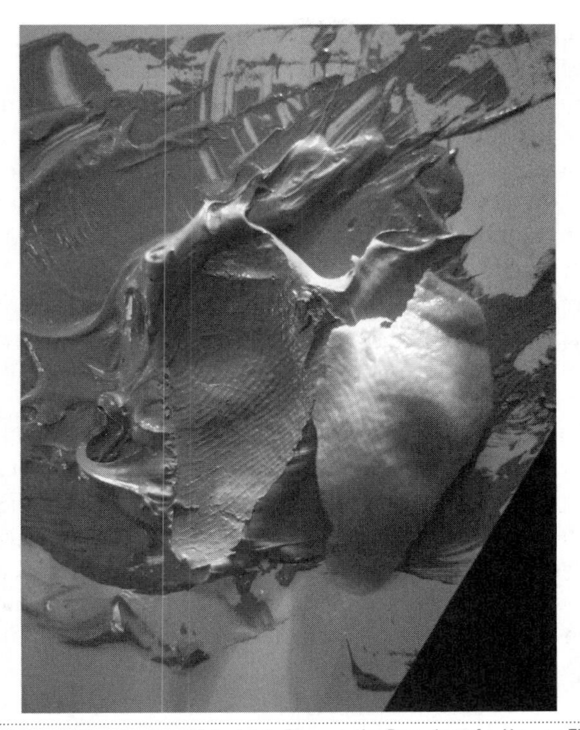

Figure 8.10 Using Silicon Casting Material to Process the Remains of a Human Fingerprint.

the print developed with fingerprint powder. After the silicone rubber has cured, the fingerprint is lifted.

- Depending on the color of the fingerprint powder that was used, white or black silicone rubber is used for contrast.

Preserving and Packaging Latent Prints

When a crime scene investigator applies physical or chemical methods of processing to latent prints, they are developing and visualizing the actual fingerprint that was left behind by an individual; they are not creating and lifting a copy of the print. The chemicals or powders will interact with or adhere to the residues left behind by the finger that touched the surface. Therefore, the recovery efforts and documentation should be the same as with any other item of physical evidence. When the print is collected and later presented within court, it does not simply represent the print left at the crime scene, it *IS* the print that was at the crime scene. This is important for a crime scene investigator to realize so that, regardless of whether or not they believe the print to be identifiable, the print should be properly processed, documented, and recovered so that the print can serve as future evidence.

The majority of times the prints recovered from crime scenes are used from a visual identification standpoint, using the actual characteristics within the print to identify the person whose print is a match to such characteristics. In this case, lifted prints can be packaged either as tape lifts adhered to print cards, or protected within an evidentiary envelope.

An object bearing friction ridge prints must be properly handled and packaged to avoid destroying the prints while in transit. Do not place items in plastic bags, or allow surfaces that contain latent prints to come in contact with or rub against the sides of the packaging materials. Mark these containers with the word "Fingerprints."

Recent advances have allowed for the use of DNA analysis with regards to latent print lifts. Some chemical processes of processing may prohibit the ability of this area of forensic analysis to be performed. If DNA analysis of print residue is anticipated to be used, the crime scene investigator is encouraged to contact their appropriate crime lab prior to utilizing chemical methods of processing so that any contamination or damage of DNA-related evidence can be avoided.

Identifying Fingerprints

Every human being carries with him from his cradle to his grave certain physical marks which do not change their character, and by which he can always be identified—and that without shade of doubt or question. These marks are his signature, his physiological autograph, so to speak, and this autograph cannot be counterfeited, nor can he disguise it or hide it away, nor can it become illegible by the wear and mutations of time. This signature is not his face—age can change that beyond recognition; it is not his hair, for that can fall out; it is not his height, for duplicates of that exist; it is not his form, for duplicates of that exist also, whereas this signature is each man's very own—there is no duplicate of it among the swarming populations of the globe! This autograph consists of the delicate lines or corrugations with which Nature marks the insides of the hands and the soles of the feet.

Mark Twain
Pudd'nhead Wilson, 1894

Fingerprint individuality, and therefore fingerprint identification, rests on four premises:

1. Friction ridges develop in their definitive form when humans are still in the womb.
2. Friction ridges remain unchanged throughout life with the exception of permanent scars.
3. Friction ridge patterns and their details are unique.
4. Ridge patterns vary within certain boundaries that allow the patterns to be classified.

The entire point of recognizing and collecting fingerprints is to identify them in order to find a suspect or identify a person. However, most people have never given much thought to the process by which fingerprint identification is actually done. When prints are found, an expert compares them with samples known to have been made by a suspect. He/she first compares overall patterns and then looks for identical ridge characteristics. When these match, they are known as **points of comparison**. The general guideline in the United States is that prints must match at 12 points of comparison before an identification is considered to be a positive match. However, there is no definitive rule on how to achieve this comparison. Current training in fingerprint comparison stresses that the quality of the print and the quality of the comparison are more important than placing emphasis on a numerical match (**Figure 8.11**).

The examiner must decide if sufficient quality and quantity of the ridge detail is present. If not, it may be concluded that there was "insufficient ridge detail to form a conclusion." The print is analyzed to determine its proper orientation, decide suitability, and then proceed to the comparison. The overall pattern and ridge flow is examined. Next, the minutiae are compared, point by point, as to type and location. Finally, pore shape, locations, numbers, and relationships, and the shape and size of edge features are compared. Any unexplained differences between known and latent during this process results in the conclusion that the known is "excluded as a source." If every compared feature is consistent with the known, and enough features are sufficiently unique when considered as a whole, the examiner makes an identification (ID). Therefore, in fingerprint identification, there are three possible conclusions that can be drawn from an analysis:

1. Insufficient ridge detail to form a conclusion
2. Print exclusion
3. Print identification

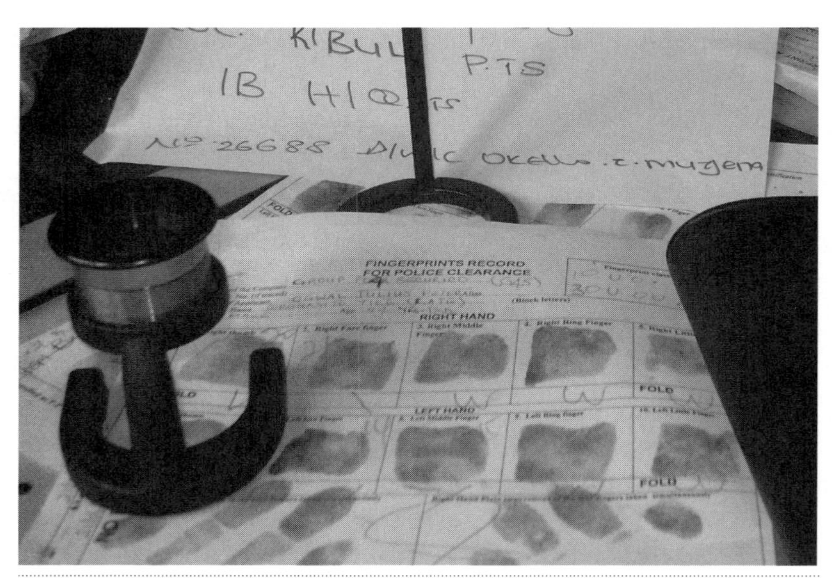

Figure 8.11 Manual Identification Using a 10-Print Card.

In law enforcement, IDs are always made by trained and often certified examiners. Sometimes inked prints may be compared with a set of inked prints on file. More commonly, the examiner compares a developed latent print to inked prints from a known person.

Fingerprint examiners are trained extensively and are required to accumulate significant experience before being entrusted with this responsibility. In addition to the general principles and approaches used, therefore, the knowledge, training, and experience are also considered. Most examiners are certified or have been declared an "expert" by a court of law.

■ Classification of Fingerprints

A classification system is necessary if large sets of fingerprint files are to be useful for criminal identification. **Classification** is a formula given to a complete set of ten fingers as they appear on a fingerprint card generally based on pattern type, ridge count, or ridge tracing. Today the **FBI National Crime Information Center–Fingerprint Classification (NCIC-FPC)** and the **Henry System** are used to classify prints.

Henry System

Developed by Sir Edward Henry, the Henry System of print classification has been used for well over a century and remains in use in many

VIEW FROM AN EXPERT

A Life of Fingerprints

"May it please the court, the claim given the front place, the claim most persistently urged, the claim most strenuously and I may even say aggressively and defiantly insisted upon by the prosecution is this—that the person whose hand left the bloodstained fingerprints upon the handle of the Indian knife is the person who committed the murder. We grant that claim." These words were written by Mark Twain (Samuel Clemens) in his 1894 book entitled, *The Tragedy of Pudd'nhead Wilson.*

Fingerprint identification has been accepted in the United States as a positive means of personal identification since around the beginning of the twentieth century. A lot has changed since the beginning; however, much has remained the same. Mark Twain also mentioned in his book that "Every human being carries with him from his cradle to his grave certain physical marks which do not change their character, and by which he can always be identified—and that without shade of doubt or question." He was speaking of the friction skin that makes up the palmer surface of the hands and the soles of the feet.

Many fingerprint examiners have quoted these passages and others from the *Pudd'nhead Wilson* book over the years. Yet, by what authority does Mark Twain speak? One would suspect that he, as any good author or fingerprint examiner would do, did his research and found many renowned people before him had done extensive study concerning the science of fingerprints.

I started my career in Fingerprint Identification in 1956. Like thousands of other teenagers I came to Washington, DC to work for the FBI. Back then, the FBI recruited people right out of high school from small towns across America. Many of us had no idea where this adventure would lead.

The FBI trained thousands of people in the classification, searching and filing of fingerprint cards. Everything was done manually and the workload was tremendous. After about five years in the Fingerprint Section, you might be selected to be trained in latent fingerprint identification.

Many police departments throughout the United States needed fingerprint examiners to fill their ranks and examiners leaving the FBI filled many of those positions. The police agencies could get people to staff their fingerprint bureaus with people who were already trained by the FBI.

Automation has caused many changes over the past several years. There is no longer the demand for thousands of fingerprint examiners to search 10-print cards at the FBI. Training in the Henry System of Fingerprint Classification is not being taught to the extent that it once was.

A police agency can now get responses to 10-print submissions in a matter of minutes as opposed to weeks or months that it used to take years ago. This has been a very positive change in the way business is done today.

(continues)

Years ago, many of the techniques used in fingerprint development were discovered by accident. The most widely used chemical treatment for developing fingerprints on paper, ninhydrin, is just one example of accidental discovery. Cyanoacrylate fuming is another example.

Today, most federal agencies and large police agencies have people who do research and development trying and testing new ways to develop latent prints on all types of surfaces. Many agencies, including the large federal agencies, have moved forward with laboratory accreditation. It was accomplished with some trepidation; however, its importance was quickly recognized.

Now, after over 100 years of recognition as the most positive means of individualization; it is the science itself that has been under heavy attack over the past few years. There are those who claim that fingerprint identification is not a proven science, or perhaps not scientific at all. Since around 1999 there have been numerous attacks against the Fingerprint community in the form of Daubert Hearings in various courts across the United States. Simply put, Daubert is concerned with the admissibility of scientific evidence in court. A Daubert motion is a motion raised before or during trial to exclude the presentation of unqualified evidence to the jury.

Fingerprint examiners today have to be better educated and have a solid understanding of what it is that leads one to form an opinion and reach a conclusion. I believe that the basics of fingerprint identification/individualization were and still are very important. These skills must now be coupled with a better understanding of the methodology that leads you to your conclusion.

Continued training and education, certification and accreditation have all been positive steps in the career of latent print examiners. To date, all of the challenges to the science have been successfully met; however, fingerprint examiners continue to face challenging cross-examination from attorneys who are keeping abreast of all of the latest information on fingerprint identification. The Internet is a wealth of information available to anyone, including defense attorneys, who wants to learn about the science of fingerprints.

Recent published accounts of "mistakes" and outright wrongdoing by a small minority of fingerprint examiners have provided fodder not only for defense attorneys, but also for those who would like to see the demise of the fingerprint science. When all is said and done, I believe that fingerprint science is extremely sound and will continue to be an effective tool for law enforcement and civilian use for many years to come.

Carmine Artone
Retired Branch Chief
Identification and Research Branch
Forensic Services Division
United States Secret Service

departments today. This system requires the complete classification of all 10 fingers of an individual in order to properly file the information. When this system was developed, it allowed for efficient searching and maintenance of large fingerprint files. However, it did not allow for manually searching for a single print. The system was built around whether or not an individual had whorl patterns present within his or her prints (primary classification) and then had a series of five extensions to the primary classification, dependent upon the type and size of the patterns present in the fingers. An example of utilizing the Henry System of classification is shown below:

Henry System of Classification

10	**I**	**1**	**R**	**-**	**r**	**2**
	S	**17**	**U**	**2 a**		

The above Henry classification is representative of the following:

Primary: 1/17
Secondary: R/U
Subsecondary/small letter group: - r/2a
Major: I/S
Key: 10
Final: 2

When trained to understand the Henry System of classification, the examiner is able to determine that there is one whorl pattern present within the hand (in the right thumb), a radial loop with a ridge count of 10 is present in the right index finger, a radial loop is present in the right ring finger, ulnar loops are present in the right middle finger, the right little finger (with ridge count of 2), the left thumb, the left index finger, and the left little finger. There are also arches present in the middle and ring finger of the left hand.

Although recognized the world over as an effective method for fingerprint filing, with the increased efficiency and affordability of computer systems, many departments are choosing to file and classify prints using a method developed by the FBI.

National Crime Information Center-Fingerprint Classification (NCIC-FPC)

This system of classification, developed by the FBI (2008), assigns a 20 character string of letters and numbers to a person's fingerprints. Every print entered into the FBI system is classified by this method, and it allows trained law enforcement personnel in the field to determine fingerprint compatibility with prints on record, along with providing an efficient and effective way for filing fingerprints. Below is an example of an NCIC-FPC.

POAA05TT19CISR58DIXX

PO: Right thumb is a plain whorl with an outer tracing.
AA: Right index finger is a plain arch.
05: Right middle finger is an ulnar loop with a ridge count of 5.
TT: Right ring finger is a tented arch.
19: Right little finger is an ulnar loop with a ridge count of 19.
CI: Left thumb is a central pocket loop whorl with an inner tracing.
SR: Left index finger is unclassifiable due to scarring.
58: Left middle finger is a radial loop with a ridge count of 8.
DI: Left ring finger is a double loop whorl with an inner tracing.
XX: Left little finger is missing (possibly amputated, or missing since birth).

ACE-V

As was discussed in Chapter 1, a systematic and thorough approach to crime scene processing means employing the scientific method within the efforts, referred to as scientific crime scene investigation. A similar methodology is utilized with reference to the comparison and identification process of latent fingerprints. David R. Ashbaugh, a scientist with the Royal Canadian Mounted Police, developed a formal method known as **ACE-V** for the scientific comparison of prints. The acronym stands for *a*nalysis, *c*omparison, *e*valuation, and *v*erification. The purpose of this comparison methodology is to either identify a print, via individualization, as having originated from the same source, or exclude impressions as having no common origin (Coppock, 2007).

Analysis

The first level of this process begins with the study of the questioned print to determine the overall print orientation, quality, shape, and ridge flow. The comparison (or known) print is analyzed in the same manner. If the information derived is found to be consistent, the analysis proceeds to the next level. If non-matching characteristics are observed, the examination is terminated, which results in an exclusion.

Comparison

If the analysis portion of the process yields sufficient information to warrant a further investigation, the next level begins with orienting the questioned and known print in the same manner and identifying a common unique point in each print to utilize as a starting point. The examination will continue from this common starting point and progress along with recognition of other areas of commonalities between the prints, with regards to ridge characteristics, beginning with the most distinctive feature identified and continuing until all of the characteristics are accounted for and there are no unexplainable variances. Differences may exist due to print quality; however, what is being compared are the print characteristics that are present, not necessarily their clarity.

Evaluation

In the event of a clear variance between the prints, following the comparison stage, an exclusion would be made. However, if the information appears consistent between the two prints, an ID can be made. Typically, this ID is based upon the degree of ridge detail. If the print is lacking in sufficient print detail, then pore distribution and ridge shapes and edges may be utilized instead or in combination.

Verification

Regardless of the conclusion reached, either exclusion or identification, another examiner re-examines the print for verification utilizing the aforementioned process. Under ideal conditions, the examiner making the identification or exclusion should be an analyst who is in no way associated with the case, or who had any significant knowledge of the case. This could impart bias to the decision process.

One important case of misidentification was the FBI's arrest of American citizen Brandon Mayfield in 2004 in connection with the Madrid bombings that killed 191 and injured over 2000. It wasn't until the Spanish government identified a different man from the fingerprint evidence that Mayfield was released (see Case in Point).

■ Automated Fingerprint Identification System

An **Automated Fingerprint Identification System (AFIS)** is an automatic pattern recognition system that consists of three fundamental stages:

1. *Data acquisition:* The fingerprint to be recognized is sensed.
2. *Feature extraction:* A machine representation (pattern) is extracted from the sensed image.
3. *Decision-making:* The representations derived from the sensed image are compared with a representation stored in the system.

CASE IN POINT

Can Fingerprints Lie?

Fingerprints do not themselves lie; however, their interpretation can certainly mislead. Brandon Mayfield, an immigration lawyer from Washington State, found this out in a very publicly humiliating and professionally damaging way in the summer of 2004. After a misidentification led to his subsequent incarceration, the FBI issued the following press release:

May 24, 2004
Statement on Brandon Mayfield Case

After the March terrorist attacks on commuter trains in Madrid, digital images of partial latent fingerprints, obtained from plastic bags that contained detonator caps, were submitted by Spanish authorities to the FBI for analysis. The submitted images were searched through the Integrated Automated Fingerprint Identification System (IAFIS). An IAFIS search compares an unknown print to a database of millions of known prints. The result of an IAFIS search produces a short list of potential matches. A trained fingerprint examiner then takes the short list of possible matches and performs an examination to determine whether the unknown print matches a known print in the database.

Using standard protocols and methodologies, FBI fingerprint examiners determined that the latent fingerprint was of value for identification purposes. This print was subsequently linked to Brandon Mayfield. That association was independently analyzed and the results were confirmed by an outside experienced fingerprint expert.

Soon after the submitted fingerprint was associated with Mr. Mayfield, Spanish authorities alerted the FBI to additional information that cast doubt on our findings. As a result, the FBI sent two fingerprint examiners to Madrid, who compared the image the FBI had been provided to the image the Spanish authorities had.

Upon review it was determined that the FBI identification was based on an image of substandard quality, which was particularly problematic because of the remarkable number of points of similarity between Mr. Mayfield's prints and the print details in the images submitted to the FBI.

The FBI's Latent Fingerprint Unit will be reviewing its current practices and will give consideration to adopting new guidelines for all examiners receiving latent print images when the original evidence is not included.

The FBI also plans to ask an international panel of fingerprint experts to review our examination in this case.

The FBI apologizes to Mr. Mayfield and his family for the hardships that this matter has caused.

Source: Information retrieved from the Federal Bureau of Investigation (FBI). (2004). Press release. Statement on Brandon Mayfield Case. Retrieved August 3, 2009, from http://www.fbi.gov/pressrel/pressrel04/mayfield052404.htm.

Different systems may use different numbers of available fingerprints (multiple impressions of a single finger or single impressions of multiple fingers) for person identification. The feature extraction stage may involve manual override and editing by experts. Image enhancement may be used for poor-quality images. Depending on whether the acquisition process is offline or online, a fingerprint may be one of three types: an inked fingerprint, a latent fingerprint, or a live-scan fingerprint.

Fingerprints no longer need to be manually matched to files. Time is often the critical factor in determining the success of a criminal investigation. The use of this computer technology not only saves time but significantly increases the accuracy match rate compared to manual comparisons. Because of this, and due to the systems becoming more affordable, AFIS is rapidly being implemented throughout law enforcement agencies.

Ten-print cards are scanned into the system. They are run against current latent prints within the system from "unknowns" The AFIS also can scan in latent prints and compare them against the 10-print cards on file. The computer assigns a percentage of probability on the matches generated. Searches can be conducted in seconds/minutes versus months for manual searches. It should be noted, however, that final determination is always left up to a professional print examiner and NOT the computer.

For AFIS to pull up candidates for a "match," an examiner must first ensure that minutia points are properly identified; sometimes these must be added or edited manually, which can be very time-consuming. However, if not performed, and minutia points are incorrectly identified, or unidentified altogether, then the chances of finding a proper match decrease dramatically.

This often tedious and problematic situation saw a dramatic improvement when in April 2009 the National Institute of Standards and Technology (NIST) released results of biometric research they had conducted with reference to Automatic Feature Extraction and Matching (AFEM). Utilizing automated fingerprint feature extraction, most of the tested print's identities were found within the top ten prints listed as possible matches. This shows a dramatic increase in efficiency of automation and bodes well for accelerating fingerprint data input and identification (Indovina, Dvorychenko, Tabassi, et al., 2009).

Today the FBI has an integrated automated fingerprint identification system in place (IAFIS). This system allows agencies to be linked together and compare/share evidence. However, not all systems are integrated. The majority of time when someone thinks of AFIS, they believe the system to be an IAFIS system, but that is not necessarily the case. An AFIS system accepts and stores input data within that system alone, and is not integrated with outside systems. Therefore, if a comparison to other prints outside of the agency's own system is to take place, the print must be e-mailed or otherwise sent digitally to an agency that has IAFIS capabilities. Agencies that are integrated, whether to the FBI IAFIS system or simply to a larger network such as several interstate crime laboratories, are able to run a print against all prints within the integrated system.

Palm Prints

The palms of the hands (as well as the soles of the feet) yield the same volar skin, and thus friction ridge skin, as that of the fingers (and toes). However, the large-scale classification of palm impressions relating to data entry or archiving is a relatively new concept. Until recently, the technology necessary to document and compare such information was not available on a large scale. Most AFIS computer databases allow only searches of fingerprints. A select few will allow for the input of and comparison of palm print impressions.

Although palm prints are relatively new for AFIS systems, latent print comparisons are not new. One of the earliest latent print identifications, possibly the first in a criminal case in this country, was a palm print identification. Palm print and footprint identifications have been part of the friction ridge identification process for many years. Palm print identifications at the Secret Service, for example, historically have been very high because of the large number of forged U.S. Treasury Checks that were processed. This was especially true before forgery and general financial fraud became mostly electronic in nature. Quite often palm prints of the side of the palm are developed under the signature area. This area of the palm is sometimes referred to as the *writers palm* because it contacts the document when a person is writing (Artone, personal communication, October 25, 2008).

Palm prints often are found during crime scene search efforts. The most commonly encountered areas of friction skin impressions typically correspond to the large padded areas of the palmer surface. As technology continues to improve, the comparison and identification efforts will improve also. A crime scene investigator should not let this deter him or her from the collection of the friction ridge evidence. To a crime scene investigator, all prints should be viewed as potentially identifiable. Such identification efforts are left up to the experience and technology of the forensic laboratory.

Preparing Fingerprint Cards

The 2003 Wisconsin Department of Justice (WIDOJ) *Physical Evidence Handbook*, suggests the following with regards to collection of inked fingerprints:

> Law enforcement personnel should strive to develop the skills necessary to take legible record ("inked") finger and palm prints (**Figure 8.12**). Absolute clarity of detail is paramount. Unless ridge detail is perfectly clear, it may be impossible to conduct comparisons against the latent prints. This can result in identifications not being made that would have been possible if the inked impression had been clearly recorded. Submissions made to the laboratory should record finger and palm prints of all persons known to have had or suspected of having had access to the item or scene. (It is especially important to secure finger and palm prints of the victim if he/she has died, since it will be nearly impossible to secure prints once the body has been interred (p. 103).

Taking Record Fingerprints

Preparing Inked Fingerprint Cards

Prepare the inking slab by placing several small dabs of ink on the surface and rolling it uniformly over the surface. Be careful not to use too much ink. The rolled out ink should be only thick enough so that when a digit is rolled across the surface, the areas where the ridges picked up the ink will appear clean. The print recorded should have good contrast with the card. Practice will allow for proper inking on a consistent basis. It is suggested that a test ink impression be made on

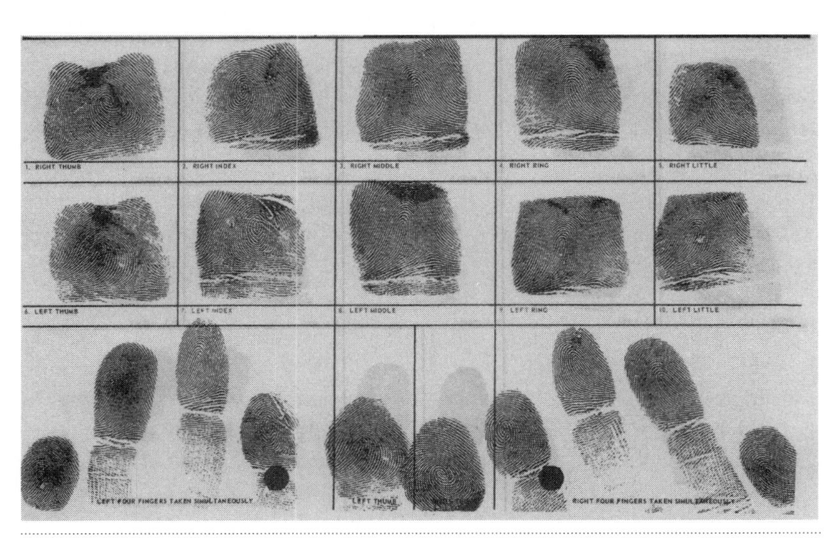

Figure 8.12 A 10-Print Fingerprint Card.

a scrap of paper to check for proper ink density. If possible, adjust the fingerprint card holder so that it is at the height of the person's elbow.

The person to be printed should wash his/her hands if excessive grease, dirt, or perspiration is present. Inked impressions should not be taken of digits having open cuts. The person being fingerprinted should be instructed not to assist with the process, but to cooperate by relaxing his/her arm so it pivots more easily. With the person positioned to the right side of the person taking the prints and slightly to his or her rear, the right hand of the person to be printed should be grasped firmly. Then, holding the four fingers back and clear of the inking slab; the thumb is inked by rolling it toward the body. Then roll the inked thumb in the designated space on the card. Repeat the process for the fingers, except rolling the fingers away from the person's body. This prevents possible drag and secures a more uniform impression. When the right hand is completed, have the person turn so he/she is standing at right angles to the card stand, with his/her back to you. Grasp the left hand and repeat the process. Do not push down on the fingers while recording them; use only enough pressure to guide and to ensure the digit does not slip. Excessive pressure will blur the impression.

To record the simultaneous impressions, re-roll ink on the slab (add more ink if necessary) and have the person wipe excess ink from his/her fingers with lint-free toweling. Simultaneous prints are not rolled; the four fingers are extended and joined. Ink and print by pressing them straight down in the appropriate block. Use only a very slight amount of pressure on the back of the person's fingers when pressing them onto the card. Repeat with the thumbs. If the inked impressions are not properly inked or recorded, retake them.

Recording Palm Prints

If palm prints are found at the crime scene then palm prints should be recorded from suspect(s) and victim(s). To do this, remove excess ink from the person's hands and then re-ink the entire palm and fingers. Palm prints are more clearly recorded by the following method:

> Secure an 8″ × 8″ plain card to a 3″ × 4″ diameter cylinder with rubber bands. Possible cylinders include a short section of plastic drain pipe or a section of the cardboard tubing on which carpet is shipped. Place the tube (with card attached) on a counter top so it can roll toward the front edge, placing it so it can make one complete revolution. After inking the palm, instruct the person to hold his/her hand perfectly flat, palm down and parallel to the counter top. Grasp the person's arm and guide the hand so the wrist is placed on the card at the bottom, ending with the tips of the fingers at the top edge of the card. As the palm draws across the card, apply very slight pressure to the back of the person's hand with

the heel of your hand (this will ensure the hollow of the palm will be recorded). The impression will be more clear if only slight pressure is applied during the procedure (only enough pressure to ensure the hand does not slip across the surface of the card).

Source: Courtesy of the Wisconsin Department of Justice (WIDOJ) (2004). *Physical Evidence Handbook* (p. 105).

All finger and palm print cards must be signed and dated by both the person and the officer. The fingerprinted person should fill out all information required on the card. This information should appear in the person's handwriting or printing.

Scientific Working Group on Friction Ridge Analysis, Study, and Technology

The **Scientific Working Group on Friction Ridge Analysis, Study, and Technology (SWGFAST)** was established in 1997 in response to a number of inconsistencies and controversies relating to fingerprint identification and technological advancement, and continues to operate through sponsorship from the FBI. Its mission is to assist the latent print community in providing the best service and product to the criminal justice system. Membership in SWGFAST comprises a diverse group of professionals within the fingerprint community. This includes not only latent print examiners from law enforcement agencies, but also defense experts, researchers, instructors, academics, managers, and others. This group's diversity provides an objective, yet varied perspective to all matters of interest to the group. The group was formed to establish consensus guidelines and standards for the forensic examination of friction ridge impressions. To date, that has been limited to the concerns of the latent print community. Although it has been realized that there are areas of common interest to other fingerprint applications (e.g., criminal history and biometrics), the existing SWGFAST guidelines were not developed with the intention of being applied to non-latent print related matters.

Chapter Summary

Fingerprints are one of the most probative types of evidence in crime scene investigation and as such, they should be searched for and gathered whenever possible. What is not looked for will not be found. What is not found cannot be identified. A crime scene investigator should consider all prints as being potentially identifiable and should call upon his or her training and experience to determine the best method for processing latent prints at a crime scene. If it is possible to collect the item and submit it to the lab for processing, it is always preferable to do so. Lacking that, processing and recovery efforts should be put to use at the crime scene.

Review Questions

1. ______________ are a textured surface, continuously corrugated with narrow ridges and found on the surface of the volar skin.
2. Friction ridge skin is characterized by hills called ______________ and valleys called ______________.
3. What are the two paradoxical truths of processing a crime scene for latent fingerprints?
4. What is the primary advantage of magnetic powder over inorganic powder?
5. What is SPR and what is it made of? What types of evidence can ALS be used to visualize?
6. List the four premises on which fingerprint individuality rests.
7. The two types of classification systems used today in the United States are ______________ and ______________.
8. Describe the issues regarding fingerprint analysis that arose during the Brandon Mayfield case.
9. What is ACE-V and how is it used in the fingerprint analysis?
10. What is SWGFAST and what is their purpose?

References

Artone. (October 25, 2008). Personal communication.

Caplan, J. (2001). *Documenting individual identity: The development of state practices in the modern world.* Princeton, NJ: Princeton University Press.

Coppock, C. A. (2007). *Contrast* (2nd ed.). Springfield, IL: Charles C Thomas Publisher, Ltd.

Girard, J. E. (2008). *Principles of environmental chemistry.* Sudbury, MA: Jones and Bartlett Publishers.

Indovina, M., Dvorychenko, V., Tabassi, E., Quinn, G., Grother, P., et al. National Institute of Standards and Technology (NISTIR). (2009). *An evaluation of automated latent fingerprint identification technologies,* NISTIR 7577. Retrieved August 1, 2009, from http://fingerprint.nist.gov/latent/NISTIR_7577_ELFT_PhaseII.pdf

James, S. H., Nordby, J. J. (2005). *Forensic science: An introduction to scientific and investigative techniques*, 2nd ed. Boca Raton, FL: CRC Press.

Lee, H. C. & Gaensslen, R. E. (2001). *Advances in fingerprint technology* (2nd ed.). Boca Raton, FL: CRC Press.

Office of the Inspector General. (2006). Special report. *A review of the FBI's handling of the Brandon Mayfield case (unclassified and redacted)*. Retrieved August 2, 2009, from http://www.usdoj.gov/oig/special/s0601/PDF_list.htm

Twain, M. (1894). *Pudd'nhead Wilson.* Hartford, CT: American Publishing Company.

U.S. Department of Justice (USDOJ), Federal Bureau of Identification (FBI). (2008). Birth of the FBI's technical laboratory 1924–1935. Retrieved August 1, 2009, from http://www.fbi.gov/libref/historic/history/birthtechlab.htm

Wisconsin Department of Justice (WIDOJ). (2004). *Physical evidence handbook* (7th ed.). Madison, WI: Wisconsin Department of Administration.

Trace Evidence

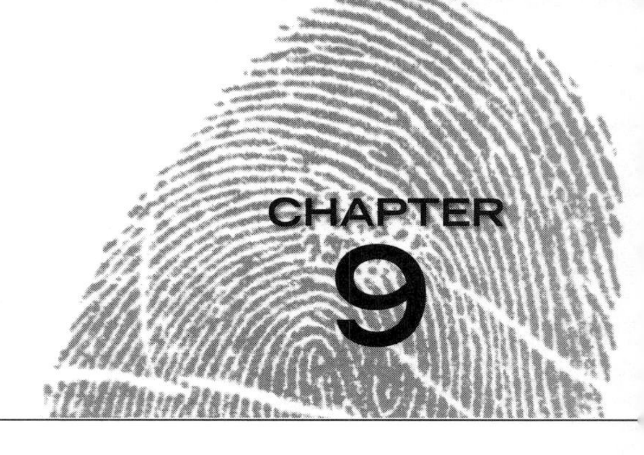

Wherever he steps, whatever he touches, whatever he leaves even unconsciously, will serve as silent witness against him.

Paul L. Kirk, 1953

LEARNING OBJECTIVES

- Understand the value of associative evidence and how it relates to trace evidence analysis
- Apply probability and statistical significance as they relate to trace evidence.
- Describe the proper methods for collection and preservation of trace evidence.
- Describe the forensic significance of glass breakage analysis.
- Know the capabilities of the PDQ database and how it is used in the investigative process.
- Differentiation between hair and fiber analysis and the forensic information possibilities relevant to each.

KEY TERMS

Associative Evidence
Concentric Fractures
Conchoidal Fractures
Cortex
Cuticle
Medulla
Primary Transfer
Probability
Product Rule
Radial Fractures
Secondary Transfer
3R Rule
Trace Evidence
Transfer Evidence

Connection of an object or person to a specific location or item is often paramount to establishing guilt or innocence. As discussed in Chapter 1, Locard's Exchange Principle states that, "whenever two objects come into contact with one another, there will be a cross-transfer of material which will occur" (Gale, 2005). This theory is the underlying premise between the documentation, collection, preservation, and identification of trace evidence. The term **trace evidence** typically refers to any evidence that is small in size, such as hairs, fibers, paint, glass, and soil, which would require microscopic analysis in order to identify it. The purpose of the microscopic analysis is to determine whether or not an association between persons, places, and things can be established, and the subsequent strength of such an association. This association would be as a result of **associative evidence**. However, sometimes enough uniqueness is present within the trace evidence to allow for an individualized identification.

Probability/Product Rule

Another way to approach the level of individualization is through association with numerous levels of class characteristics. This association brings to light a mathematical concept known as **probability**. Probability is most simply defined as the frequency with which an event will occur, sometimes referred to as the *odds of occurrence*. For instance, if one were to flip a coin 50 times, the probability of the coin landing heads down will be 25 in 50. The concept of probability is important to comprehend from a forensic aspect because it is this frequency of occurrence that an examiner is depending upon to limit potential suspect items or individuals, and to explain the odds of such occurrences to the potential jury.

The concept of probability can be taken a step further when considering all aspects of probability associated with an event in an effort to determine a statistical probability for the event or item in question. For instance, if all eye witnesses agree that a white male suspect over the age of 30 committed a crime, then utilizing the **product rule** would lend statistical support to such matters, and explain the frequency of such an event occurring. The product rule is when the frequency of independently occurring variables are multiplied together to obtain an overall frequency of occurrence for the event or item. **Table 9.1** shows the product rule applied to the case example given above.

TABLE 9.1 Utilizing the Product Rule

Demographic	Statistic	Percentage	Equation
Number of people in the United States (est)	295,734,134		
Population that is white	241,614,787	81.70%	(241,624,787 ÷ 295,734,134)
Population that is white male	118,705,344	49.13%	(118,705,344 ÷ 241,614,787)
Population of white males over age 30	42,401,549	35.72%	(42,401,549 ÷118,705,344)

Using the product rule, 0.817 × 0.4913 × 0.3572 = 0.1433, or 14.33%, or 1 in 7 persons could have been involved in the commission of the crime.

Source: Data from the Central Intelligence Agency. (2006). The World Factbook. Retrieved August 4, 2009, from https://www.cia.gov/library/publications/the-world-factbook/

Most evidence is not truly individualistic; instead, it is looked at as being individual in nature because the probability of a match between any other item is so remote (so large a number). However, the establishment of probabilities relating to evidence is typically the duties of the criminalist who analyzes or interprets the evidence, rather than for the crime scene investigator. Criminalists will determine probability and utilize the product rule during their analysis, interpretation, and eventually within their testimony. They will do this in order to attempt to explain the probability of a person's DNA matching that of the evidence presented (i.e., 1 in 1,476,565,780), or the frequency of another carpet fiber matching the one discovered at the crime scene (1 in 7,896). Their utilization of such statistical analysis does not by any means define an absolute but instead presents the trier of fact with an educated extrapolation of a number that a reasonable person would infer as being sufficient to believe that the probability of a more likely match being present is extremely remote.

Collection of Trace Evidence

Trace evidence is collected in a variety of ways, including with the use of a forceps, tweezers, by hand, tape lift, or even vacuuming. As discussed in Chapter 8 with regards to fingerprints, if possible, the entire item containing the suspected evidence should be collected and preserved for later analysis. If conditions do not allow for this, then on-scene steps must be taken to properly collect and preserve the trace evidence (**Figure 9.1**).

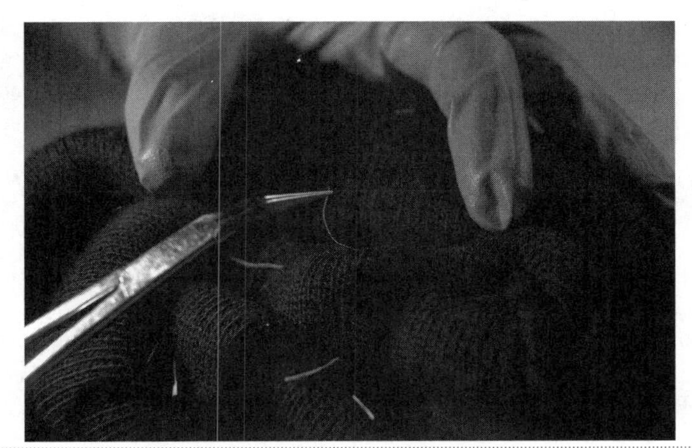

Figure 9.1 Collection of Trace Evidence from Crime Scene Evidence.

CASE IN POINT

The Product Rule at Work

While investigating a cold hit-and-run accident, a crime scene investigator located a piece of plastic that appeared to be from a turn signal of a vehicle. The plastic did not match the struck vehicle and was believed to have belonged to the vehicle that fled the scene. The plastic piece had a number stamped on it, along with the Chrysler logo. This led the investigator to contact a local Chrysler dealer's parts department, which was able to give the investigator information that the plastic piece belonged to a 2003 Dodge Ram pickup truck. The dealership was confident in this because that part was only placed into that specific year and model. The crime scene investigator knew that the vehicle that fled the scene was a white colored vehicle due to paint transfer found on the red Chevy S-10 that was struck.

The crime scene investigator contacted local body shops and asked that if they had anyone who showed up with a 2003 white Dodge Ram pickup truck with front right end damage, that they call the investigating officer. Within one week a body shop called to report that an individual was requesting body work be done on the front of the damaged vehicle that had resulted from "accidentally striking their garage." The crime scene investigator responded to the body shop and noted red transfer paint on the damaged portion of the vehicle. Photographs were taken, and eventually hit-and-run charges were filed against the registered owner for the accident.

In court, the evidence was presented and the prosecution brought up the probability of a Dodge Ram pickup truck being the vehicle involved. It was determined through contact with Chrysler that there were 1196, 2003 Dodge pickup trucks sold within 100 miles of the accident scene. Of those, 257 were white. This meant that the probability of the fleeing vehicle having been a 2003 white Dodge Ram pickup truck was 21.48%. The prosecution took this a step further and looked at the number of registered vehicles within a 100 mile radius, through information derived from the Department of Motor Vehicles. It was determined that there were 175,765 validly registered vehicles within the defined area. Utilizing the product rule, the frequency of a 2003 white Dodge pickup being the vehicle involved in the incident was explained as follows:

2003 Dodge pickup (1196 out of 120,125)	0. 99%
2003 white colored Dodge pickup (257 out of 1196)	21.48%

Therefore, $0.0099 \times 0.2148 = 0.0021$, or 0.21%, or 1 in 500 vehicles on the road were 2003 white, Dodge Ram pickups.

This was a very large number and the jury took this as evidence of there being a very remote chance that another vehicle matching the description of the defendant's vehicle also sustained front end damage consistent with the crash.

When collecting trace evidence, the crime scene investigator must document and collect not only the questioned samples but they must also collect known samples for comparison purposes. For example, if a suspect is located in a burglary case involving entry through a broken glass window, the crime scene investigator should document and collect trace glass evidence from the clothing, or submit the clothing to the lab for trace glass analysis. The crime scene investigator must also collect known samples of the glass from the broken window for comparative analysis. Failure to collect the known/comparison sample will almost always result in the forensic lab declining the ability to conduct an analysis. The crime scene investigator is attempting to show that Locard's Theory of Exchange is alive and well, and that the broken glass from the window can be used to show transfer and thus connectivity to the suspect, and vice versa.

It is important for crime scene investigators to understand the mechanisms of primary and secondary transfer. As trace evidence can be transferred during the commission of a crime, it can also be transferred during the search process. Not only can investigators inadvertently pick up hairs and fibers, but they can be inadvertently deposited at the crime scene. The following are considerations at the crime scene:

- Elimination hair and/or fiber samples may need to be obtained from personnel conducting the search.
- Prioritize the order of evidence collection. Collect large items first and then proceed to the trace evidence. Use caution when walking the crime scene.
- Once the trace evidence is collected via vacuuming, taping, or tweezing, take blood samples, remove bullets, dust for fingerprints, and so on.
- Processing the crime scene for fingerprints prior to trace evidence collection is not recommended because it can inadvertently transfer trace evidence onto the clothing of the technicians, can move trace evidence, and/or contaminate trace evidence with dusting powder.

Glass Evidence

Because glass is so prevalent, breaks easily, and when fragmented has a tendency to adhere to clothing and body surfaces, it is frequently encountered within the context of crime scene investigation as **transfer evidence**.

Fracture Analysis

Examination of glass that has been fractured can lead the crime scene investigator to determine which direction the impact originated. This can be useful in determining whether or not a window was broken from the inside or outside (e.g., in attempting to identify whether or not a window was broken to cover up an employee theft and disguise it as a burglary). To make this determination, the crime scene investigator must understand what happens when glass fractures. First of all, a crime scene investigator can observe the **radial fractures**, those originating from the point of impact and moving away from that point, and other fractures that appear to make a typically broken series of concentric circles around the impact point, known as **concentric fractures**. Next, the crime scene investigator can observe the edges of the glass and will see the glass will show characteristics referred to as **conchoidal fractures** (**Figure 9.2**). These stress marks are shaped like arches that are perpendicular to one side of the glass surface, and curved nearly parallel to the opposite glass surface. This is telling to a crime scene investigator because for radial fractures, the perpendicular edge is always opposite the direction from which the force was applied to the glass surface. In concentric fractures, the perpendicular end always faces the direction to which the force was applied to the glass surface. One way to remember this is the **3R rule**: Radial fractures form a Right angle at the Reverse side to which force was applied.

A crime scene investigator can also determine glass breakage sequence through careful analysis of present characteristics (**Figure 9.3**). By taking

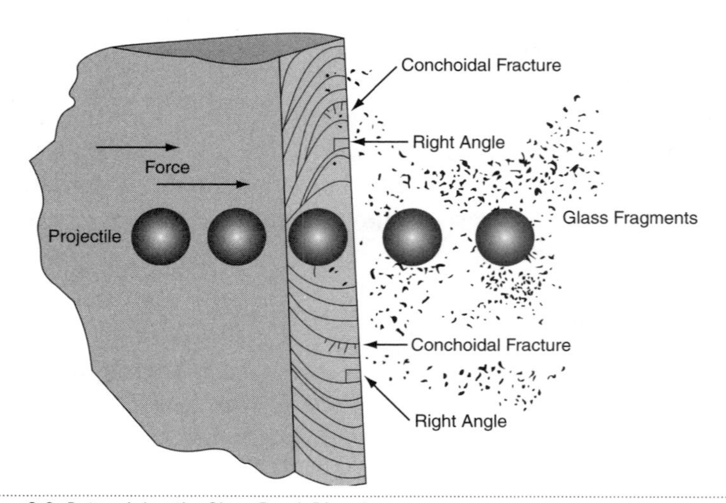

Figure 9.2 Determining the Glass Break Direction.

Source: Courtesy of Ellie Bruchez, University of Wisconsin–Platteville.

note that some cracks reach a termination point at their intersection with other fractures, the crime scene investigator can conclude that fractures that terminate occurred after those that do not. This will give an idea as to the glass break sequence. Impact direction can be gauged if enough of the initial impact point is still present. At the point of impact, a crime scene investigator will note that there will be evidence of a cone-like core ejection from the exit side of the glass. An example of when this would be used is in determining which bullet hole was made first within a window pane. If a suspect was shooting out of a glass picture window at police, and the police were shooting into the picture window from outside, a court will want to know who fired the first shot. Careful analysis of the glass fractures will help to determine this fact.

Collecting and Packaging Glass Evidence

In order for glass evidence to serve an evidentiary purpose, it must be properly collected and preserved. The gathering of glass evidence at the crime scene and from a suspect or victim must be thorough if the forensic lab is to have a chance to attempt analysis that would individualize the glass fragments as being from a common source. This requires that the crime scene investigator have a working knowledge as to what kinds of information can be derived from glass evidence.

Firstly, glass evidence is typically viewed as class evidence and is only rarely able to be individualized. This can result in both inclusions and exclusions of persons and glass surfaces, which although not unique, is certainly advantageous within the context of a criminal investigation.

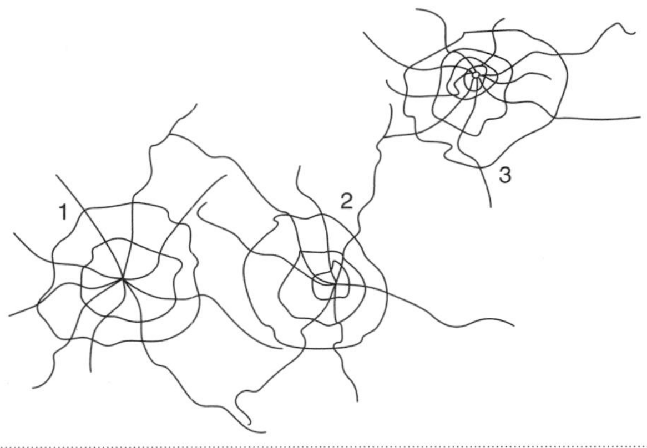

Figure 9.3 Determining the Glass Break Sequence.

Source: Courtesy of Ellie Bruchez, University of Wisconsin–Platteville.

Depending upon the crime, and the scene of the crime, it may be necessary for the crime scene investigator to attempt to collect all of the glass evidence located at the scene. In other cases it may only be necessary for the crime scene investigator to collect a known sample of the glass. For instance, in a vehicle versus pedestrian hit-and-run accident, a crime scene investigator may find it necessary to recover all related headlight and reflector lens glass so that if a vehicle is later identified, the pieces can be puzzled together to show the uniqueness and thus include the vehicle as a suspect vehicle. Simply collecting a known sample of the glass at the scene would only result in glass analysis being able to determine that the suspect vehicle had glass components consistent with the glass found at the scene, which would typically be the case for any such make and model vehicle having the same automotive parts. This would be enough to include the vehicle as a possible suspect vehicle but not sufficiently individualize it, as would the puzzling together of all related glass and showing the corresponding match.

However, when an individual fit is not likely or necessary, the crime scene investigator should collect a reference sample of broken glass found at the scene of the crime, and then properly collect and package the suspect or victim's clothes for comparative analysis. The reference sample should be collected as near the point of impact as possible. Approximately one square inch of material is all that will be required for forensic analysis relating to this comparison. Therefore, there is no need to collect an abundance of glass from the scene.

The fragment(s) should be packaged in solid containers, which will avoid further breaking the glass while in transit, and also not allow the glass to protrude from the packaging and injure personnel handling the evidence. If there are suspect or victim clothes that will be submitted to the lab for glass analysis, they should be wrapped individually in paper to ensure that the trace evidence is not lost in transport, and then packaged in a porous material such as paper bags or cardboard boxes. When the item is wrapped in the paper, the item should be folded within the paper in such a way that no surface of the clothing touches another surface of the same clothing, but instead has paper between them. This will ensure that there is no transfer or contamination of trace evidence as well as properly contain the trace evidence. The crime scene investigator should not attempt to shake the clothes to dislodge and thus collect the glass particles but should instead leave the discovery, collection, and analysis of glass evidence to forensic lab personnel.

If direction of force analysis is to be determined by the forensic lab, the crime scene investigator should attempt to collect as much of the glass evidence at the scene as possible. Also, if possible, the crime scene investigator should note which side of the glass was interior and which was exterior. If unknown, dirt, residue, etc. may be used to assist with the determination.

Paint Evidence

Paint chips and fragments of other protective coatings such as varnishes, sealers, lacquers, enamels, and plastics are frequently recovered at scenes of burglaries, hit-and-run vehicles and scenes, forced entries, and other crime scenes. A determination of common origin is possible in cases where irregularly shaped adjoining edges of paint chips can be physically joined to form a fracture match. However, the value of a single-layered paint chip or paint smear should not be overlooked.

Paint Evidence Analysis

The International Forensic Automotive Paint Data Query (PDQ) is a searchable database of chemical and color information of original automotive paints. Developed by Forensic Laboratory Services (FLS), the database represents 30 years of accumulated information provided by automotive companies and samples of vehicles submitted by other forensic laboratories or police. The PDQ contains information about the make, model, year, and assembly plant for many vehicles.

The PDQ is used by forensic laboratories around the world to assist with criminal investigations requiring vehicle identification. FLS maintains the database so that forensic laboratories can provide timely and effective support to police investigations. Police agencies do not access the PDQ directly.

The PDQ is one of the most collaborative ventures to be found in forensic science. Accredited users are given a free copy of the PDQ in exchange for submitting 60 new automotive paint samples per year. The paint samples are collected from vehicles at body shops or junk yards as well as from automobile manufacturers. Contributors to the PDQ include the Royal Canadian Mounted Police (RCMP), provincial forensic laboratories in Ontario and Quebec, 40 American forensic laboratories, and police agencies in 21 other countries.

Paint samples provided by automotive manufacturers are analyzed by their chemical composition. The chemical components and proportions are coded into the database. An automotive paint job usually consists of four layers. These known paint samples are then available for comparison against paint samples taken from a crime scene or from suspect vehicles.

The PDQ team samples each paint layer with scalpels and microscopes to determine the spectra and chemical composition of the paint chips. Once unknown vehicles are matched in the database, police can use the possible make, model, and year information to search for the vehicle involved in the criminal activity, most often a hit-and-run.

The PDQ serves to narrow the search for vehicles to enable further investigation. Once a suspect vehicle is located, the comparison of paint from the crime scene with paint of the suspect vehicle can be conducted to match the vehicle more conclusively to the crime.

Collecting and Packaging Paint Evidence

The following procedures are recommended for recovery of paint samples (WIDOJ, 2004):

A. Recover, package, and seal all paint samples separately.
B. Recover known paint samples from areas immediately adjacent to the damaged area. The hoods, trunks, and fenders of vehicles may not be painted at the same location or with the same paint used on the body. Therefore, it is of utmost importance that a known paint sample be taken from the exact part of the vehicle upon which the damage occurred. In hit-and-run investigations, the known paint samples should be taken near the point of impact, but should not be taken from areas of corrosion, such as the rocker panels.
C. When tool marks exist on a damaged object, recover paint samples from areas immediately adjacent to tool marks without mutilating the tool mark.

   Tape a clean sheet of paper (do not use envelopes) to the object in the manner shown, forming a pocket. Mark the paper for identification. Scrape the questioned paint into the pocket formed by the paper. It is important to use a new, disposable scalpel blade or razor blade for each sample to avoid contamination. Some razor blades are coated with oil to prevent rusting. Therefore, all razor blades should be thoroughly cleaned with a clean cloth or tissue just before they are used. Carefully remove the paper from the object and fold each edge toward the center so that the packet is completely closed. Insert each packet into a separate new, clean bottle, together with the scalpel blade or razor blade used in recovering the sample. Seal and label according to instructions previously recommended.

D. When areas of paint are missing from sheet metal parts of vehicles or doors and windows of residences and businesses, consideration should be given to bringing the entire part to the laboratory for possible fracture match analysis.

E. Avoid use of any container that would permit loss or contamination of contents, especially envelopes and plastic petri dishes, because the manufacturer's seal is not leak proof.

F. Use a new, clean scalpel blade or razor blade for each sample recovered; enclose in the bottle with the sample.

G. Do not use gummed tape to recover paint samples because it interferes with the chemical analysis.

■ Soil Evidence

Soil is a generic term for any disintegrated surface material, natural or man-made, which lies on or near the earth's surface (Saferstein, 2007). This is important for a crime scene investigator to realize because forensic analysis of soil not only relates to such naturally occurring components as rocks, vegetation, pollen, minerals, and animal matter, but it also includes the identification and examination of such man-made components as paint particles, glass, asphalt, concrete, and other items present that may assist in characterizing the soil as being unique to a specific or definable location.

The evidentiary value of soil has its foundation in its prevalence at crime scenes and the ease at which it is transferred between the scene and the suspect. Thus, soil or the dried remains of mud found adhered to a suspect's clothing or footwear, when compared to soil samples collected from the crime scene, may provide an associative link between a suspect or an object and the crime scene. As with most physical evidence, soil analysis is comparative in nature. This means that soil that is located in the possession of a suspect must be carefully collected in order to be compared to soil samples from the crime scene and its surroundings. However, even if the scene of a crime has not been determined, the crime scene investigator should not overlook the evidentiary value of soil. For instance, a forensic geologist who is educated in the local geology may be capable of using her or his knowledge along with geological maps, to direct law enforcement to the possible vicinity where the soil most likely originated from, and thus provide a possibility as to the area where the crime was committed.

Sand or soil is encountered in many types of investigations and should not be overlooked by the investigating officer. The following items frequently have soil related to the crime scene adhering to their surfaces: footwear, clothing, tool containers, vehicle operating pedals, under-carriages, or wheel wells. Soil found on the floor of a vehicle also may become valuable evidence. Most soil adhering to such objects is representative of the upper one-fourth inch of surface soil from which it originated and may be associated with its source if proper known samples are recovered.

Most times soil can be distinguished through a gross examination, primarily based upon coloration. It is estimated that there are

Collection of Soil Evidence

A. Photograph impressions (tire, footwear, etc.) that have patterns containing discernible class and/or individual characteristics.
B. Do not recover soil samples until proper photographs and casts have been made.
C. Obtain samples consisting of three (3) tablespoonfuls of soil from the top 1/4″ of soil from within the impression. Collect a soil sample from each impression that has a different color or texture (dirt, clay, sand, etc.).
D. The tablespoon should be cleaned after each area has been sampled to avoid cross-contamination of samples.
E. The soil samples should be recovered using a systematic method.
F. Prepare a diagram of the area showing the point from which each sample was recovered and submit copy of diagram. On the diagram, show distance and direction of each sample recovered. The diagram should be oriented with fixed objects in the area such as buildings, utility pole, etc.
G. If soil samples can be immediately transported to the Laboratory, put samples in individual clean glass containers and seal.
H. If samples cannot be transported to the Laboratory immediately, to prevent the growth of mold, spread the samples on separate sheets of clean paper to air dry for at least 24 hours. When dry, package and seal. Do not mix or contaminate samples.
I. If suspect shoes have been recovered and soil is present on the shoes, recover soil sample from the area of the footwear impression that corresponds to the location of the soil on recovered shoes. Example: clump of soil adhering to arch of shoe.
J. When excavating grave sites, remember soil could be adhering to the tools used by the suspects or their clothing. Therefore, a core sample should be taken from as close to the grave as possible and to a depth equal or greater than the grave itself. The core sample should be kept intact and packaged so the layer structure is not altered or damaged during drying, handling or shipping.

Source: Courtesy of the Wisconsin Department of Justice (WIDOJ). (2004). *Physical evidence handbook*. Madison, WI: Wisconsin Department of Justice.

approximately 1100 different soil colors; thus, comparison of soil color is a logical first step in soil analysis. A crime scene investigator should note, however, that soil is darker when it is wet and, therefore, color comparisons should always be made when the samples are dry.

Collecting and Packaging Soil Evidence

Because soil variation is the fundamental consideration in soil analysis, establishing such variation must be considered when gathering soil specimens. It is suggested that standard/reference samples of soils are collected at various intervals within a 100-meter radius of the crime scene, in addition to collection at the site of the crime, for comparison to the questioned soil.

Hair Evidence

Hairs, which are composed primarily of the protein keratin, can be defined as slender outgrowths of the skin of mammals. Each species of animal possesses hair with characteristic length, color, shape, root appearance, and internal microscopic features that distinguish one animal from another (**Figure 9.4**). Considerable variability also exists in the types of hairs that are found on the body of an animal. In humans, hairs found on the head, pubic region, arms, legs, and other body areas have characteristics that can determine their origin. On animals, hair types include coarse outer hairs or guard hairs, the finer

Figure 9.4 Hair and Fiber Comparison Under a Microscope.

fur hairs, tactile hairs such as whiskers, and other hairs that originate from the tail and mane of an animal.

Because hairs can be transferred during physical contact, their presence can associate a suspect to a victim or a suspect/victim to a crime scene (**Figure 9.5**). The types of hair recovered and the condition and number of hairs found all impact on their value as evidence in a criminal investigation. Comparison of the microscopic characteristics of questioned hairs to known hair samples helps determine whether a transfer may have occurred.

Hair is composed of three primary sections or layers. The outermost layer is known as the **cuticle**. The cuticle contains the scaly protective layer that covers the shaft of the hair. Each species has identifiable cuticle characteristics. The innermost region is known as the **medulla**. The medulla in humans is amorphous and lacks visible cellular material, whereas the medulla of other species is often seen as cellular in nature, some displaying characteristics similar to a bead of pearls. The region between the medulla and cuticle is known as the **cortex**. The cortex contains the pigment cells that are responsible for imparting hair color characteristics (**Figure 9.6**).

Hair is present on many different regions of the body. Each region, such as the head, pubic area, chest, axillae, and limbs, has hairs with microscopical characteristics attributable to that region. Although it is possible to identify a hair as originating from a particular body area, the

Figure 9.5 Photograph of Hair/Fur Evidence at a Crime Scene.

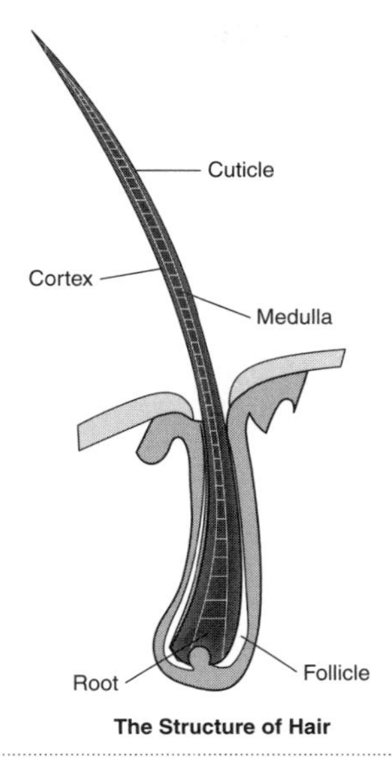

Figure 9.6 The Structure of Hair.

Source: Courtesy of Erica Lawler, University of Wisconsin–Platteville.

regions of the body that are primarily used in forensic comparisons are the head and pubic areas. As hairs undergo a cyclical growth (anagen) and resting phase (telogen), the visible microscopic characteristics are sufficient to determine the phase of growth of the hair.

During the anagen phase, the hair is actively growing, and materials are deposited in the hair shaft by cells found in the follicle. Metabolically active and dividing cells above and around the dermal papilla of the follicle grow upward during this phase, to form the major components of the hair: the medulla, cortex, cuticle, and accompanying root sheath. In the telogen phase, the follicle is dormant or resting. The transition period between the anagen and telogen phases is referred to as the catagen phase.

Hairs are routinely lost during the telogen phase and often become a primary source of evidentiary material. An example of this natural shedding process can be seen when one combs through the hairs on the head. It is common for hairs of this type to be transferred to another individual or to an object during physical contact. Hairs also can become dislodged from the body while they are in an actively growing state, such as by pulling or by striking with an object. The microscopical appearance of the root area will allow for the determination of the growth phase.

The basic morphology of human hairs is shared by each individual in the population, but the arrangement, distribution, and appearance of individual microscopic characteristics within different regions of hair routinely allow a skilled hair examiner to differentiate hairs between individuals. An analogy would be the ability of an individual to recognize the face of a friend or relative in a crowd even though each person in the crowd possesses ears, eyes, a nose, and a mouth.

Animal Hairs

Animal hairs discovered on items of physical evidence can link a suspect or location to a crime of violence. A victim placed in a vehicle or held at a location where animals are routinely found often have animal hairs transferred to the victim's clothing. Cat or dog hairs can be found on the adhesive portions of ransom and extortion notes prepared by pet owners. The transfer of pet hairs to the victim or crime scene may also occur when the suspect is a pet owner and has animal hairs on his or her clothing when the contact occurs. This is referred to as a *secondary transfer* of trace material (discussed below).

When an animal hair is found, it is identified to a particular type of animal and microscopically compared with a known hair sample from either an animal hair reference collection or a specific animal. If the questioned hair exhibits the same microscopic characteristics as the known hairs, it is concluded that the hair is consistent with originating from that animal. It is noted, however, that animal hairs do not possess enough individual microscopic characteristics to be associated with a particular animal to the exclusion of other similar animals.

The collection of a suitable known animal hair standard is necessary before a meaningful comparison can be conducted. Because hairs can vary widely in color and length on different areas of the body of an animal, hairs should be collected from each area. While a minimum number of hairs is difficult to determine, good judgment should be used in collecting enough hairs to represent the various types and colors of hairs found on the animal. The sample should contain full-length hairs and should include combings as well as pluckings. If the animal is not available for sample collection, a brush or comb used for the animal may be substituted. Sometimes hair samples collected from a dog or cat bed may be useful when actual samples from the animal cannot be obtained.

In crimes where personal contact has occurred, especially if there was physical force, hair and fibers are frequently found as evidence. A cross transfer of hair and/or fibers between a victim and an assailant can provide supportive evidence of an association. In addition, hair recovered from the scene may serve to associate an individual with the scene. Fibers recovered from the clothing of the victim, suspect, and crime scene can be compared to known textile materials to determine possible sources of origin.

If a hair is determined to be of human origin and is deemed probative to a case, DNA analyses may be performed on the root (if present) of the hair. If DNA is obtained from a questioned hair root, this DNA can be compared to DNA from a standard blood sample from an individual. Identification may be the result.

Body Area Determination

The body area from which a hair originated can be determined by general morphology. Length, shape, size, color, stiffness, curliness, and microscopic appearance all contribute to the determination of body area. Pigmentation and medullar appearance also influence body area identification. Hairs that exhibit microscopic characteristics shared by different anatomical areas are often referred to as *body hairs*. These include hairs found on the upper legs, lower abdomen, and back. Because there is a wide range of interpersonal variation in head and pubic hairs, the majority of work in forensics has been in comparing and differentiating hairs from the head and pubic regions.

Racial Determination

A human hair can be associated with a particular racial group based on established models for each group. Forensic examiners differentiate between hairs of Caucasoid (European ancestry), Mongoloid (Asian ancestry), and Negroid (African ancestry) origin, all of which exhibit microscopic characteristics that distinguish one racial group from another. Head hairs are generally considered best for determining race, although hairs from other body areas can be useful. Racial determination from the microscopic examination of head hairs from infants, however, can be difficult, and hairs from individuals of mixed racial ancestry may possess microscopic characteristics attributed to more than one racial group. The identification of race is most useful as an investigative tool, but it can also be an associative tool when an individual's hairs exhibit unusual racial characteristics.

Age and Sex

The age of an individual cannot be determined definitively by a microscopic examination; however, the microscopic appearance of certain human hairs, such as those of infants and elderly individuals, may provide a general indication of age. The hairs of infants, for example, are generally finer and less distinctive in microscopic appearance.

As individuals age, hair can undergo pigment loss and changes in the configuration of the hair shaft to become much finer and more variable in diameter.

Although the sex of an individual is difficult to determine from microscopic examination, longer, treated hairs are more frequently encountered in female individuals. Sex can be determined from a forcibly removed hair (with tissue), but this is not routinely done. Definitive determination of sex can be accomplished through the staining of sex chromatin in the cells found in the follicular tissue, but nuclear DNA and mitochondrial DNA (mtDNA) tests will provide more specific information regarding the possible origin of the hair.

Significance of Hair Evidence

The significance of hair examination results is dependent on the method of evidence collection used at the crime scene, the evidence processing techniques employed, the methodology of the hair examination process, and the experience of the hair examiner. Head hairs and pubic hairs are routinely held as more significant than hairs from other body areas.

Collection of Hair and Fiber Standards

It is necessary to obtain standard hair and fiber samples from all possible sources (suspect, victim, and scene) for comparison with questioned hairs and fibers. DNA analysis on hair roots has replaced microscopic hair comparisons. Pubic and head hair standards are still necessary for determining which foreign, questioned hairs may be subjected to DNA analysis. Due to the ease of head hair transfer and potential limited probative value, DNA analysis on hairs will be limited.

Collection of Questioned Hair and Fibers

Pubic Hair Combings

These are usually collected in sexual assault cases and from homicide victims.

Place a piece of paper under the pubic region of the individual and comb through the entire pubic area to dislodge any foreign hairs or other material that may be present. Place the used comb onto the paper and fold the paper around the comb, being careful not to lose any of the dislodged evidence. Place the wrapped comb in a clean, properly-labeled envelope, and seal.

Other Recovered Questioned Hairs and Fibers

Separately package the hairs and fibers collected from different persons and different locations.

Suggestions for Collection of Hair and Fiber Standards

Head Hair Standards

Obtain at least fifty (50) head hairs by cutting them at the skin surface. These hairs should be collected from various areas of the head such as the crown, sides, front, and back to assure that all shades of color and texture have been adequately sampled. The quantity of hairs obtained from a deceased individual should be doubled and the hairs should be pulled. Place the hairs in a clean, properly labeled envelope, and seal.

Pubic Hair Standards

Obtain at least twenty (20) pubic hairs by cutting them at the skin surface. The hairs should be collected from various areas within the pubic region. If this is a deceased individual, double the number of hairs to be collected and pull them. Place the hairs in a clean, properly labeled envelope, and seal.

Known Fibers

Known fibers should be obtained from all possible sources (clothing, drapes, rugs, etc.).

Submit the suspected source in total if possible. Place the source in a clean, properly-labeled paper bag, and seal. If it is not feasible to submit the source in total, a sufficient quantity should be taken to ensure that each color and type of fiber involved has been sampled. Place the fibers in a clean, properly-labeled envelope or glass jar, and seal.

Source: Courtesy of the Wisconsin Department of Justice (WIDOJ). (2004). *Physical evidence handbook*. Madison, WI: Wisconsin Department of Justice.

When the amount of evidence is very small, extreme care should be exercised so that the material is not contaminated or inadvertently lost. The hair or fiber should be placed on a piece of clean white paper and the paper should be tightly folded around the hair or fiber. Place the paper packet in a clean, properly-labeled envelope or glass jar, and seal.

Collecting and Packaging Hair Evidence

Submit individual hairs and fibers in clean paper or in an envelope with sealed corners. The primary paper or envelope should be placed inside a secondary sealed envelope with all corners taped. Many times individual hairs identified on items of clothing are not removed or secured. These hairs may move or be lost, so it is recommended that they be removed and placed in an envelope (first noting where they were removed).

If a floor surface is vacuumed, the debris should be placed on a white sheet of paper (8 × 11 inches) and folded at the corners. This paper should be placed in a heat-sealed or re-sealable plastic bag.

Hair Evidence Analysis

When evidence is received by the forensic laboratory, the case is assigned to an examiner. The examiner will read the incoming communication to determine the nature of the offense, the names of the suspects and victims, and the types of requested examinations. It is important to have an understanding of the offense to help determine the course of action in the laboratory. Contact between family members or friends would be treated differently, as the transfer of trace evidence would be more likely in these cases.

Typically, there are several processes by which debris is collected. This is conducted through a combination of picking, scraping, and sometimes taping. Whereas taping is considered by some laboratories to be a preferred technique, it can be time-consuming and tedious, and may present a storage problem for the tape collections. Vacuuming is not recommended for clothing items. The clothing of a victim is processed in a room other than the room in which the suspect's clothing is processed. The collected debris is placed in pillboxes and examined with a stereo-binocular microscope. Hairs are then mounted on glass microscope slides for identification and comparison purposes.

When a questioned hair exhibits the same microscopic characteristics as the known hairs of an individual, the hair could have originated from that individual (see Figure 9.5). If the questioned hair is microscopically dissimilar to the known hair standard, it cannot be associated with the individual. Different people generally have different hair characteristics, but differences in microscopic characteristics also can be the result of time and alteration. Known hair samples should be collected from individuals as soon as possible to the date of the crime. As time passes, microscopic characteristics can change, and the individual may alter the color with dyes.

Certain case situations affect the significance of identifying hairs. When a family member may be involved in a crime, the location, number, and condition (forcibly pulled or burnt, for example) of recovered hairs may be important. The involvement of the victim's associates, including dates, coworkers, and other people who may have logical contact with or access to the victim and/or crime scene is an additional consideration in hair examinations. Situations involving strangers have greatest significance when hair associations have been made.

■ Fiber Evidence

A *fiber* is the smallest unit of a textile material that has a length many times greater than its diameter. Fibers can occur naturally as plant and animal fibers, but they can be man-made also. A fiber can be spun with other fibers to form a yarn that can be woven or knitted to form a fabric. The type and length of fiber used, the type of spinning method, and the type of fabric construction all affect the transfer of fibers and the significance of fiber associations. This becomes very important when there is a possibility of fiber transfer between a suspect and a victim during the commission of a crime.

As discussed previously, fibers are considered a form of trace evidence that can be transferred from the clothing of a suspect to the clothing of a victim during the commission of a crime. Fibers also can transfer from a fabric source such as a carpet, bed, or furniture at a crime scene. These transfers can either be direct (primary) or indirect (secondary). A **primary transfer** occurs when a fiber is transferred from a fabric directly onto a victim's clothing, whereas a **secondary transfer** occurs when already transferred fibers on the clothing of a suspect transfer to the clothing of a victim. An understanding of the mechanics of primary and secondary transfer is important when reconstructing the events of a crime.

When two people come in contact or when contact occurs with an item from the crime scene, the possibility exists that a fiber transfer will take place. This does not mean that a fiber transfer will always take place. Certain types of fabric do not shed well (donor garments), and some fabrics do not hold fibers well (recipient garments). The construction and fiber composition of the fabric, the duration and force of contact, and the condition of the garment with regard to damage are important considerations.

Another important consideration is the length of time between the actual physical contact and the collection of clothing items from the suspect or victim. If the victim is immobile, very little fiber loss will take place, whereas the suspect's clothing will lose transferred fibers quickly. The likelihood of finding transferred fibers on the clothing of the suspect a day after the alleged contact may be remote, depending on the subsequent use or handling of that clothing.

Natural Fibers

Many different natural fibers originating from plants and animals are used in the production of fabric. Cotton fibers are the plant fibers most commonly used in textile materials, with the type of cotton, fiber length, and degree of twist contributing to the diversity of these fibers. Processing techniques and color applications also influence the value of cotton fiber identifications.

Other plant fibers used in the production of textile materials include flax (linen), ramie, sisal, jute, hemp, kapok, and coir. The identification of less common plant fibers at a crime scene or on the clothing of a suspect or victim would have increased significance.

The animal fiber most frequently used in the production of textile materials is wool, and the most common wool fibers originate from sheep. The end use of sheep's wool often dictates the fineness or coarseness of woolen fibers: Finer woolen fibers are used in the production of clothing, whereas coarser fibers are found in carpet. Fiber diameter and degree of scale protrusion of the fibers are other important characteristics. Although sheep's wool is most common, woolen fibers from other animals also may be found. These include camel, alpaca, cashmere, mohair, and others. The identification of less common animal fibers at a crime scene or on the clothing of a suspect or victim would have increased significance.

Man-Made Fibers

More than half of all fibers used in the production of textile materials are man-made. Some man-made fibers originate from natural materials such as cotton or wood; others originate from synthetic materials. Polyester and nylon fibers are the most commonly encountered man-made fibers, followed by acrylics, rayon, and acetates. There are also many other less common man-made fibers. The amount of production of a particular man-made fiber and its end use influence the degree of rarity of a given fiber.

The shape of a man-made fiber can determines the value placed on that fiber. The cross section of a man-made fiber can be manufacturer-specific: Some cross sections are more common than others, and some shapes may only be produced for a short period of time. Unusual cross sections encountered through examination can add increased significance to a fiber association.

Fiber Color

Color influences the value given to a particular fiber identification. Often several dyes are used to give a fiber a desired color. Individual fibers can be colored prior to being spun into yarns. Yarns can be dyed, and fabrics made from them can be dyed. Color can be applied to the surface of fabric also, as found in printed fabrics. How color is applied and absorbed along the length of the fiber are important comparison characteristics. Color-fading and discoloration also can lend increased value to a fiber association.

Fiber Evidence Analysis

Whenever a fiber found on the clothing of a victim matches the known fibers of a suspect's clothing, it can be a significant event. Matching dyed synthetic fibers or dyed natural fibers can be very meaningful, whereas the matching of common fibers such as white cotton or blue denim cotton would be less significant. In some situations, however, the presence of white cotton or blue denim cotton may still have some meaning in resolving the truth of an issue. The discovery of cross transfers and multiple fiber transfers between the suspect's clothing and the victim's clothing dramatically increases the likelihood that these two individuals had physical contact.

It is argued that the large volume of fabric produced reduces the significance of any fiber association discovered in a criminal case. It can never be stated with certainty that a fiber originated from a particular garment because other garments were likely produced using the same fiber type and color. The inability to positively associate a fiber with a particular garment to the exclusion of all other garments, however, does not mean that the fiber association is without value.

Another important consideration is coincidence. When fibers that match the clothing fibers of the suspect are found on the clothing of a victim, two conclusions may be drawn: The fibers originated from the suspect, or the fibers originated from another fabric source that not only was composed of fibers of the exact type and color, but was also in a position to contribute those fibers through primary or secondary contact. The likelihood of encountering identical fibers from the environment of a homicide victim (i.e., from his or her residence or friends) is extremely remote **(Figure 9.7)**.

Figure 9.7 Scaled Photo of Fiber Evidence.

Collecting and Packaging Fiber Evidence

Collection and preservation of fiber evidence is quite similar to the methodology for collection of hair evidence. Reference and elimination samples must be collected, and each must be properly documented and packaged, as mentioned previously in regards to hair evidence.

Chapter Summary

Edmond Locard dedicated his life to the microscopic analysis of trace evidence in an effort to prove that one's actions could be traced back to a particular item or location. It is this study that the forensic value of trace evidence is based upon. Utilizing the product rule, one is able to determine the frequency for the occurrence of an event and the probability that the item of trace evidence in question is related to a particular location, item, or person. While most all trace evidence is not individualistic in nature, this does not minimize its forensic importance. Such matters are up to the analyst who is performing the analysis of the evidence. What is imperative is that the crime scene investigator knows the proper method for documenting, collecting, and preserving the various types of trace evidence to ensure their ability to undergo proper forensic analysis.

Review Questions

1. ______________ is the frequency of an event occurring.
2. The ______________ is when the frequency of independently occurring variables are multiplied together to obtain an overall frequency of occurrence for the event or item.
3. The 3R rule states: ____
____
____.
4. It is important for crime scene investigators to understand the mechanisms of primary and secondary transfer. As trace evidence can be transferred during ______________, it also can be transferred during ______________.
5. Which type of trace evidence samples need to be collected from the personnel conducting the search and collection of trace evidence?
6. What is PDQ and how is it involved with CSI?
7. ______________ is a generic term for any disintegrated surface material, natural or man-made, which lies on or near the earth's surface.

8. Most soil can be distinguished through a gross examination, primarily based upon ______________.
9. Hairs are routinely lost during the ______________ phase and often become a primary source of evidentiary material.
10. Polyester, nylon acrylics, rayons, and acetates are all examples of ______________.

Case Studies

Research the case of Wayne Williams and determine how it relates to the topics presented in this chapter. Discuss how trace evidence was used within that case, and how statistical analysis assisted crime scene personnel with limiting the possible suspect pool.

References

Gale, T. (2005). Locard's Exchange Principle. *World of Forensic Science.* Retrieved November 14, 2009, from http://www.encyclopedia.com/doc/1G2-3448300354.html

Kirk, P. L. (1953). *Crime Investigation* (p. 4). New York: Interscience.

International Association of Identification (IAI). (2008). *Welcome to the IAI!* Retrieved September 3, 2008 from http://www.theiai.org

Saferstein, R. (2007). *Criminalistics: An introduction to forensic science,* 9th ed. Upper Saddle River, NJ: Pearson Prentice Hall.

U.S. Department of Justice (USDOJ), Federal Bureau of Investigation (FBI). (2008). Birth of the technical laboratory 1924–1935. Retrieved September 2, 2008, from http://www.fbi.gov/libref/historic/history/birthtechlab.htm

Wisconsin Department of Justice (WIDOJ). (2004). *Physical evidence handbook* (7th ed.). Madison, WI: Wisconsin Department of Administration.

Blood and Biological Evidence

A man is suspected of a crime months perhaps after it has been committed. His linen or clothes are examined and brownish stains discovered upon them. Are they blood stains, or mud stains, or rust stains, or fruit stains, or what are they? That is a question which has puzzled many an expert, and why? Because there was no reliable test. Now we have the Sherlock Holmes test, and there will no longer be any difficulty.

Sir Arthur Conan Doyle (1859–1930)
Sherlock Holmes, in *Study in Scarlet*, 1887

▸▸ LEARNING OBJECTIVES

- Identify the necessary steps to address universal precautions where biological fluids are concerned.
- Differentiate between nuclear and mitochondrial DNA and the types of evidence each can yield.
- Understand the types and differences between presumptive tests and confirmatory tests for biological fluids.
- Explain the purpose of CODIS and how it is used within the law enforcement community.

▸▸ KEY TERMS

Complimentary Base Pairing
Criminalistics Light-Imaging Unit (CLU)
Combined DNA Index System (CODIS)
Deoxyribonucleic Acid (DNA)
Luminol
Mitochondrial DNA (mtDNA)
Polymer
Sexual Assault Evidence Collection Kit (SAE Kit)

■ Biological Evidence

In addition to being dynamic and unique, crime scenes also present hazards to crime scene personnel, as discussed in Chapter 5. One of the greatest hazards to personnel comes in the form of biological fluids. Crime scenes frequently involve such biological evidence as blood, saliva, seminal fluid, vaginal secretions, urine, and feces. As discussed in Chapter 5, it is imperative that personnel take proper universal precautions in such instances for two reasons: to protect themselves from the hazards inherent with such evidence, and also to prevent the contamination of the evidence from the crime scene investigators themselves. Once precautions have been taken, then crime scene personnel must set about the task of attempting to identify, interpret, document, collect, and preserve the biological evidence. This chapter considers the recognition, collection, and preservation of some of these commonly encountered forms of biological evidence.

Blood Evidence

Is it blood? When a crime scene investigator encounters material at a crime scene that he or she believes may be blood, the following questions must be answered:

- Is it blood?
- Is it human blood?
- If the blood is of human origin, how closely can it be associated with a particular individual?

These questions are answered through a variety of methods. The first step is to test to see whether or not the item in question is in fact blood.

Presumptive Test for Blood

Crime scenes may often contain minute quantities of blood that, because of their small size, may not be readily noticed. Presumptive blood tests are initial screening tools that indicate, but are not specific for, the identification of blood. A positive result can result from the presence of human or animal blood and a variety of plant materials that contain the chemical peroxidase (e.g., horseradish, potatoes, etc.). The most common crime scene presumptive tests for blood include leucomalachite green, phenolphthalein, and tetramethylbenzidine (TMB). TMB is a presumptive test for blood that is based on the peroxidase-like activity of hemoglobin. It is also the most common lab presumptive test performed on suspected blood. TMB testing is easily conducted in the field, making use of the chemical TMB, distilled water, and 3% hydrogen peroxide. TMB also will yield false-positives on stains that visually resemble blood, such as rust and iodine. This is why it is only a presumptive test and the stain must be confirmed as being human blood in a forensic laboratory.

In some cases, presumptive blood tests are used as a search method. They can also help to differentiate blood from the presence of rust, chocolate, tar, or other brownish colored stains that are often incorrectly assumed to be blood. **Luminol** is an example of a presumptive search technique that results in chemiluminescence as a result of the chemical reaction occurring between the reagent and the biological stain. It is fast acting and must be documented in darkness, using photography in an expeditious manner. Most times the chemiluminescence will not be present for more than 30 or 40 seconds.

Utilizing a presumptive test can also assist the crime scene investigator in identifying priority areas or evidence that must be tended to prior to other crime scene processing activities. It is extremely important to collect and preserve biological evidence early in the investigative process in order to prevent its loss or deterioration.

Once the item or area has been identified as being presumptive positive for blood, it is up to the forensic laboratory to conduct confirmatory testing that will identify whether or not it is truly blood, and if it is, whether or not it is of human origin. If the blood is of human origin, then it may become necessary to identify whether or not the blood can be sourced back to a particular individual.

Collection of Blood Evidence

In crimes of violence, blood is usually found in the form of dried stains. Blood frequently can be identified and genetically compared to blood standards from individuals if a sufficient quantity is properly collected and submitted. In addition to identification and comparison testing, the shape and pattern of the bloodstaining may provide information concerning how the blood was deposited. (Documentation of bloodstain patterns is covered in greater depth in Chapter 11).

Record the following information:

- Physical state (fluid, moist, dry)
- Amount present (few drops, small pool, etc.)
- Shape (smear, round drops)
- Exact location in relation to fixed objects
- Pattern of stains (all in one spot, trail)
- Atmosphere conditions (temperature, humidity)
- Date and time of observation
- Take scaled and unscaled photographs of stains

Liquid Blood

If wet blood or a blood pool of fluid blood is present, collection should be made in the following manner:

- Wear gloves, mask, and eye protection while soaking up samples.
- Using cotton swabs, soak the suspected blood onto the swabs.
- Continue collecting the stain until it is either completely collected or until five swabs have been saturated.
- Make sure to avoid contamination of swabs.

- It is suggested to change gloves frequently and to change them immediately if contamination of gloves occurs.
- Allow the swabs to dry in place or place them on a nonporous surface like a glass microscope slide and allow to thoroughly air dry.
- Package the dried swabs in a paper container (i.e., white slide box, envelope, paper bag, etc.). Use separate containers for each area recovered. Label the paper container with a "biohazard" sign.
- Properly label and seal each container.
- Select an unstained area adjacent to the suspected bloodstain and collect a sample from this area as before. This sample will serve as a control. Package, label, and seal this control separately from the stained material.

Dried or Moist Bloodstain Recovery

If the stained object is transportable, submit the item intact. If the suspected blood is still moist, allow it to thoroughly air dry in a well ventilated but draft-free area prior to packaging. Label the area with "biohazard" signs. Package the item in a clean paper container, and then seal and label it. If it is impractical to submit the bloodstained item to the laboratory or it is not possible to cut or remove a portion of the stained and unstained area of the item, then collect in the following manner:

- Wear gloves, mask, and eye protection while collecting samples.
- Moisten a sufficient number of cotton swabs to collect the stain. It is better to underestimate the amount of swabs required as additional swabs can always be used.
- Wet the swabs using distilled water, or clean tap water if distilled water is not available.
- Do not allow the swabs to come into contact with any other object.
- Gently swab the stain with the moistened swabs until the swabs thoroughly absorb the blood and are a dark reddish brown coloration.
- Continue collecting the stain until it is either completely collected or five swabs have been saturated.
- Allow the stained swabs to thoroughly air dry either directly on the stained object or on a clean glass microscope slide. To avoid contamination, swabs can be dried by making a small perforation in a pillbox and placing the swab into the perforation. This will allow the swab to dry without contacting any other surfaces.
- Package, label, and seal the air-dried swabs in a paper container (i.e., white slide box, envelope, etc.). Label with a "biohazard" sign.

- Select an unstained area adjacent to the suspected bloodstain and collect a sample from this area as before. This sample will serve as a control. Package, label, and seal this control separately from the stained material.

Deoxyribonucleic Acid (DNA)

There was a time when the closest that one could get to identifying blood to a person was through blood type, which was class evidence and not at all individualistic. In 1901 Dr. Karl Lansteiner discovered that blood could be grouped into four different categories, called *types*:, A, B, AB, and O blood types. This finding intrigued Dr. Leone Lattes and in 1915 he devised a procedure for determining the blood group of a dried bloodstain. This offered forensic significance for the analysis of crime scene-related blood. However, again, it only offered the ability of investigators to isolate the evidence to a particular group but not to an individual. It was not until 1984 when Sir Alec Jeffreys made the revolutionary discovery of *deoxyribonucleic acid (DNA) fingerprinting* that the ability to isolate a specific individual (or individuals in the case of identical twins) was possible.

DNA Structure

Deoxyribonucleic acid (**DNA**) is the genetic building block for all living organisms, including some viruses. **Figure 10.1** shows the double helix structure of DNA. Virtually every cell contains DNA. DNA does

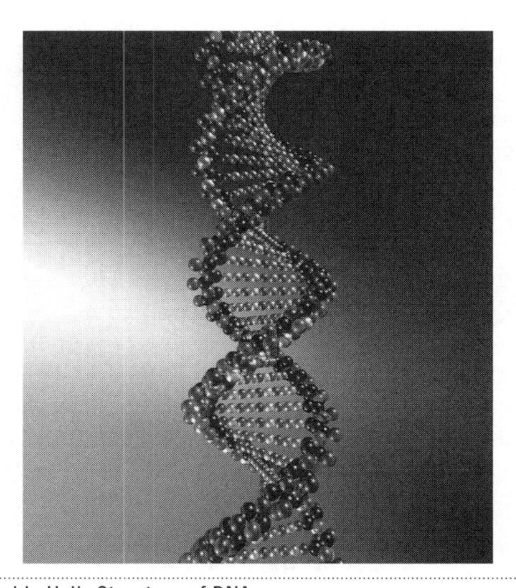

Figure 10.1 The Double Helix Structure of DNA.

not change throughout a person's life. With the exception of red blood cells, nearly all of the cells in the human body contain nuclear DNA that is housed within the nuclei of each cell. In blood, however, only white blood cells have a nucleus. Red blood cells shed their nucleus during their early formation.

Nucleic acid is typically extracted as source material for most crime scene DNA. Each nucleated cell has a single nucleus, and each nucleus holds 23 pairs of chromosomes. At conception, each parent contributes one half of these 23 pairs, creating a completely distinct set of chromosomes in the offspring. A person therefore inherits half of his or her genetic material (DNA) from each parent. However, what is important to understand about heredity is that a person inherits

Understanding DNA Evidence

"DNA fingerprinting" or DNA typing (profiling) as it is now known, was first described in 1985 by the English geneticist, Alec Jeffreys. Dr. Jeffreys found that certain regions of DNA contained DNA sequences that were repeated over and over again next to each other. He also discovered that the number of repeated sections present in a sample could differ from individual to individual. By developing a technique to examine the length variation of these DNA repeat sequences, Dr. Jeffreys created the ability to perform human identity tests.

These DNA repeat regions became known as VNTRs, which stands for variable number of tandem repeats. The technique used by Dr. Jeffreys to examine the VNTRs was called restriction fragment length polymorphism (RFLP) because it involved the use of a restriction enzyme to cut the regions of DNA surrounding the VNTRs. This RFLP method was first used to help in an English immigration case and shortly thereafter to solve a double homicide case. Since that time, human identity testing using DNA typing methods has been widespread. The past 15 years have seen tremendous growth in the use of DNA evidence in crime scene investigations as well as paternity testing. Today over 150 public forensic laboratories and several dozen private paternity testing laboratories conduct hundreds of thousands of DNA tests annually in the United States. In addition, most countries in Europe and Asia have forensic DNA programs. The number of laboratories around the world conducting DNA testing will continue to grow as the technique gains in popularity within the law enforcement community.

DNA evidence is playing a larger role than ever before in criminal cases throughout the country, both to convict the guilty and to exonerate those wrongly accused or convicted. This increased role places greater importance on the ability of victim service providers to understand the potential significance of DNA evidence in their clients' cases.

Source: Information retrieved from National Institute of Justice and Office for Victims of Crime. (May 2001). *Understanding DNA Evidence: A Guide for Victim Service Providers.* Brochure.

the information required to produce his or her characteristics in the form of DNA, but they do not inherit the characteristics, only the information to produce them.

DNA is a **polymer**, which means it is a molecule that is formed by linking together a series of repeating units known as *nucleotides*. A nucleotide is composed of a sugar molecule, a phosphorous-containing group, and a nitrogen-containing molecule called a *base*. There are only four types of bases associated with DNA: adenine (A), cytosine (C), guanine (G), and thymine (T). There is no specified length of the DNA strand; in fact, a DNA strand can be composed of a long chain having millions of bases.

The unique feature of a DNA molecule is that it is composed of two DNA strands coiled into a *double helix*. The only way the bases (A, C, G, T) on each strand can be properly aligned is in specific paired combinations. G always pairs with C, and A always pairs with T. This is known as **complimentary base pairing**. Although A-T and G-C pairs are always required, there are no restrictions on how the bases are to be sequenced on a DNA strand. Any base can follow another on a DNA strand. Therefore, the possibilities of sequencing are staggering. In fact, the average human chromosome has DNA containing 100 million base pairs. In the human body, all of the chromosomes taken together contain approximately 3 billion base pairs.

During Alec Jeffreys' process towards discovery of DNA fingerprinting, he found that portions of the DNA molecule contain sequences of letters that are repeated numerous times. What is important to understand is that all humans have the same type of repeats, but there is tremendous variation in the number of repeats in each individual. It is this variation that gives each person his or her biological equivalent of a fingerprint.

Mitochondrial DNA

Mitochondrial DNA (mtDNA) is found outside the nucleus of the cell and is inherited solely from the mother. Each cell contains one nucleus yet hundreds to thousands of mitochondria. Therefore, there are hundreds to thousands of mtDNA copies in a human cell compared to just one set of nuclear DNA in that same cell.

This is useful in situations where nuclear DNA is significantly degraded, such as in charred remains, or in cases where nuclear DNA may be present in small quantity (i.e., a hair shaft). In situations where authorities cannot obtain a reference sample from an individual who may be long deceased or missing, an mtDNA reference sample can be obtained from any maternally-related relative. However, mtDNA is

CASE IN POINT

DNA Can Both Exonerate and Convict

In 1985, Steven Avery was convicted in the State of Wisconsin for sexual assault, attempted murder, and false imprisonment. On July 29, 1985, a woman was jogging on a Lake Michigan beach when a man attacked and sexually assaulted her. Three days later, the victim identified Avery from a police lineup. He was sentenced to 32 years. On September 11, 2003, Avery was exonerated as a result of efforts by the Innocence Project, due to re-analysis of DNA evidence that matched a hair found at the crime scene to Gregory Allen, a man already serving a 60-year sentence for a similar sexual assault. Contributing causes to Avery's original conviction were eyewitness misidentification and unreliable science at the time of analysis.

DNA had set Avery free, but it was also to put him away for life. In November of 2005, while in the midst of a civil case that he had filed regarding the false rape conviction, Avery was charged with the murder of Teresa Halbach. DNA evidence at the scene of the brutal murder tied Avery to the homicide event. He was found guilty of murder and of illegally possessing a firearm. Avery was sentenced to life in prison.

not as unique as nuclear DNA. All individuals of the same maternal lineage will be indistinguishable by mtDNA analysis.

Value of DNA Evidence

DNA is a powerful investigative tool because, with the exception of identical twins and other multiples, no two people have the same DNA. Therefore, DNA evidence collected from a crime scene can be linked to a suspect or can eliminate a suspect from suspicion. During a sexual assault, for example, biological evidence such as hair, skin cells, semen, or blood can be left on the victim's body or other parts of the crime scene. Properly collected DNA can be compared with known samples to place a suspect at the scene of the crime. In addition, if no suspect exists, a DNA profile from crime scene evidence can be entered into the FBI's Combined DNA Index System (CODIS) to identify a suspect anywhere in the United States or to link serial crimes to each other.

The effective use of DNA as evidence also may require the collection and analysis of elimination samples to determine the exact source of the DNA. Elimination samples may be taken from anyone who had lawful access to the crime scene and may have left biological material. When investigating a rape case, for example, it may be necessary to obtain an elimination sample from everyone who had consensual intercourse with the victim within 72 hours of the alleged assault to account for all of the DNA found on the victim or at the crime scene.

What Is CODIS?

The **Combined DNA Index System (CODIS)** is an electronic database of DNA profiles administered through the Federal Bureau of Investigation (FBI). The system lets federal, state, and local crime labs share and compare DNA profiles. All 50 states have enacted legislation which requires offenders convicted of particular crimes to submit DNA samples for inclusion within the CODIS database. Through CODIS, investigators match DNA from crime scenes with convicted offenders and with other crime scenes using computer software, just as fingerprints are matched through automated fingerprint identification systems.

CODIS Uses Two Indices

1. The Convicted Offender Index, which contains profiles of convicted offenders.
2. The Forensic Index, which contains DNA profiles from crime scene evidence.

The real strength of CODIS lies in solving no-suspect cases. If DNA evidence entered into CODIS matches someone in the offender index, a warrant can be obtained authorizing the collection of a sample from that offender to test for a match. If the profile match is in the forensic index, the system allows investigators—even in different jurisdictions—to exchange information about their respective cases.

From National Institute of Justice (NIJ). (July 2003). DNA: What law enforcement officers should know. *NIJ Journal, 249,* 10–15.

Comparing DNA profiles from the evidence with elimination samples may help clarify the results.

Basics of DNA Typing

Only one-tenth of a single percent of DNA (about 3 million bases) differs from one person to the next. Scientists can use these variable regions to generate a DNA profile of an individual, using samples from blood, bone, hair, and other body tissues and products.

In criminal cases, this generally involves obtaining samples from crime-scene evidence and a suspect, extracting the DNA, and analyzing it for the presence of a set of specific DNA regions (markers). If the sample profiles don't match, the person did not contribute the DNA at the crime scene. If the patterns match, the suspect may have contributed the evidence sample. DNA from crime scenes also can be compared to profiles stored in a database.

DNA analysis is a powerful tool because each person's DNA is unique (with the exception of identical twins and multiples). Therefore, DNA evidence collected from a crime scene can implicate or

RIPPED FROM THE HEADLINES

CODIS: Only as Good as the Information

In September of 2009 the State of Wisconsin found that dependency upon a database to provide evidence and potentially solve crimes can be misleading. Police investigating the 21 year serial murder actions of Walter E. Ellis, found that running a DNA sample against the state and national database and then discounting suspects when there is no "hit" can have catastrophic consequences. Police were investigating Ellis for a string of homicides occurring on Milwaukee's North Side between 1986 and 2007. Eventually, through the course of investigative efforts, Ellis was developed as a prime suspect. A toothbrush was used to compare his DNA to that found in the various cases under investigation. The comparison resulted in a match.

This flummoxed investigators because they had run the DNA profile found on 9 murdered women against the Wisconsin state databank of 125,000 people, and 6 million in the national CODIS databank but did not get any hits. A Wisconsin law enacted in 2000 required that people convicted of a felony provide DNA samples. Ellis had been incarcerated within a Wisconsin State prison from 1998–2001. This led law enforcement to believe that the suspect was not in prison and had not provided a genetic sample in any state, and that Ellis could not then be their suspect. However, numerous other clues and a great deal of other evidence pointed in Ellis' direction. Therefore, police executed a high-risk search warrant to collect DNA evidence from Ellis' home, even at the risk of tipping off Ellis that police were investigating him. A recovered toothbrush was used to make the comparison.

It is believed that a fellow prisoner, posing as Walter E. Ellis in 2001, gave a DNA sample for him, to Department of Corrections personnel, thus keeping Ellis out of a statewide and national database that would otherwise have linked him to the previous deaths, and could have possibly avoided the subsequent slaying following his incarceration. One of the victims, Ouithreaun Stokes, was killed in 2007, six years after Ellis' DNA should have been archived within the DNA database.

Through the course of investigating the failure, it was discovered that in excess of 12,000 felons convicted from 2000–2009 were missing from the state (and thus the national) database. The samples appeared to be missing for a variety of reasons, not all of which was due to inmate impersonation. Many of the missing samples were from convicted felons incarcerated near the time the law took effect and were simply missed in the rush to collect DNA from them all. Other felons who were not incarcerated, never had samples collected. DNA collection should have been a condition of being released from probation, but for some reason it did not occur.

eliminate a suspect, similar to the use of fingerprints. It also can analyze unidentified remains through comparisons with DNA from relatives. Additionally, when evidence from one crime scene is compared with evidence from another using CODIS, those crime scenes can be linked to the same perpetrator locally, statewide, and nationally.

DNA is also a powerful tool because when biological evidence from crime scenes is collected and stored properly, forensically valuable DNA can be found on evidence that may be decades old. Therefore, old cases that were previously thought unsolvable may contain valuable DNA evidence capable of identifying the perpetrator.

Since its introduction in the mid-1980s, DNA typing has revolutionized forensic science and the ability of law enforcement to match perpetrators with crime scenes. Thousands of cases have been closed and innocent suspects freed with guilty ones punished because of the power of a silent biological witness at the crime scene.

The study of DNA is often challenging. The reader is directed to the President's DNA Initiative: Advancing Justice Through DNA Technology to answer questions about such matters (http://www.dna.gov).

Evidence Collection

Contamination can take place if someone sneezes or coughs over the evidence or touches his or her hair, nose, or other part of the body and then touches the area containing the sample to be tested. DNA left at a crime scene also is subject to environmental contamination. Exposure to bacteria, heat, light, moisture, and mold can speed up the degradation (or erosion) of DNA. As a result, not all DNA evidence yields usable profiles.

Crime scene personnel should not drink, eat, litter, smoke, or do anything else that might compromise the crime scene. They should remember that valuable DNA evidence may be present even though it is not visible. For example, because evidence could be on a telephone mouth- or earpiece, investigators should use their own police radios or cellular phones instead of a telephone located at the crime scene.

To further avoid compromising evidence, any movement or relocation of potential evidence should be avoided. Personnel should move evidence only if it will otherwise be lost or destroyed. Potential evidence can become contaminated when DNA from another source gets mixed with samples gathered for a specific case. In those situations, laboratory analysts have to request samples from all persons with access to the crime scene, including officers and anyone who had physical possession of the evidence while it was being recovered, processed, and examined.

Maintaining a precise chain of custody of all DNA materials collected for testing is critical, as it may at some point become an issue in court. Every action officers take at a crime scene must be fully documented. Improvements in analysis and interpretation of physical

RIPPED FROM THE HEADLINES

Contaminated DNA Swabs Solve the Case of the "Phantom of Heilbronn"

In March of 2009, German police detectives announced that they believed evidence linking 14 murders together, to a supposed serial killer, was in fact a result of contaminated DNA swabs. The killer, referred to as the "Phantom of Heilbronn", had been perplexing German police for almost two years.

The killer was an oddity for a number of reasons. One, she was determined through DNA analysis to be female. Also, DNA had linked her to a total of 40 crimes (including the 14 homicides), ranging from stolen motor vehicles to drug violations. It was not until early in 2009 that police discovered a red flag that signaled that something was amiss. While attempting to establish the identity of a male asylum seeker's charred remains, they found that the fingerprints of the individual contained the killer's female DNA. Perplexed as to how this would be the case, investigators performed the test again, with a separate cotton swap, and no trace of the killer's DNA was discovered.

This led investigators to believe that the DNA connection between all of the supposed serial cases could be connected to a single innocent industry employee, most likely employed in the packaging stage of the swab manufacturing process. This individual, if not adhering to proper protocol could have affected the swaps through sneezing, coughing, etc. Although swabs used for DNA collection are sterilized, sterilization removes fungi, bacteria, and viruses, but it will not destroy DNA.

As a result, Germany's Federal Criminal Police are now investigating at what point in the manufacturing process the contamination occurred. The female donor of the DNA has not been identified, and the 40 criminal cases are now back to stage one.

evidence recovered from crime scenes continue to develop. Properly documented and preserved DNA evidence will be given increased weight in court, so it is extremely important that an officer's approach to gathering evidence be objective, thorough, and thoughtful.

■ Elimination Samples

The DNA of several individuals may be present at a crime scene. So, officers must ensure that technicians collect the victim's DNA along with the DNA of anyone else who may have been present at the scene. These *elimination samples* help determine if the evidence is from a suspect or another person. The types of elimination samples to be collected depend on the details of the crime, but they are generally samples of blood or saliva. For example, in a residential burglary where the suspect may have sipped from a glass of water, DNA samples should be obtained from every person who had access to the crime scene both before and after the burglary. The laboratory will then compare these samples with the saliva found on the glass to determine if the saliva contains probative evidence.

In homicide cases, the victim's DNA should be obtained from the medical examiner at the autopsy, even if the body is badly decomposed. This process may help to identify an unknown victim or to distinguish between the victim's DNA and other DNA found at the crime scene.

Contamination and Preservation

DNA evidence can become contaminated when DNA from another source gets mixed with DNA relevant to the case. For this reason, investigators and laboratory personnel should always wear disposable gloves, use clean instruments, and avoid touching other objects, including their own bodies, when handling evidence.

Environmental factors, such as heat and humidity, can accelerate the degradation of DNA. For example, wet or moist evidence that is packaged in plastic will provide a growth environment for bacteria that can destroy DNA evidence.

Therefore, biological evidence should be thoroughly air dried, packaged in paper, and properly labeled. Handled in this manner,

Safeguard DNA Evidence and Yourself

Biological material may contain hazardous pathogens, such as the hepatitis A or B virus, which can lead to potentially lethal diseases. At the same time, such material can easily become contaminated. To protect both the integrity of the evidence and the health and safety of law enforcement personnel, crime scene personnel should:

- Wear gloves and change them often.
- Use disposable instruments or clean them thoroughly before and after handling each sample.
- Avoid touching any area where DNA might exist.
- Avoid talking, sneezing, or coughing over evidence.
- Avoid touching one's own nose, mouth, and face when collecting and packaging evidence.
- Air-dry evidence thoroughly before packaging.
- Put evidence into new paper bags or envelopes. Do not place evidence in plastic bags or use staples.

Source: Information retrieved from National Institute of Justice (NIJ). (July 2003). DNA: What law enforcement officers should know. *NIJ Journal, 249,* 10–15.

DNA can be stored for years without risk of extensive degradation, even at room temperature. For long-term storage issues, contact the local crime laboratory.

Other Biological Fluids

In addition to blood at crime scenes there are numerous other biological fluids that may be collected and identified in order to explain what occurred at the scene of the crime. Most will yield the ability to identify DNA from them also. Other fluids that are present at crime scenes, and for which identification is sometimes necessary include but are not limited to: semen, saliva, vomitus, perspiration, urine, and fecal matter.

DNA Now and in the Future

Broader Implementation of the CODIS Database

States will continue to enact legislation requiring DNA samples from more offenders, resulting in more crimes being solved and increased cooperation among the states. Procedures for making international matches are expected to be developed, especially with Great Britain, which has a well-developed convicted felon database. Increased use of automated laboratory procedures and computerized analyses should improve departmental efficiency and accuracy of handling evidence. Although these timesaving approaches are not expected to replace human judgments in the final review of data, automation of many of the more routine aspects of analysis is expected to result in significant cost savings.

Portable Devices Capable of DNA Analysis

These devices, plus other advances in communications technology, may permit DNA evidence to be analyzed closer to the crime scene.

Remote Links to Databases and Other Criminal Justice Information Sources

Prompt determinations of the DNA profile at the crime scene could speed up identification of a suspect or eliminate innocent persons from being considered suspects. Such forecasts of the future are somewhat uncertain. However, the fact that private laboratories, federal agencies, and universities are aggressively researching these and other new technologies raises expectations that more sophisticated innovations will be developed.

Even with the latest innovations, DNA testing alone cannot provide absolute answers in every case. The prosecutor, defense counsel, judge, and law enforcement should confer on the need for such testing on a case-by-case basis.

Source: Information retrieved from National Institute of Justice (NIJ). (July 2003). DNA: What law enforcement officers should know. *NIJ Journal, 249*, 10–15.

Semen

In 1935, Kutscher and Wolbergs discovered that human semen contains uniquely high levels of seminal acid phosphatase (SAP) compared with other body fluids and plant tissues. SAP is produced by the prostate and is therefore found in both animals and humans. However, it is 20 to 400 times more concentrated within humans than in any other body fluid (Saferstein, 2007). This is the scientific basis for the presumptive ID of semen. Many methods for the presumptive ID of semen have been devised.

One method is utilizing an ALS because seminal fluid contains an acriflavine, which fluoresces bluish-white when exposed to the light from an ALS, with a proper barrier filter. However, this test is not sufficient to presumptively identify semen, as there are numerous other substances that have the same fluorescence. These include: fabric softeners, toothpaste, cosmetics, sweat, and urine. This is a good starting point, however. There are presumptive field test kits to screen for SAP that are more isolating in nature. This test is performed by promoting a color reaction within the sample. In forensic laboratories, alphanaphthyl phosphate is the preferred substrate and Brentamine Fast Blue the color developer. However, Brentamine Fast Blue B is potentially carcinogenic, so liberal applications made directly onto items of clothing or areas of flooring is not recommended.

Confirmatory tests for the presence of semen can only be made at the forensic laboratory. Microscopic identification of spermatozoa is one type of confirmatory test for semen. At the lab, the criminalist will stain the prepared sample using a "Christmas tree" stain test, that stains the head of spermatozoa red and their tails (flagella) green. This then provides unambiguous proof that the stain in question contains semen.

Another method to confirm the presence of semen in a sample is for the forensic lab to perform what is known as a *p30 test*. This test is performed when no sperm are found to be present within the sample (i.e., vasectomized male). P30 is a prostate-specific antigen produced in the human male. There have been cases where p30 was able to be detected in samples stored at room temperature for up to 10 years. In fact, semen cells can be located within an oral sample for up to 24 hours, and inside of the vagina for up to 72 hours. This provides a finite window of collection opportunities for a crime scene investigator (Saferstein, 2007).

Procedure for the Collection of Seminal Stains

Where a sexual offense has occurred, stains may be found on clothing, bedding, rags, upholstery, or other objects. Seminal stains can be helpful

Using CLU Fluorescence to Detect Semen Stains

The advent of DNA technology and databases has made semen stains found at the scene of a sexual assault the most valuable piece of evidence. The problem is that the semen stains must first be located and sampled.

The conventional method—fluorescence detection—illuminates the crime scene with light from a high-intensity lamp while an investigator views the area through optical filter glasses. This method has a number of drawbacks. Although semen fluoresces, the light it emits is weak compared to surrounding room light, thereby hindering detection. If the crime scene is outdoors, investigators must wait until nightfall to use the technique. If the crime scene is indoors, investigators must turn off all lights and black out the windows to maximize the method's effectiveness. This takes time and effort and increases the possibility that investigators will contaminate the area.

Moreover, when blacking out a room, many other substances besides semen fluoresce, such as food spills and animal urine. In order to complete their search in a reasonable amount of time, investigators often collect all questionable fluorescing materials. Detecting and documenting semen stains thus become the task of technicians back at the crime lab.

It would be best to photograph potential evidence at the crime scene. However, setting up a camera is time-consuming, and investigators often do not have enough time for this step. If the police do photograph evidence at a crime scene, there is no guarantee of any evidentiary value until the film is developed.

The use of a **Criminalistics Light-Imaging Unit (CLU)** at the crime scene offers significant improvements over conventional approaches. CLU is a multispectral imaging system that uses various colors of light to view the substance or structure being examined. It can locate body fluids at crime scenes under normal lighting conditions. By using a strobe lamp, signal processing, and improved optics, CLU rejects surrounding light and thereby improves both the sensitivity and specificity of the area being viewed. CLU is five times more sensitive than current fluorescing methods. CLU allows investigators to find fluorescing evidence under normal lighting conditions and to easily view and highlight images of suspected evidence at the crime scene. Furthermore, CLU greatly reduces the chances of crime scene contamination.

Source: Information retrieved from the National Institute of Justice (NIJ). (July 2003). Without a trace? Advances in detecting trace evidence. *NIJ Journal*, *249*, 2–3.

in establishing whether or not an alleged sexual act occurred and can also provide information concerning the man who contributed the semen.

Carefully recover all suspected stained material, including the clothing worn by the victim and the suspect at the time of the offense. Each item of evidence should be packaged separately, labeled, and sealed. Air dry all damp stains in a well-ventilated but draft-free area.

Clean paper should be spread under the item to catch any debris that may be dislodged during the drying process. Package, label, and seal each item along with the paper upon which the item dried. Use only paper containers for packaging (i.e., paper bags). If the suspected seminal stain is on an object that cannot be transported, collect utilizing swabs and distilled water, as with blood evidence.

Standard blood samples are normally used for comparison purposes, rather than collecting semen standards for comparison.

Saliva

Humans produce 1 to 1.3 liters of saliva per day. Its primary purpose is to aid in the initial stages of digestion by lubricating food for ease of swallowing and to begin the process of digestion. However, no test is specific for saliva. Presumptive tests will test for presence of amylase (found in saliva, perspiration, semen, vaginal secretions, and breast milk). However, amylase is found in 50 times higher concentration within saliva. There have been cases where activity has been detected for up to approximately 28 months (Saferstein, 2007).

Saliva stains are not usually evident from a visual examination. However, certain types of evidence frequently contain traces of saliva (e.g., cigarette butts, gummed surfaces of envelopes, stamps, bite marks, areas where oral contact may have occurred, etc.) and sometimes the amount of saliva present is sufficient to determine the DNA type of the individual who is the source of the saliva.

Procedure for the Collection of Evidential Forms of Saliva

Easily transportable objects such as individual cigarette butts and envelopes should be placed in a paper container (i.e., paper bag or envelopes) and the container should be properly labeled and sealed. If transporting the object is not practical, such as in the case of bite marks on the body of sexual assault victims, then the saliva can be collected as follows:

- Moisten a cotton swab with distilled water.
- Shake the swab to eliminate excess water.
- Gently swab the suspected saliva stain. Using a dry swab go over the stained area to absorb any remaining moisture.
- Allow the swabs to thoroughly air dry prior to packaging, labeling, and sealing in a paper envelope. Air drying can be accomplished by making a perforation in the center of a pillbox, inserting the swab into the perforation and allowing the swab to air dry.
- Select an unstained area and collect as before. Package, label, and seal separately from the stained material. This swab will serve as a control.

Standard blood samples are normally used for comparison purposes, rather than collecting saliva standards for comparison.

Other Types of Biological Evidence

Urine

This is a rarely identified biological fluid. The presumptive test for urine by the forensic laboratory tests for the presence of creatinine, which is found in heavy concentrations in urine. It is also found, however, in high concentrations in sweat. There is no confirmatory test that is regularly utilized.

Vomitus

There is no presumptive or confirmatory test for vomitus. However, the material can be tested to determine food stuff and to compare pH levels.

Vaginal Secretions

In order to meet the elements of sexual assault, often it is necessary to prove that an object was placed within the vagina. In these cases it is necessary for the forensic laboratory to make use of microscopy in order to identify fluids and cells associated with vaginal secretions.

Vaginal secretions in the form of a foreign DNA (DNA that did not originate from the individual swabbed) can sometimes be attributed to another individual when the penis of a suspected sexual perpetrator is swabbed at the time of apprehension. The sample is collected by wetting a cotton swab with distilled water and swabbing the external area of the penis. This type of analysis is most successful when the perpetrator is apprehended shortly after the alleged occurrence of sexual activity, generally within twenty-four hours and prior to bathing. The outer area of condoms also can yield this type of DNA.

■ Preservation of Dried Biological Evidence

The ideal way to preserve biological evidence is to freeze it. This can become impractical with large amounts of evidence. Evidence with dried biological stains can be stored in a temperature-controlled room, which is maintained at normal room temperature or colder. Large fluctuations in temperature should be avoided. When biological evidence is returned after processing by the DNA unit of the crime laboratory, it frequently will contain a manila envelope labeled "DNA packet." This packet contains cuttings of stains and extracts of those stains. This packet needs to be frozen. If this packet is included with the evidence, it will be noted on the return release form and the evidence will be labeled "Biological Evidence Enclosed, Please Remove and Freeze."

Sexual Assault

In a rape case, investigators may need to collect and analyze the DNA of every consensual sexual partner the victim had up to four days prior to the assault. Testing can eliminate those partners as potential sources of DNA suspected to be from the rapist. A sample also should be taken from the victim. It is important to approach the victim with extreme sensitivity and to explain fully why the request is being made. A qualified victim advocate or forensic nurse examiner can be a great help.

In sexual assault cases, it is especially important that victims are told why they should not change clothes, shower, or wash any part of their body after an assault. Depending on the nature of the assault, semen may be found on bedding or clothing, or in the anal, oral, or vaginal region. Saliva found on an area where the victim was bitten or licked may contain valuable DNA. If the victim scratched the assailant, skin cells containing the attacker's DNA may sometimes be present under the victim's fingernails. Victims should be referred to a hospital where an exam will be conducted by a physician or sexual assault nurse examiner.

In all cases it is essential to have the victim(s) examined by a medical professional as soon as possible after the assault and before the affected areas (pubic area, vagina, rectum, etc.) or clothing are washed or cleaned.

Evidence on or inside a victim's body should be collected by a physician or sexual assault nurse examiner. A medical examination should be conducted immediately after the assault to treat any injuries, test for sexually transmitted diseases, and collect forensic evidence, such as fingernail scrapings and hair. Typically, the vaginal cavity, mouth, anus, or other parts of the body that may have come into contact with the assailant are examined.

The examiner also should take a reference sample of blood or saliva from the victim to serve as a control standard. Reference samples of the victim's head and pubic hair may be collected if hair analysis is required. A control standard is used to compare known DNA from the victim with that of other DNA evidence found at the crime scene to determine possible suspect(s).

Most forensic laboratories will have a **Sexual Assault Evidence Collection Kit (SAE kit)** available that can assist the crime scene investigator and attending medical professional in properly collecting the specimens required by the laboratory. This kit can be used to collect appropriate samples from both male and female sexual assault victims and suspects.

Each kit gives specific collection methods for various types of evidence that must be collected and documented. These include:

- Clothing
- Pubic hair combings
- Vaginal swabs (four) and smear (one)
- Cervical swabs (two) and smear (one)
- Rectal swabs (two) and smear (one)
- Oral swabs (two) and smear (one)
- Pubic hair standards
- Penile swab (one)
- Buccal cell standard (cheek or mouthwash)
- Fingernail scraping (if indicated)
- Bite marks (if indicated)
- Toxicology specimens

The clothing of the victim and suspect(s) is the next most important type of evidence. Articles of clothing worn by the victim (and suspect, if possible) should be submitted to the laboratory for examination, as there may be seminal stains, blood stains, foreign hairs and fibers, or other trace evidence adhering to the clothing. In addition, items at the crime scene may provide important evidence that associates either the victim, the suspect, or both to the scene.

The recommended procedure for the collection and preservation of clothing is as follows:

- Clothing of the victim must be kept separate from that of the suspect at all times.
- Clothing worn at the time of or immediately after the offense must be recovered and preserved. This includes undergarments, handkerchiefs, sanitary napkins and/or tampons (only if used during or after the offense).
- Garments should be handled as little as possible to avoid the loss of trace evidence.

Additional Consideration for Processing the Scene

Sexual assault crimes involve physical contact between perpetrator and victim. This contact results in the transfer of materials, such as hairs, fibers, and body fluids, particularly seminal fluid and saliva. In many cases, the perpetrator is not known to the victim and the assault occurs in seclusion or with no witnesses. As a result, there are some additional considerations that a crime scene investigator should take in order to ensure the proper documentation and collection of related evidence.

The crime scene investigator should attempt to recover articles such as handkerchiefs, rags, tissues, etc., that may have been used as a wipe after ejaculation. They should also recover and submit any articles that may have become stained during the offense or might have foreign hairs present (bedding, rugs, sofa cushions, etc.).

When condoms are recovered in suspected sexual assault cases, they should be placed in a glass specimen jar and frozen until submitted to the crime laboratory.

Chapter Summary

The identification, documentation, collection, and preservation of biological evidence are of paramount concern in establishing the facts of a case. It is important that a crime scene investigator realize that there are correct ways to go about such collection efforts, and that there are scientific considerations that pertain to the evidence being properly preserved. If the samples are not properly collected and preserved, subsequent analysis and identification will be nearly impossible.

Review Questions

1. What is TMB, and how is it used in relation to biological evidence?
2. ______________ is an example of a presumptive search technique that results in chemiluminescence as a result of the chemical reaction occurring between the reagent and the biological stain.
3. Before DNA analysis was used on blood evidence, ______________ was used to determine the class characteristics of A, B, AB, or O.
4. Who was the discoverer of DNA fingerprinting?
5. ______________ is the building block for the human body.
6. DNA is a molecule that is formed by linking together a series of repeating units, this is also known as a ______________.
7. Complimentary base pairing refers to the fact that ______________ always pairs with ______________, and ______________ always pairs with ______________.
8. What is mtDNA?
9. What is CODIS and how is it used?
10. Explain what is meant by a "Christmas tree" stain.
11. What is an SAE kit and how is it used?

Case Studies

1. Look up the name Anastasia Manahan, who is also known as Anna Anderson. Explain how the use of DNA technology helped to uncover this imposter and set the historical record straight.
2. Look up the name William Gregory and explain how the use of DNA evidence applied to his case and to discovering the truth. Be thorough in the summary.

References

Caplan, J. (2001). *Documenting individual identity: The development of state practices in the modern world.* Princeton, NJ: Princeton University Press.

Doyle, A. C. (1887). *A Study in Scarlet.* London, United Kingdom: Ward Lock & Co.

Federal Bureau of Identification (FBI). (2008). *Birth of the FBI's technical laboratory.* Retrieved September 2, 2008 from http://www.fbi.gov/libref/historic/history/birthtechlab.htm

Fisher, B. A. J. (2004). *Techniques of crime scene investigation* (7th ed.). Boca Raton, FL: CRC Press.

Fish, J. T., Miller, L. S., & Braswell M. C. (2007). *Crime scene investigation.* Newark, NJ: LexisNexis Group.

Girard, J. E. (2008). Principles of environmental chemistry. Sudbury, MA: Jones and Bartlett Publishers.

Kirk, P. L. (1974). *Crime investigation* (2nd ed.). New York: John Wiley & Sons.

International Association of Identification (IAI). (2008). *Welcome to the IAI!* Retrieved September 3, 2008 from http://www.theiai.org

Lee, H. C., Palmbach, T., Miller, M. T. (2001). *Henry Lee's crime scene handbook.* San Diego: Academic Press.

Muehlberger, C. W. (1955, May–June). Col. Calvin Hooker Goddard 1891–1955. *The Journal of Criminal Law, Criminology, and Police Science, 46,* 103–104.

National Institute of Justice and Office for Victims of Crime (May 2001). *Understanding DNA Evidence: A Guide for Victim Service Providers.* Brochure.

National Institute of Justice (NIJ). (July 2003). DNA: What law enforcement officers should know. *NIJ Journal, 249,* 10–15.

National Institute of Justice (NIJ). (July 2003). Without a trace? Advances in detecting trace evidence. *NIJ Journal, 249,* 2–3.

Radziki, J. (1960). *Slady Krwi w Praktyce Sledczej (Bloodstain Prints in Practice of Technology).* Warsaw, Poland: Bibliotecka Kryminalistyczna.

Saferstein, R. (2007). *Criminalistics: An introduction to forensic science* (9th ed.). Upper Saddle River, NJ. Pearson Prentice Hall.

Bloodstain Pattern Analysis

Art in the blood is liable to take the strangest forms.

Sir Arthur Conan Doyle (1859–1930)
Sherlock Holmes, *The Adventure of the Greek Interpreter*

LEARNING OBJECTIVES

- Explain the forensic significance of bloodstain pattern analysis.
- Understand how the properties of human blood allow bloodshed events to be interpreted.
- Know how impact angle, surface type, and drop height affect a resulting stain, and how they are analyzed.
- Differentiate between the primary types of bloodstain patterns found at crime scenes, and what can be surmised from their discovery.
- List the correct way to document bloodstain evidence.

KEY TERMS

Area of Convergence
Area of Origin
Bloodstain Pattern Analysis (BPA)
Cast-Off Patterns
Directionality
Expirated Patterns
Passive Bloodstain
Projected Patterns
Swipe
Terminal Velocity
Transfer Stain
Viscosity
Void Pattern
Wipe

Bloodstain pattern analysis (BPA), is the science of examining and interpreting blood present at a bloodshed event in order to determine what events occurred, in what order, and who possibly left the stains. As with any area of forensic science, this discipline seeks to define the facts surrounding the incident in question. The volume of blood, patterns present, shape characteristics, dispersion, and number of bloodstains present within a scene all play a role in the evaluation and analysis. The training and experience of the crime scene investigator will determine the accuracy of the analysis. BPA is not the answer to all of the questions concerning what happened at the scene of a crime; it is instead a tool within the greater toolbox of crime scene investigation.

History of Bloodstain Pattern Analysis

Much like track impression evidence, the historical origin and skill development of bloodstain analysis most likely dates from the earliest of mankind's hunting efforts. Paleolithic art document the skill of early human hunters and show the use of blood tracks to locate prey. Biblical passages relate bloodstains with injury and with mortality. So, it is reasonable to assume that humans have been analyzing bloodstain patterns for well over 4,000 years. Investigators have been attempting to interpret bloodstains for as long as there have been criminal investigations; however, bloodstain analysis as a distinct area of forensic science and crime scene investigation appears to be a more modern occurrence.

The early 1900s saw several prominent scholars and scientists researching and experimenting with blood dynamics and properties of human blood. While numerous individuals recognized the importance of such matters, historically BPA as a forensic discipline is credited to Dr. Paul Leland Kirk of the University of California at Berkeley. In his book entitled *Crime Investigation* (1952, 1974), Kirk's chapter, "Blood: Physical Investigation," explains the benefits of and application of BPA within criminal investigations. Kirk put his teachings into practice when submitting an affidavit of his examination of bloodstain evidence and findings in the case of the *State of Ohio v. Samuel H. Sheppard*, in 1955. In this case, Dr. Kirk testified as to the consideration of drying times for blood and pattern evaluation in an effort to explain the events that had occurred at the Sheppard residence, leading to the death of his wife.

This case and Dr. Kirk's textbook are considered an impetus for the modern forensic study of bloodstain pattern analysis. Afterwards, the number of authors postulating on such matters began to increase in a dramatic fashion. One such author was Herbert Leon MacDonell, whose research recreated bloodstains observed at crime scenes. His initial book was entitled, *Flight Characteristics of Human Blood and Stain Patterns* (1971). While numerous texts have been published on the topic since, MacDonell is credited for reawakening the discipline and providing the stimulus for professional organizations such as the International Association of Bloodstain Pattern Analysts (IABPA) and International Association of Identification (IAI) to develop and/or begin to offer training and certification in this field. As a result of this professional and academic growth, the discipline as a whole has gained far greater acceptance within courtrooms across the United States.

Properties of Human Blood

A normal healthy adult has approximately 4.5 to 6.0 liters (L) of blood in his/her body. This is important for a crime scene investigator to be aware of, because if there appears to be a volume of blood as large as this, then either the evidence suggests that the injury that resulted in the bloodshed was inconsistent with the sustenance of life or there is blood present from several individuals that has resulted in the volume observed.

Blood will act in a predictable manner when subjected to external forces. The surface tension of blood is slightly less than that of water. Blood falls in a spheroid configuration, not "tear drop" shape as typically drawn and portrayed. The volume of a typical or average drop of blood has been reported to be approximately 0.05 milliliters (mL), with an average diameter of 4.56 mm (while in the air) (Saferstein, 2007). Molecules of blood are mutually attracted to one another, this is called **viscosity**. The more viscous a fluid, the more slowly it flows. Blood is approximately six times more viscous than water and has a specific gravity slightly higher than water.

Bloodstains dry from their perimeter inward toward their center. As the distal margins of the stain dry, they become more resistant to disruption or removal than the stain's still wet interior. As a general guide, a minimum of one minute is required for sufficient drying to create a perimeter stain. Because the drying time of blood is dependent upon the volume of blood deposited, the surface on which the blood is deposited, and the environmental conditions to which it is subjected, this interval may be experimentally determined at the discretion of the blood pattern analysis expert (BPAE).

All of the aforementioned characteristics of blood allow for its subsequent interpretation within the context of crime scene investigation.

Bloodstain Pattern Interpretation

The ability to interpret a bloodshed event should be viewed as a forensic tool that assists the crime scene investigator to better understand what took place and what could not have taken place during a bloodshed event. Information obtained may assist in apprehending a suspect, corroborating a witness' statements, assisting in interrogating a suspect, allow for reconstruction of event, and possibly exonerate an accused.

Through the careful examination of the physical evidence (bloodstains), the investigator will be able to deduce specific information with regards to the events that occurred during the bloodshed event. The crime scene investigator will observe three primary areas pertaining to the bloodstains in order to assemble their findings. The crime scene investigator will look at the size, the shape, and the distribution of the bloodstains. Each of these, separate and in combination, will result in information specific to the event.

Bloodstain Shape

Shape is one of the chief sources of information that a bloodstain possesses. Careful analysis of bloodstain shape will allow the crime scene investigator to determine the direction from which the blood originated. This can be extremely beneficial in helping to understand what occurred at the scene of a crime. In order to determine **directionality** or the direction that blood was traveling, a crime scene investigator need only look at the shape of the bloodstain. The narrow end of an elongated bloodstain will typically point in the direction of travel (**Figure 11.1**). For many stains, particularly those deposited on non-porous surfaces, it may be possible to determine the direction a drop was moving by a comparison of the edge characteristics in the two extreme margins of the stain. That distal margin that displays lesser disruption with respect to its opposing margin was deposited first. The margin that displays the greater disruption was deposited second. In this way, a bloodstain can be said to have a directionality that tracks along the path of its long axis and in the direction that encounters its edge of lesser disruption first and the edge of greater disruption next (**Figure 11.2**).

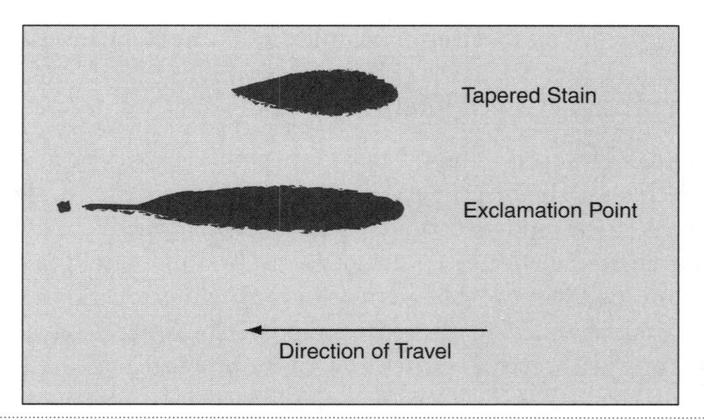

Figure 11.1 Determining Blood Spatter Directionality.

Source: Courtesy of Erica Lawler, University of Wisconsin–Platteville.

Figure 11.2 Example of Blood Spatter Directionality.

However, the information derived from a bloodstain's shape does not stop there. When a source of blood is impacted or otherwise subjected to sufficient force, blood droplets will be dispersed, and the resulting blood droplets will be projected upon target surfaces at various impact angles. If the crime scene investigator was to draw an imaginary line through the long axis of each bloodstain in the direction opposite of observed travel they would arrive at a two-dimensional point known as an **area of convergence (Figure 11.3)**. It is at this point that an event occurred (i.e., impact) that led to the subsequent dispersal of the blood. It must be remembered that this is only a two-dimensional explanation (X and Y axis) and does not determine how far away from the area that the blood event originated; it instead gives an area in which to determine such information. For instance, once the area of convergence within a room has been determined, in order to determine the origin distance from the wall or floor within the area of that room, a third dimension must be considered.

By establishing the impact angles of representative bloodstains and projecting their trajectories back to a common axis (Z) extended at 90 degrees from the intersection of the X and Y axis, an approximate location of where the blood source was when it was impacted may be established. This area is known as the **area of origin (Figure 11.4)**.

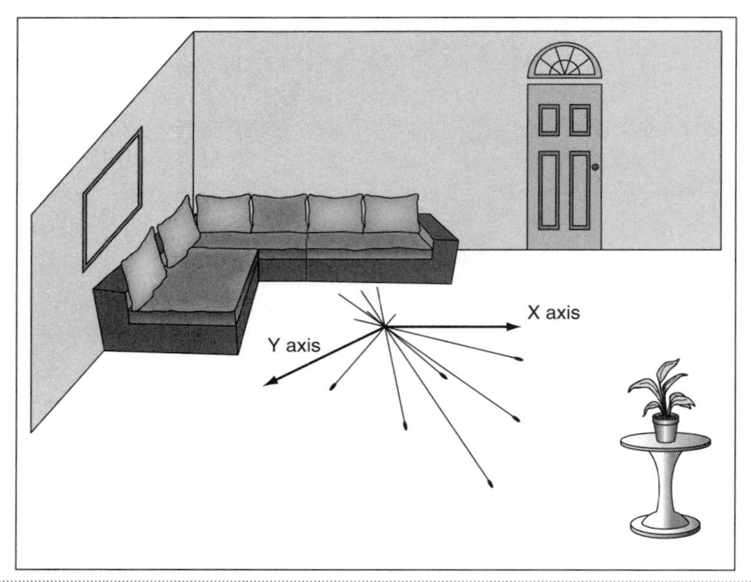

Figure 11.3 Determining Area of Convergence from Blood Spatter Evidence.

Source: Courtesy of Ellie Bruchez, University of Wisconsin–Platteville.

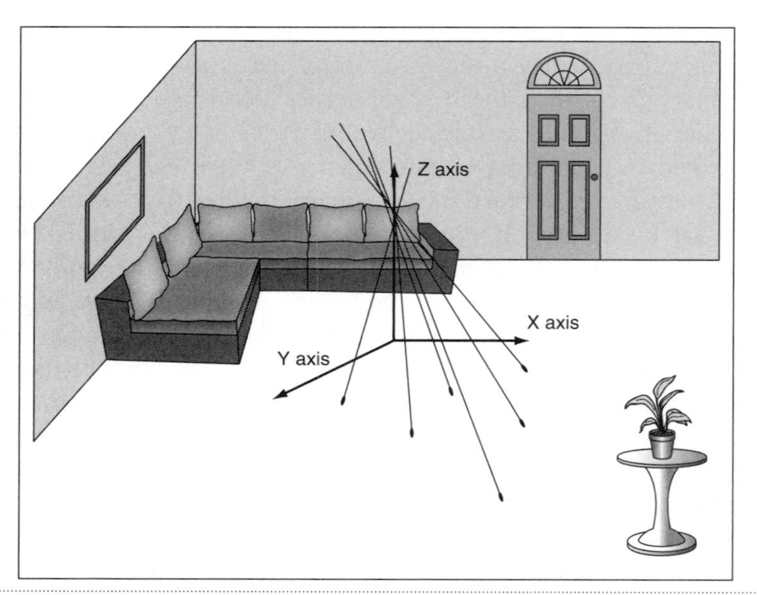

Figure 11.4 Determining Area of Origin for Blood Spatter.

Source: Courtesy of Ellie Bruchez, University of Wisconsin–Platteville.

However, in order to locate this, the examined blood must be analyzed to determine its angle of impact.

Angle of Impact

If blood impacts a target surface at 90 degrees, the resulting bloodstain generally will be circular in shape. Blood droplets that strike a target at an angle less than 90 degrees will create elliptical bloodstains (**Figure 11.5**). This is helpful in understanding the approximate origin of a bloodstain; however, a crime scene investigator is able to gain a much closer approximation when he or she realizes that a mathematical relationship exists between the width and length of a bloodstain, which will allow for the calculation of the angle of impact for the original spherical drop of blood.

To calculate this angle, a crime scene investigator must first measure the width and length of the bloodstain to be analyzed. The width of the bloodstain is then divided by its length. When measuring the length of a blood stain it is important not to include in this measurement the tail portion of the stain. Inclusion of these trailing edge characteristics in the measurement of length may result in the under estimation of the impact angle of a stain (**Figure 11.6**).

Next, with the help of a scientific calculator, the crime scene investigator should take the resulting answer from the previous computation and utilize the function key, arc sin, sin-1, or inverse sin function, and the corresponding angle of impact will be displayed. A quick test

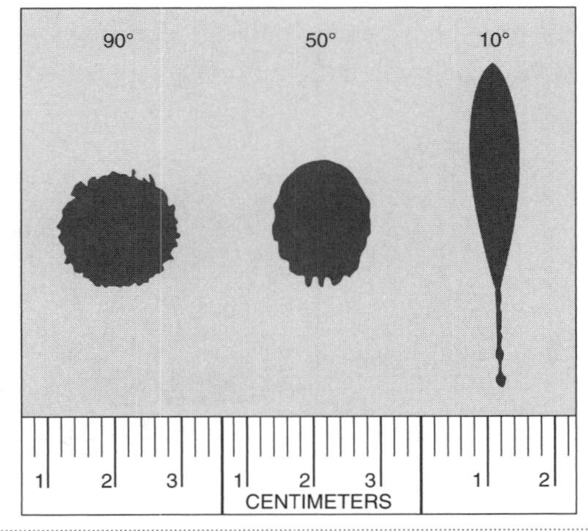

Figure 11.5 Determining Blood Spatter Angle of Impact (Variance Between 90 degree and Others).

Source: Courtesy of Erica Lawler, University of Wisconsin–Platteville.

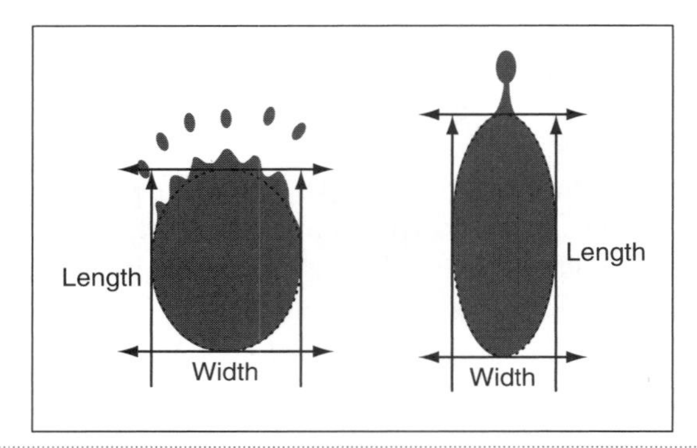

Figure 11.6 Measuring Bloodstain Dimensions to Determine Angle of Impact.

Source: Courtesy of Erica Lawler, University of Wisconsin-Platteville.

of this will show the earlier summation of a circular bloodstain to be true. For a circular bloodstain, the width and length are equal and thus the sin is 1.0, which corresponds to an impact angle of 90 degrees.

Stringing a Scene

After establishing the angle of impact for each of the bloodstains, the three-dimensional origin of the bloodstain pattern can be determined. One method is to place strings at the base of each bloodstain and project these strings back to the axis that has been extended 90 degrees up or away from the area of convergence. This is accomplished by placing a protractor on each string and then lifting the string until it corresponds with the previously determined impact angle. The string is then secured to the axis (**Figure 11.7**). This is repeated for each analyzed bloodstain. It is important to remember that this is not exact. In practical terms it is used to determine whether a victim was standing, lying down, or sitting in a chair when the blood was spattered (**Figure 11.8**).

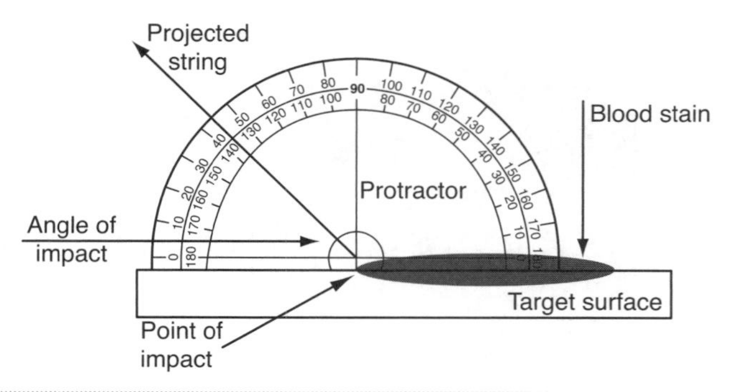

Figure 11.7 Determining Blood Spatter Angle of Impact Using Stringing.

Source: Courtesy of Ellie Bruchez, University of Wisconsin–Platteville.

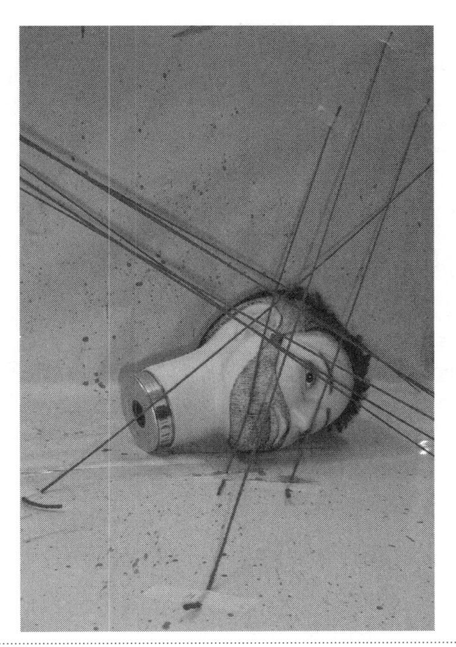

Figure 11.8 Example of Stringing a Crime Scene.

Common sense and quality observations will often resolve the question of where someone was when injuries were inflicted. For instance, if no blood spatter appears on a table top or chair seat, but spatter associated with a gunshot is found on the underside of the table and chair, the obvious conclusion is that the victim was on or near the floor when shot.

Effect of Height on Bloodstain Size

A blood drop falling through the air will increase its velocity until the force of air resistance that opposes the drop is equal to the force of the downward gravitational pull. At this point the droplet is said to have achieved its **terminal velocity**. Blood therefore falls in a predictable manner and a typical droplet of blood that has a volume of 0.05 mL will then produce bloodstains of increasing diameters when dropped from increasing increments of height onto a smooth, hard surface. The measured diameters will range from 13.0 to 21.5 mm over a dropping range of 6 inches to 7 feet (Saferstein, 2007) (**Figure 11.9**). Drops that fall distances greater than 7 feet, however, will not produce stains with any significant or even perceptible increases in their diameter, as the droplets have reached their terminal velocity and cannot impact the target surface with any greater force.

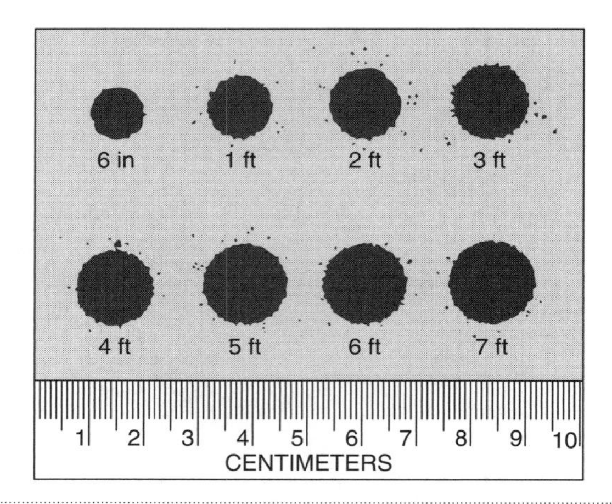

Figure 11.9 Impact of Height on Diameter of Blood Evidence.

Source: Courtesy of Erica Lawler, University of Wisconsin–Platteville.

Types of Bloodstains

All bloodstains left at a crime scene can be grouped into one of four categories: passive, transfer, projected/impact, and miscellaneous (**Table 11.1**). This grouping was first suggested by Jozef Radziki (1960) and is for the large part still in use today.

Passive Bloodstains

Passive bloodstains consist of clots, drips, flows, and blood pools. They are not related to a specific action within the context of the scene with regards to bloodshed violence. Instead, they are the aftermath associated with this violence, as the resulting bloodshed begins to move and cure due to environmental and gravitational forces.

Drip (Blood in Blood)

Drip patterns result from blood dripping into blood. Drip patterns generally display satellite staining that results from the deposition of small droplets that are ejected from the blood pool upon its being struck by a falling drop. Because these ejected blood droplets generally follow arching trajectories, they strike the surface at or near 90 degrees and result in circular to near-circular stains. Additionally, because they are generally slow moving, the satellite stains surrounding the central pool of a drip pattern often display an observable thickness because their droplets lack the energy necessary to fully spread themselves over the target surface upon striking it.

TABLE 11.1 Categories of Bloodstains

Passive	Transfer	Projected/Impact	Miscellaneous
Drips	Pattern	Arterial spurt/gush	Voids
Flow	Wipe	Cast-off	Fly/insect spots
Clot	Swipe	Expiratory	Skeletonized
Pool		Spatter (impact)	

The satellite stains in the lateral margins of a drip pattern on a vertical surface often display evidence of gravity effects. Such satellite stains display directionality toward the horizontal surface because the droplets deposited were on the downward flight path of their travel arch. Also, the size of the satellite stains on a vertical surface will generally decrease as the distance from the blood pool that resulted in the drip pattern increases.

Single drops of blood will produce small spatters around the parent stain as a result of striking a rough surface. This is referred to as *secondary* or *satellite* spatter. When multiple free-falling drops of blood are produced from a stationary source onto a horizontal surface, particular drip patterns will result that are large and irregular in shape, with a satellite spatter.

Flow Patterns

Flow patterns are a change in the shape and direction of a bloodstain due to the influence of gravity or movement of the surface and are generally observed in circumstances where a volume of liquid blood moves freely along a downward path.

Flow patterns also may display one or more long, narrow stains that may or may not originate from a distinct bloodstain pattern. Often the terminal end of this narrow stain will show a thickening that represents the residual blood volume whose weight was not sufficient to continue its flow. These narrow stains always will extend in the direction of gravity's influence on the object as it rested at the time of blood flow.

A flow pattern that does not extend downward on the object or surface as observed indicates that the object or surface was moving during blood flow or that it was moved after blood flow had ceased.

Clot

Clotted blood is blood that has undergone the physiological process of fibrin formation that congeals the solid elements of blood into a gelatinous mass. Blood that is shed from the body in sufficient volume to remain liquid for an extended period of time may begin to display clot formation. The minimum time necessary for clot formation to begin generally ranges from 3 to 15 minutes. This time course is dependent on many factors including the volume of blood deposited, the surface on which it is deposited, and the environmental conditions to which it is subjected.

Large blood pools that collect on nonporous surfaces may remain liquid for a sufficient period of time to display clot formation. Blood pools that undergo clot formation generally display large gelatinous masses with serum collections in the peripheral margins of the blood pool.

Individual stains that display clotted material may be produced when already clotted blood is subjected to disruption by a force or action. Such stains may display a central region of dense staining surrounded by a ring of less dense staining. This "bull's eye" or "fried egg" appearance is the result of the deposition of gelatinous clotted material with subsequent radial diffusion of the remaining liquid serum in porous material or continued clot retraction on nonporous surfaces.

Clotted material in the stain(s) of a bloodstain pattern may indicate that the blood that created the pattern was shed and remained undisturbed for a period of time sufficient for clot formation. In this way, the presence of clots within a bloodstain pattern may indicate that some period of time elapsed between the event(s) that resulted in bloodshed and clot formation.

It may not be possible to determine the actual interval of time between the blood's shedding and clot formation, however, unless ancillary information is available concerning the volume of bloodshed and the environmental conditions to which it was subjected. As a result, while the presence of clotted blood within the stains of a bloodstain pattern may indicate a passage of time, care must be taken in any attempt to refine the estimate of such an elapsed interval. At the discretion of the crime scene investigator, this interval may be experimentally determined.

Pooling

This pattern occurs when gravitational forces pull blood to the lowest possible level and there is an accumulation of blood that develops, sometimes referred to as a *blood basin*. This type of pattern allows the crime scene investigator to gauge how much blood could possibly have been lost from the victim, whether or not there was any movement of the victim from the area where the pooling occurred, and a time frame (due to clotting and drying of the blood within the pool).

Transfer Bloodstains

Transfer stain patterns (sometimes referred to as *contact* patterns) are generally indistinct stains that can be of virtually any size or shape. The physical appearance of such stains is generally mottled with numerous variations in their color and/or density. The shape of a contact/transfer pattern may retain some of the physical characteristics of the object that created it. In this way, the shape of a contact/transfer pattern may suggest the object that created it through the recognizable patent image **(Figure 11.10)**.

The direction of motion of a transfer pattern may be determined from the general density distribution of the stain. Generally, a transfer pattern is heaviest in that portion of the stain deposited first, and its density decreases as blood is removed from the object.

- Feathered margins may be created as an object gradually comes into or loses contact with a target surface. Generally, feathering displays a gradual decrease in density in the direction of an object's movement. As a result, the trailing edge of a contact/transfer pattern may display lightened staining with respect to the leading edge.

Figure 11.10 Photograph of Blood Evidence at a Crime Scene.

- The presence of distinct front boundaries deposited along the margins of a contact/transfer pattern may allow a direction of movement to be determined. These front boundaries are the result of blood that collects along the leading edge of an object. Because these front boundaries are left where an object loses contact with a surface, they are most numerous in that portion of the stain deposited later.

Swipe

A **swipe** occurs when a bloodied surface rubs across a non-bloodied one. This is beneficial to the crime scene investigator as it will usually give some indication as to movement within the scene. There will be an initial point of impact that leads towards a feathered or disappearing edge. Movement is towards the feathered edge.

Wipe

A **wipe** occurs when a non-bloodied surface moves through/across a stationary one. This type of pattern is also able to give the crime scene investigator an idea as to movement at the crime scene, just as with a swipe pattern.

Pattern Transfer

When an object wet with blood comes into contact with an unstained object or secondary surface, a *blood transfer pattern* occurs. A recognizable image of all, or a portion, of the original surface may be observed in the pattern (i.e., footwear impression, fingerprint).

Projected Bloodstain Patterns

Projected patterns are produced by blood released under pressure such as arterial spurting. These patterns generally result from volumes of blood larger than those that produce passive drop stains or other dynamic patterns (e.g., impact patterns).

If the trajectory of a bloodstream strikes an intervening target, a relatively large central stain is created that is surrounded by numerous spines of varying lengths. The appearance of this peripheral staining may be of value in assessing whether a volume of blood struck an object or surface forcefully. Projected blood streams that do not strike a vertical surface continue along individual parabolic flight paths until they deposit on a horizontal surface.

Cast-Off

Cast-off patterns are those created when blood is released or thrown from a blood-bearing object in motion. They can be the result of two basic actions: (1) arc cast-off patterns created when blood is released from an object through the influence of centrifugal acceleration and (2) the instances of cessation, or stop-action, when blood is thrown from an object when the object's motion is abruptly stopped.

Arc Cast-off Patterns. The individual stains of an arc cast-off pattern are generally distributed in a linear configuration and may be suitable for approximating the plane in space through which the blood-bearing object was moved. It may be possible to determine the object's direction of travel based on the shapes of the individual stains within the pattern. The presence of an arc cast-off pattern on a piece of clothing or other evidence item may indicate its association with a blow or other source of cast-off action.

Cessation/Stop-Action Cast-Off Patterns. The individual stains of a cessation cast-off pattern are generally smaller in size than those resulting from passive blood drops and can be a range of shapes. Cessation cast-off patterns generally do not display the linear arrangement of individual stains characteristic of arc cast-off patterns. Because cessation cast-off patterns are created as a blood-bearing object is abruptly stopped, such patterns could be produced by one or more of the impacts received by a blood source. If the blood source receiving the multiple blows is stationary, cessation cast-off patterns may be present within the context of other bloodstain pattern types such as impact patterns. In such circumstances, care must be taken in the selection of individual stains for impact site reconstruction to avoid use of stains of cessation cast-off origin.

Determination of Number of Blows

The number of cast-off patterns present may be used to estimate the minimum number of blows received by a given blood source. Because a blow applied to a blood source may not result in the creation of a cast-off pattern, however, it may not be possible to determine with certainty the specific number of blows received.

Handedness Determinations

Handedness determinations may be possible when the locations of the cast-off patterns are limited by the relative positions of the individuals and objects within the scene itself. In the absence of either ancillary information or spatial constraints at the scene, the appearance and location of cast-off patterns is not generally sufficient information on which to base a determination of the handedness of an individual wielding an object.

Not Produced from Initial Impact

Spatter and castoff patterns are created with subsequent blows to the same general area where a wound has occurred and blood has accumulated. As the object of the injury is swung away and towards the body centrifugal force is generated. If this force is great enough to overcome the adhesive force that holds the blood to the object, blood will be flung from the object and form a castoff bloodstain pattern. These are typically linear in distribution and frequently are larger in size than impact blood spatters.

Spatter

When a bloodshed event occurs at a crime scene, typically it will result in a spatter pattern. There is no "L" in the word. Those new to the field of bloodstain pattern analysis are constantly reminded that in BPA *spatter* is investigated and documented, not *splatter*. Some of this is semantics, as "splatter" is a verb, and "spatter" can be a verb, noun, and adjective. Truthfully, spatter is a more worthwhile and more descriptive term.

Blood spatter is defined as a random distribution of bloodstains that vary in size and may be produced by a variety of mechanisms. The pattern is created when sufficient force is available to overcome the surface tension of the blood. A crime scene investigator should be aware that a single small stain does not constitute a spatter pattern. Recognition of spatter patterns is important in interpreting the action and events that took place at the scene of the crime. Spatter will vary considerably between gunshot, beating, and stabbing events, and recognition of such may assist the crime scene investigator in determining the mechanism that resulted in the bloodshed. Oftentimes recognition of spatter may allow the determination of an area or location of the origin of the blood source. Recalling Locard's Exchange Principle, if blood spatter is found on a suspect's clothing, it may place them at the scene of a violent altercation.

Impact Patterns

Impact patterns are bloodstain patterns created by a blow or some other force that results in the random distribution of smaller drops of blood. They are therefore technically a subcategory of spatter. Impact patterns may be used to determine the point where the force encountered the blood source (i.e., the impact site).

While an impact pattern may display characteristics indicative of the nature of the blow or force that created it, determinations of the specific object and/or details of the impacting event generally require ancillary information. This information may include investigative information (e.g., crime scene reports, etc.), reconstruction of the point (area) of convergence, point (area) of origin, and/or experimentation. In some circumstances, however, this ancillary information may prove to be insufficient for such determinations.

Low Velocity Impact Spatter. Stains relating to a low velocity impact are considered to involve force that is equivalent to the earth's normal gravitational pull, up to a force of 5 feet per second (Bevel & Gardner, 2002). The resulting stain is typically large and has a diameter greater than 4 millimeters (mm). Such bloodstains are usually associated with blood dripping from an injury or weapon.

Medium Velocity Impact Spatter. According to Bevel and Gardner (2002), blood stains associated with this group are typically associated with a pattern and have diameters ranging from 1 to 4 mm; they may be smaller or larger depending on force of impact and quantity of available exposed blood. The force necessary to produce medium velocity impact spatter is suggested to be between 5 and 25 feet per second. It must be remembered by the crime scene investigator, however, that the first blow rarely produces spatters. Bloodstains produced from this impact group usually involve those associated with beating and stabbing events. Cast-off stains will sometimes be grouped into this category as well.

High Velocity Impact Spatter. These stains typically have an associated diameter of less than 1 millimeter and are associated with a pattern. The force necessary to produce these stains is generally in excess of 100 feet per second (Bevel & Gardner, 2002). Stains associated with this category usually involve those stemming from a gunshot, explosion, or power tool and high-speed machinery injuries.

Stains associated with gunshots may produce minute spatters of blood that may be less than 0.1 mm in diameter. This will create a very distinct "misting" effect that is not seen in beatings, stabbings, or satellite spatter created by blood dripping into blood. These stains also may exhibit spatter ranging in size from less than 0.1 mm up to several mm. Size range is dependent on the quantity of available blood, the caliber of the weapon, the location and number of shots, and factors such as hair, clothing, etc. When associated with an entrance wound, it is referred to as *back spatter* or *blowback*. This spatter may be found on the weapon and the shooter, especially hand. When associated with an exit wound, it is referred to as *forward spatter*.

Arterial Spurt

When an artery is breached, blood is projected from it in varying amounts. This pattern can vary in size from large gushing or spurting patterns to very small spray types of patterns.

Projected/impact patterns produced by arterial spurting sometimes may be identified by their forceful appearance as an inverted V-shape pattern. Because the inverted V-shape (if present) is the result of the differential intravascular blood pressure that cycles as the heart contracts and relaxes, it may be possible to identify those portions of the patterns created by less forceful heartbeats.

Expirated

Expirated patterns are created when blood is blown out of the nose, mouth, or wound as a result of air pressure and/or airflow and often display numerous, relatively small stain sizes that may vary in shape.

To differentiate between the individual stains of an expirated or impact pattern, serological tests for the presence of amylase may be performed. Additionally, a histological search for squamous epithelial cells may be performed. The presence of bubble-rings within the individual stains may also be indicative of an expirated pattern.

Identification of expirated patterns must be augmented with ancillary information such as photographs, emergency room and autopsy reports, etc. Should such documentation exist and fail to support the existence of a source for an expirated pattern, this information may be used to support the conclusion that a specific stain pattern cannot be of expiratory origin.

As a result of trauma, blood will often accumulate in the lungs, sinuses, and airway passages of the victim. In a living victim, this blood will be forcefully expelled from the nose and/or mouth in order to free the airways. These stains may appear diluted due to saliva or nasal secretions, and they may contain tiny air bubbles also.

Miscellaneous Bloodstains

Void

If there is an item or individual between the area of impact and the surrounding surfaces, the object will create a **void pattern** upon the surrounding wall or items. A void is an area that is absent of any projected blood. This void pattern often can be used to determine the size and shape of the interfering object.

Fly/Insect Spots

This category is included because often the activity of flies within a crime scene can present the crime scene investigator with confusing information about the bloodstain pattern analysis, incorrectly appearing as spatter. Flies may regurgitate, excrete, or track minute amounts of blood around a scene or onto items within the scene, which can appear to be blood spatter but in fact is not a result of the actual bloodshed event.

Skeletonized

When the center of a dried bloodstain flakes away and leaves a visible outer rim, the result is referred to as a *skeletonized* stain. Another type of skeletonized bloodstain occurs when the central area of a partially dry bloodstain is altered by contact or a wiping motion.

Stains are created when a partially dried bloodstain is disrupted by any subsequent action. Such stains retain their outer dried peripheral outlines, but generally the central area will have been partially or completely removed. Because drying time is dependent upon the surface on which blood is deposited and the environmental conditions to which it is then subjected, estimates of drying times must be experimentally determined.

Target Surface Considerations

No matter how far a drop of blood falls, it will not break into smaller droplets or spatters unless something disrupts the surface tension and cohesive properties of the droplet. In general, a hard, smooth, nonporous surface such as glass will create little if any spatter. However, in contrast, a rough surface texture such as untreated wood or concrete will create a significant amount of spatter.

Documentation of Bloodstain Evidence

Document the size, shape, and distribution of stains and patterns. Pictures with and without a scale of reference are required in order to accurately re-create the events that occurred to cause the bloodshed. Photographs, sketches, video, notes all should be done. Overall, midrange, and close-up photos are suggested in order for the entire event to be understood. If only close-up photos are taken, later the analyst will not have any way to comprehend the overall pattern and distribution of the blood. Photos and sketches should be completed in such a manner as to allow a third party to utilize the documentation to place the bloodstain patterns and articles of evidence back in their original locations. This process is typically referred to as *reconstruction*. This reconstruction can sometimes involve experimentation as to the timing and manner of blood stain distribution. In order for this to occur, a crime scene investigator or BPAE must attempt to recreate every aspect of the crime scene environment in order to determine the age of bloodstains, clots, and patterns. Such recreation and replication will be necessary in order for associated analytical findings to be admissible within court. For this reason, it is suggested that such experiments be left up to an experienced pattern analyst.

Certification in Blood Pattern Analysis

In an effort to ensure scientific validity and court acceptance of BPA as a forensic discipline, the IAI has specified specific requirements for those who would seek certification within the field of BPA. According to the organization's Web site (http://theiai.org/certifications/bloodstain/requirements.php), they are as follows:

A minimum of forty (40) hours of education in an approved workshop providing theory, study and practice as follows:

- Flight characteristics and stain patterns
- Examination and identification of bloodstain evidence
- Documentation of blood stains and patterns

This forty (40) hour course must include the following:

- Flight characteristics and stain patterns
- Oral and/or visual presentation of physical activity of blood droplets illustrating blood as fluid being acted upon by motion or force. Past research, treatise or other reference materials for the student.
- Laboratory exercises that document bloodstains and standards by previous research. Exercises must include but not be limited to the following:

Falling Blood

- Spot size related to distance fallen and blood volume variations.

Surface Considerations

- texture: smooth, rough, porous
- flat or angular surfaces
- drying time

Impact Angle Determinations

Blood in motion striking a horizontal surface or vertical target. Study of spot size, shape and spatter characteristics related to:

- distance fallen
- speed of travel
- direction of travel

Increased Blood Volumes

- stain shape and satellite spatter related to volume and distance
- radial patterns
- dripped
- splashed

Flow Patterns (horizontal/vertical)

- accumulative/passive
- secondary, altered

Projected Bloodstains and Patterns

- Static blood affected by impact of object (shoe, hand, etc.)
- Spurted, gushed, expectorated, respirated

(continues)

Certification in Blood Pattern Analysis (continued)

- Cast-off: arc of the swing, number of swings or patterns, weapon variations
- Forceful Impact Spatter Patterns
  - Blunt force, explosive (gunshot)
  - Size, shape, and distribution of spots related to degree of force and distance traveled

Transfer Stains and Impression Patterns

- Secondary targets, intermediate or intervening objects, images and impressions, wipe and swipe patterns (hair, skin, cloth, etc.)

Other Topics Suggested (not required)

- Correlation of stain and patterns to the scene surroundings and/or body trauma
- Serological/DNA considerations
- Chemical and light source absorption and enhancement
- Blood detection/collection techniques
- Stain pattern reconstruction

Source: Courtesy of the International Association of Identification and the Bloodstain Certification Board (2009).

Chapter Summary

Bloodstain pattern analysis is a contribution to the process of crime scene investigation. It may not necessarily be conclusive by itself. When combined with photographs, sketches, witness statements, and other crime scene evidence, BPA may help to provide for a clearer explanation of the events and ordering of events within a crime scene. When conducted properly, BPA is considered to be a science because the patterns and interpretation of those patterns can be replicated under controlled circumstances.

Review Questions

1. ______________ is the science of examining and interpreting blood present at a bloodshed event in order to determine what events occurred, in what order, and who possibly left the stains.
2. What were MacDonell's contributions to BPA?
3. About how much blood is present in a normal, healthy, adult body?
4. Identify the three factors whereby an investigator may deduce bloodstain information.
5. The point at which the impact event occurred that led to the subsequent dispersal of the blood is known as the ______________.

6. Explain how to calculate the angle of impact of a blood droplet.
7. If the length of a blood droplet is 3.2 cm and the width is 1.8 cm, what was the angle of impact?
8. What angle of impact produces circular blood droplets?
   a. 30°
   b. 45°
   c. 65°
   d. 90°
9. The point at which the force of air resistance that opposes the drop is equal to the force of the downward gravitational pull is known as ______________.
10. Single drops of blood will produce small spatters around the parent stain as a result of striking a rough surface. This is referred to as ______________.
11. Explain the difference between a swipe and a wipe.
12. A ______________ pattern is formed when there is an item or individual between the area of impact and the surrounding surfaces.

Case Studies

Research *State of Ohio v. Samuel H. Sheppard* (1955), and explain how the science of blood pattern analysis aided in the interpretation of crime scene evidence related to this case. Be specific with regards to impact angle, size, distribution, etc., associated with the homicide case in question.

References

Bevel, T. & Gardner, R. M. (2002). *Bloodstain pattern analysis with an introduction to crime scene reconstruction* (3rd ed.). Boca Raton, FL: CRC Press.

Caplan, J. (2001). Documenting Individual Identity: The Development of State Practices in the Modern World. Princeton, NJ: Princeton University Press.

Doyle, A. C. (1892). The Adventure of the Greek Interpreter. In the Adventures of Sherlock Holmes. *Strand Magazine.*

Federal Bureau of Identification (FBI). (2008). Birth of the FBI's technical laboratory. Retrieved September 2, 2008, from http://www.fbi.gov/libref/historic/history/birthtechlab.htm

Fish, J. T., Miller, L. S., & Braswell, M. C. (2007). *Crime scene investigation.* Newark, NJ: LexisNexis Group.

Fisher, B. A. J. (2004). *Techniques of crime scene investigation* (7th ed.). Boca Raton, FL: CRC Press.

Girard, J. E. (2008). *Principles of environmental chemistry*. Sudbury MA: Jones and Bartlett Publishers.

Kirk, P. L. (1974). *Crime investigation* (2nd ed.). New York: John Wiley & Sons.

International Association of Identification (IAI). (2009). *Welcome to the IAI!* Retrieved August 5, 2009, from http://www.theiai.org.

Lee, H. C., Palmbach, T., Miller, M. T. (2001). *Henry Lee's crime scene handbook*. San Diego: Academic Press.

MacDonnell, H. L. (1971). *Flight characteristics of human blood and stain characteristics*. Washington, DC: National Institute of Law Enforcement and Criminal Justice.

Muehlberger, C. W. (1955, May–June). Col. Calvin Hooker Goddard 1891–1955. *The Journal of Criminal Law, Criminology, and Police Science*, *46*(1), 103–104.

National Institute of Justice and Office for Victims of Crime. (May 2001). *Understanding DNA Evidence: A Guide for Victim Service Providers*. Brochure.

Radziki, J. (1960). *Slady Krwi w Praktyce Sledczej (Bloodstain prints in practice of technology)*. Warsaw, Poland: Bibliotecka Kryminalistyczna.

Saferstein, R. (2007). *Criminalistics: An introduction to forensic science*, 9th ed. Upper Saddle River, NJ: Pearson Prentice Hall.

State of Ohio v. Samuel H. Sheppard, 1955.

Impression Evidence

There is no branch of detective science which is so important and so much neglected as the art of tracing footsteps.

Sir Arthur Conan Doyle (1859–1930)
Sherlock Holmes, *A Study in Scarlet*, 1887

LEARNING OBJECTIVES

- Understand how impression evidence documentation, collection, and preservation differs from other physical evidence.
- Know what information can be derived from the proper processing of impression-related evidence.
- Differentiate between compression and striated forms of impression evidence.
- Know how to perform SICAR and how it can assist with the interpretation of impression evidence.

KEY TERMS

Compression Evidence
Electrostatic Lifting Device (ELD)
Ligature
Negative Impression
Shoeprint Image Capture and Retrieval (SICAR)
Striated Evidence

What Is Impression Evidence?

Minute imperfections on a large variety of objects such as tools, footwear, tires, and so on produce markings due to their normal (and sometimes unusual) usage. These markings are often characteristic of the type of tool or object used. In many instances, very small and sometimes microscopically unique markings are left that can be traced directly to the object or instrument in question. Two types of marks are made: compression or scraping/striated evidence.

Compression evidence are those marks left when an instrument is in some way pushed or forced into a material capable of picking up an impression of the tool. These would include shoe, tire, bite mark, impressions left by a hammer hitting a piece of wood, breech mark impressions on shell casings, etc.

Scraping or **striated evidence** are those marks produced by a combination of pressure and sliding contact by the tool, which result in microscopic striations imparted to the surface onto which the tool was worked. These include fired bullet striations, or a screwdriver blade dragged over a surface.

Comparative Examination

Comparative examination is the method by which impression-type evidence is studied. Marks left at the crime scene (or castings of the mark) are compared with test markings made by the tool or object in question. Through careful and often tedious examination of the known and questioned evidence, a determination can be made as to whether or not a particular item was responsible for a specific mark. Often the item/items used to strangle or bind a person can be matched through impression evidence comparison.

Types of Impression Evidence

There are numerous types of impressions that are left at crime scenes or upon victims of crimes. Some examples of impressions typically located during crime scene investigation include:

- Footwear
- Tire
- Bitemarks
- Tools
- Ligature/binding
- Fabric

Each is important in its own way and has a specific methodology associated with its documentation, collection, and preservation.

Footwear Impression Evidence

The history of using tracks and footwear as a form of identification is far older than that of fingerprints. Since humans first began hunting animals for food the benefit of tracks was obvious. Through keen observation, humans learned to identify their various prey animals and animal behaviors through the patterns made by the animal's pads and claws/hooves. Being able to discern the direction of travel, the number of animals present, and even whether the animal was injured due to differences in the gait features present increased the success rate of an efficient hunt. However, such knowledge did not simply apply to the gathering of food. Eventually, humans recognized the significance of such information in helping to solve criminal cases as well.

Footwear identification as a criminal justice application has its first recorded application in Scotland in 1786 in a case that involved the murder of a young girl (Hamm, 1989). An astute investigator noticed some footwear impressions that appeared to be boots, and noted from the markings that they appeared newly patched and had a great deal of nails present. The investigator made a crude casting of the impressions. The investigator, understanding that the majority of homicides involve a known person to the victim, compared the cast impression with boots of individuals who attended the victim's funeral. A suspect was found and his footwear was taken for examination. A further comparison of the suspect's footwear to the casting taken on scene led to his conviction for murder.

As was evidenced early in criminal investigative history, footwear impressions provide an important link between the criminal and the place where the crime occurred. Footwear evidence is viewed as proof that an event occurred or a person was present at a given place. Neil Armstrong's footwear impressions on the Moon are some of the most famous prints that are pointed to as evidence that an event occurred. Although footwear impressions are present at many crime scenes, the quantity found is far less than those that are actually present. It goes without question that there are very few crime scenes where the suspect does not walk to or from the scene of the crime. However, although it has been reported that 30% of all crime scenes contain useful shoeprint marks (Alexandre, 1996), very few agencies actually lift footwear evidence, and even fewer submit the gathered evidence for analysis.

Are we missing something? Is valuable evidence being overlooked due to a lack of training and technology with regards to footwear impression collection and analysis? Or is it simply misinformation and uninformed personnel?

The entire footwear recovery process can be time-consuming and extremely labor-intensive. The number of footwear evidence found at a scene may be so large that the sheer volume makes the entire process of recovery overwhelming and not at all desirable or cost efficient. However, surely this was the case with fingerprints until the age of computerization and a linked database such as AFIS. So, what is the real reason that footwear evidence is not given as much attention as other areas of forensic evidence collection?

It seems that there is a gap between the usefulness and evidentiary value of footwear impressions and gathering that type of impression evidence. If it could be determined that this was because of a lack of training or technology, misinformation, or skewed views relating to footwear evidence, then the positive conclusion of and response to this gap could provide useful information that may lead to an increase in evidence gathering, which in turn might result in successful prosecutions and increased case closures.

Many hold the belief that footwear evidence is purely circumstantial in nature. While a shoeprint does not itself provide the same level of conclusive evidence as a fingerprint, it can act as valuable corroborating evidence. In many cases, footwear impressions can reveal the type, make, model, description, and approximated or precise size of the footwear of the person who wore them. In addition, the location of footwear impressions at the scene can often help in reconstructing the crime. Footwear impressions at crime scenes can be used to corroborate or refute information provided by witnesses of suspects (Bodziak, 2000). Therefore, this evidence should not be neglected or disregarded because it is considered only circumstantial in nature. It can still prove extremely valuable in supporting or refuting the facts of the case.

The average investigator could list many possible reasons why fingerprints would be left at a crime scene and their probable locations. The same investigators are not usually as knowledgeable about footwear impressions, however. Yet with every step that is taken, whether on soil, snow, concrete, a tile floor, carpeting, glass, a wooden window sill, a piece of paper, a bank counter, or on countless other objects and materials, a representation of the characteristics of the shoe sole can be impressed against and retained by that surface, either visible or in latent form that can later be visualized.

"Because of the direct physical contact under the weight of the wearer, there is no doubt that some type of interaction between the shoe and the substrate occurs with each and every step" (Bodziak, 2000, p. 2). An example of this phenomenon can be found in **Figure 12.1**.

Where Can Footwear Impressions Be Found?

Footwear marks might not seem like an obvious source of evidence, particularly as they can be difficult to see and may be present along with footwear marks unrelated to the crime. But they can be unique and, if discovered, can provide evidence that a particular shoe was present at the scene of the crime.

Perhaps it would be easier to locate footwear evidence if the individual attempting to collect the impressions understood the basic premise behind footwear. Three phenomena occur when someone

Figure 12.1 Scaled Photograph of a Dusty Footwear Impression Left at a Crime Scene.

walks across a surface: static charges are created; the surface deforms; and/or there is an exchange of materials between the shoe and the surface (Bodziak, 2000, p. 6–8). Footwear marks are hence either two- or three-dimensional depending on the surface.

In soft surfaces, such as mud or snow, a permanent three-dimensional impression is formed (**Figure 12.2**) But on hard surfaces, such as carpet or tiles, any impression is temporary, as the surface will quickly return to its original shape. However, often there will be a two-dimensional residue of marks on hard surfaces due to the transfer of trace materials. Static charges picked up by footwear facilitate this transfer.

It is imperative that crime scene personnel be aware of the full importance of footwear impressions as physical evidence. When securing the crime scene area, they must preserve all forms of evidence, including footwear impressions. Before beginning the crime scene search, careful thought should be given to understanding what occurred at the crime scene, how footwear impression evidence would be relevant and could contribute to the proof of facts, and what areas of the crime scene might be the most logical to check first for footwear impressions. Then a careful and aggressive search for the footwear impressions should begin. "What is not looked for will not be found!" (Bodziak, 2000, p. 2).

Why Is Footwear Impression Evidence Overlooked?

According to Dwane S. Hilderbrand's book entitled, *Footwear, The Missed Evidence* (1999), there are two major reasons why footwear

Figure 12.2 Scaled Photograph of a Footwear Impression in Dirt.

evidence is overlooked. The first is a lack of training and education in the proper search, collection, and preservation methods, and the second is that the evidence is often undervalued or misunderstood. However, with the proper education in footwear evidence, both of these concerns can be overcome. This type of evidence has great evidentiary value when collected and preserved in the correct manner. As mentioned earlier, footwear evidence can reveal the type of shoe, the make, description, and, in some cases, approximate or precise size. When a crime scene is searched and documented in the correct manner, footwear evidence can either provide or assist in determining the number of suspects, their path, their involvement, and the events that occurred during the crime (Hilderbrand, 1999).

William J. Bodziak, retired FBI forensic examiner and author of *Footwear Impression Evidence*, believes that there are a number of reasons that finding footwear impressions at a crime scene is often difficult or why the evidence is simply overlooked. These include: (a) not believing that visible and/or latent footwear impressions will be found at the crime scene and not aggressively searching for them; (b) incomplete searches of the scene, possibly due in part to the inability to determine the exact points of entry and exit. Searches may be incomplete because of the lack of proper knowledge of the ways footwear impressions can occur and how they can be found; (c) arrival at the scene after other persons have trampled over the impression evidence and failure to look for footwear impressions in those areas; (d) the combination of shoe and surface characteristics not being conducive to the production of footwear impression evidence; (e) the impressions were intentionally destroyed by the suspect, and; (f) weather destroyed exterior impressions.

The presence of footwear evidence is almost guaranteed at any crime scene involving crimes against person and property. This evidence probably stands a better chance of being present than fingerprints, yet latent footwear evidence is considered to be low on the list of laboratory examinations. Footwear evidence, like fingerprints, only has to be properly searched for and recovered to be of value (Hamm, 1989). If the submission of impression evidence is low in incidence, it is because individuals responsible for the processing of the scenes are not making an effort to obtain this meaningful type of evidence. The reasons for first response officers failing to make an effort to protect scenes and recover footwear evidence can probably be traced to a lack of training in the value of footwear impressions or not being comfortable with the methods of how to properly recover the impression evidence.

Is Footwear Evidence Undervalued or Misunderstood?

Many times, courts have ruled that footwear impression evidence falls under reliable technical or specialized knowledge (such as in *Daubert v Merrell Dow Pharmaceuticals, 509 U.S. (1993)* and *Kumho Tire Co. v. Carmichael (97-1709) 526 U.S. 137 (1999)*) and is generally admissible expert testimony in a criminal case. However, although footwear impression evidence has been given its credit in our court systems, the subject itself has not been widely publicized.

Another reason that the value of footwear impression identification may be minimized is because it is viewed as only circumstantial in nature. While an item of footwear can be positively identified to a crime scene, who was wearing the shoe at the time the impression was made may be called into question. This same argument could be applied to the science of firearm identification. The bullet can be positively matched to the weapon, but who fired the weapon is not able to be discerned from the ballistic analysis. And yet, firearm examination is seen as reliable and there is great emphasis placed on it.

In many instances, footwear impressions can be positively identified as having been made by a specific shoe to the exclusion of all other shoes. This identification is based on a physical match of random individual characteristics the shoe has acquired with those respective features present in the impression. The identification is as strong as that of fingerprints, tool marks, or typewritten impressions (Bodziak, 2000).

For anyone who has not had the success of locating many footwear impressions at the scene of the crime and is not yet convinced that there are more footwear impressions present than those that are blatantly obvious, two examples are offered. A recent study conducted in several jurisdictions in Switzerland disclosed that footwear impressions were located at approximately 35% of all crime scenes. The crimes investigated in those areas consisted primarily of burglaries (Girod, 1997). Another fellow examiner reported that in the past three years, after re-emphasizing to crime-scene personnel and teaching the basics of locating and recovering footwear impressions, the percentage of cases in which footwear impression evidence was now being submitted to their laboratory had increased from less than 5% to approximately 60% (Kelderman, M., personal communication to William Bodziak, Environmental Sciences Research Ltd., Auckland, New Zealand, March 1997).

It is important that the value of footwear impression evidence be known to officers in the field and that they know how to recover it. For this reason, it is sometimes necessary to re-train them in the basics. It is necessary to reiterate the importance of effective identification, collection. and maintenance of evidence samples.

Reviewing the Basics

The location of footwear impression evidence, primarily on ground surfaces, makes it sometimes difficult or inconvenient to find, particularly if the impressions are latent or nearly invisible. Specialized lighting techniques are often required along with an aggressive effort to find the impressions (**Figures 12.3** and **12.4**).Many crime scene technicians have had little or no experience in searching for these types of impressions. In other instances, they simply may not have dedicated the time and effort needed to search for this evidence. In addition, the footwear impressions of the subject that may be mixed with those of other persons, such as paramedics and investigators who may have arrived at the scene ahead of the crime scene technician, does not necessarily mean that the suspect's were destroyed, yet many technicians are discouraged by such tracked over areas and mistakenly abandon any efforts to look for impressions.

There are four basic methods of recording footwear impressions at the crime scene: (1) photography; (2) documentation/sketching; (3) casting, and; (4) lifting (Hilderbrand, 1999).

When a footwear mark is found, distance photographs should be taken first to show the position of the mark in relation to its surroundings.

Figure 12.3 Showing How to Utilize Oblique Lighting to Capture the Details of a Footwear Impression.

Figure 12.4 Footwear Impression Photographed Utilizing Oblique Lighting.

The mark is then photographed close-up, with and without a scale. It is important to ensure that the scale is on the same plane as the impression. This is especially important so that the photograph can be enlarged to the actual shoe size. Just as with all crime scenes, the footwear evidence should be thoroughly documented in both the written report and the crime scene sketch.

It is obviously preferable to remove any evidence that has footwear marks on them, and take them into the lab for analysis. However, when this is not possible, an impression of the mark must be taken at the scene. Three-dimensional casting is an important part of footwear impression examination. A cast can reveal much more than a photograph, as minor details that may not show up in a photograph have a greater chance of being captured in a cast. Two-dimensional marks are recovered by lifting them onto another surface. This enables greater contrast between the mark and its background, resulting in more detail. Lifting is typically done through the use of gelatin lifters, an electrostatic lifting device, or adhesive lifters.

Gel Lifting of Footwear Evidence

Gel lifters contain a thick, nonaggressive, low-adhesive gelatin layer that permits the lifting of traces from almost every surface, including porous material such as paper or cardboard. Lifted impressions can be taken along for photography or closer examination. Each kind of lifter can be cut to size using scissors, and can be marked with the appropriate case information. All lifters are protected by a cover of transparent polyester film. There are several brands and types but the typical varieties come in black, white, and transparent.

Surfaces from which shoeprints can be lifted using gel lifters include all smooth and hard surfaces such as floor coverings, painted wood, paper, table tops, etc. They are especially useful in the case of **negative impressions,** when an item of footwear has left its pattern within a dusty area, essentially removing the dust that was present (**Figure 12.5**). Even if the footprints do not show up on initial lighting of the surface and were not visible upon lifting, they may show up under oblique lighting of the lifter surface in a dark room (after removing the polyester cover sheet). Lifters with no apparent prints in normal light can then show a highly detailed image.

Footwear impressions can be secured using gelatin lifters both before and after applying fingerprint powders. When the use of appropriate lighting reveals latently present impressions, it is not advisable to use powders prior to recovering the impression with a gelatin lifter. When no impressions are found, it is often useful to search or develop

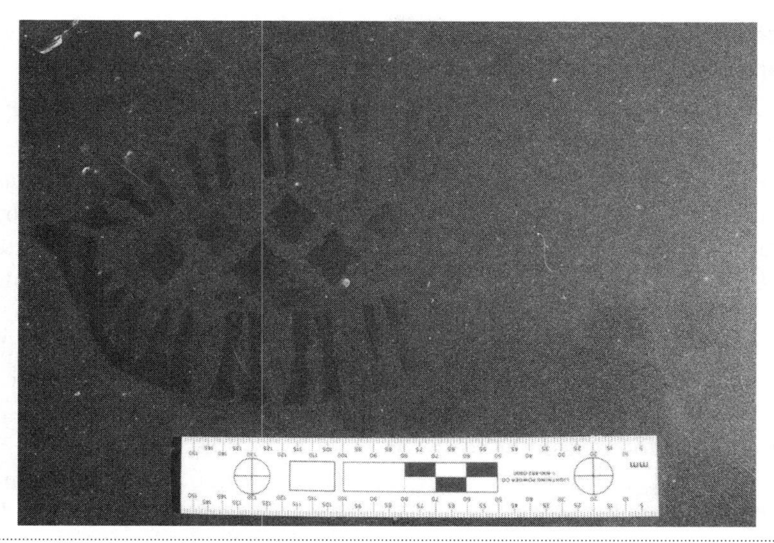

Figure 12.5 Footwear Impression Made in Dust (A Negative Impression).

footwear impressions using fingerprint powders, and then recover the developed impression with a gelatin lifter.

Electrostatic Lifting

Another and more exotic approach to lifting and preserving dust impressions involves the use of a portable **electrostatic lifting device (ELD)**. The principle employed is similar to creating an electrostatic charge on a comb and using the comb to lift small pieces of tissue paper. The device operates by charging a lifting film that has been placed over a surface bearing a dust print impression. During operation, the electrostatically-charged lifting film is drawn down to the surface, and the dust particles in the impression are attracted to the lifting film. The construction of the lifting film allows it to store the electrostatic charge, thus retaining the dust particles after the power supply has been disconnected. The impression-bearing film is then viewed with an oblique light source to search for any impressions that may have been recovered. Dust impressions may be enhanced through chemical development also.

The power supply generates either a low-powered electric field of direct voltage, or by generating an impulsive electric field of high voltage. The stronger the electric field, the stronger the device's ability to attract dust impressions to the lifting film. This equipment is designed for lifting of dry dust impressions. It works poorly on wet origin impressions.

Casting Impressions

When shoe marks are impressed into soft earth at a crime scene, their preservation is best accomplished by photography and casting. Class I dental stone, a form of gypsum, is widely recommended for making casts of shoe impressions (Bodziak, 2000). Dental stone, like plaster of paris, is a form of gypsum, but provides a superior result. Dental stone is much stronger and, therefore, does not require reinforcement material to be placed in the cast during pouring. Most importantly, dental stone is more durable and harder than plaster of paris, which means it can be cleaned in the laboratory with virtually no loss or erosion of forensic detail from the surface.

Casting is conducted to give lifelike reproduction of the original impression. This is of great forensic benefit because it will: (a) back-up the scene photograph; (b) provide for tangible three-dimensional evidence; (c) eliminate focus or scale problems; (d) reproduce microscopic characteristics; and; (e) reproduce characteristics of the side of outsoles and midsoles of the shoe that are not usually reproduced or evident in photographs (Hilderbrand, 1999).

Chemical Development of Impressions

A number of chemicals can be used to develop and enhance footwear impressions made with blood. In areas where a bloody footwear impression is very faint or where the subject has tracked through blood and left a trail of bloody impressions, chemical enhancement can visualize latent or nearly invisible blood impressions (Saferstein, 2001). The investigator may adapt any one of several chemical formulas for use in processing footwear evidence. A few examples include: luminol, amido black, and leucocrystal violet.

Computerized Footwear Databases

Crime scene footwear classification systems can be utilized to look for possible suspects, to get models and brand names for crime scene impressions, and to link crime scenes together (Mikkonen, Suominen, & Heinonen, 1996). Shoe impression or reference files are not new; they are maintained worldwide by such bodies as the U.S. FBI, the Metropolitan Police Service in London, the National Bureau of Investigation in Finland, and the Bundeskriminalamt in Germany. Without such computerized assistance, time and resources are expended in compiling information about a single incident or enquiry that at its conclusion may be discarded (Ashley, 1996).

New computer software may be able to assist the forensic scientist in making shoeprint comparisons. One example is the automated shoeprint identification system developed in England by Foster and Freeman Ltd., called **Shoeprint Image Capture and Retrieval (SICAR)**, incorporates multiple databases to search known and unknown footwear files for comparison against footwear specimens (Bodziak, 2000). Using this system, an impression from a crime scene can be compared to a reference database to find out what type of shoe caused the imprint. That same impression also can be searched in the suspect and crime databases to reveal if it matches the shoes of a person who has been in custody or the shoeprints left behind at another crime scene. When matches are made during the searching process, the images are displayed side by side on the computer screen (Saferstein, 2001).

In Europe, footwear evidence is the second most collected item of evidence after fingerprints. This high regard for footwear evidence is due, in part, to the fact that police agencies in several countries have created national databases for footwear evidence: United Kingdom, SICAR; Germany, BILDDATEBANK and SCHARS; the Netherlands, REVEZO and TRIS; Switzerland, TDP; and Finland, NBISPDB. SHOE is established in Australia, as well. This capability does not yet exist in the United States (Mankevich, 2002).

Currently, a detective cannot determine if the shoe print under investigation by his/her agency is also appearing at crime scenes in other jurisdictions. The detective also has limited options on how to effectively search for the brand name and manufacturer of the shoe print evidence.

In the past, detectives have resorted to flipping over numerous shoes in shoe stores in the hopes of blindly stumbling upon the correct pair of shoes and matching the crime shoe prints. Flipping through the pages of magazines and catalogs, and surfing online Internet sites are other time-consuming, rarely productive means often used to recognize the brand of shoe prints.

A computerized database of outsole and footwear impression images on a central server would allow case detectives and laboratory analysts the most productive means to search for and recognize footwear's brand and manufacturer without the labor-intensive efforts required in the past.

Footwear impression evidence has repeatedly exhibited its value as a means to provide a criminal investigation with information related to (and often, the actual identification of) the crime perpetrator. One needs only recall the preflight activities of the September 11 terrorist group, as they "blended in" and operated for a length of time out of diverse locations such as Florida, Virginia, Massachusetts, Maryland, and Arizona prior to the bombings, to see how associative evidence could help identify and possibly be used to thwart criminal enterprises.

An example of footwear evidence having a direct relation to the Homeland Security issue is Richard Reid's attempt to destroy American Airlines Flight #63 with a bomb that had been inserted in his shoe. On January 25, 2002, an image of the outsole of Reid's shoe was posted on the Wanted Page as Case #2/2002, by the FBI Laboratory, asking if anyone knows the "brand name or manufacturer of the enclosed shoe." Using its SICAR system, Maryland State Police was able to provide the match as "an 'Ashburn' brand of hiking boot" to the FBI Laboratory. The only other law enforcement agency to respond to the request for brand/manufacturer recognition was the National Bureau of Investigation in Vantaa, Finland.

The ability to quickly recognize the brand and manufacturer of footwear impressions is invaluable to detectives during any criminal investigation. Once the brand of shoe prints appearing at a crime scene is known, detectives can focus their investigation upon suspects who conform to the marketing and brand "aura" of the footwear such as age, gender, stature, recreational pursuits, aesthetic, stylistic, and fashion preferences. A recognized footwear brand can be specified during requests for search warrants. It used to be that when there was footwear evidence the detectives would serve a search warrant and take all of the shoes in the house. However, if they were able to know the make, model, and/or brand ahead of time, as specified in the search warrant, they only need to take those shoes that are similar to the outsole pattern identified.

Collection of Footwear Impression Evidence

Footwear impressions may yield information as to the type, make, and approximate size of the shoe or boot. In some cases, conclusive identification can be made, linking a particular shoe to the crime scene impression. The likelihood of conclusive identification is a function of several factors, but it is primarily governed by the amount of wear sustained by the outsole, the ability of the receiving surface to resolve the fine detail of the impression, and the fineness of the material by which the impression is deposited.

A footwear impression may be either positive or negative, that is, the impression may be the result of dust or other material being deposited on a clean surface, or it may be the result of the outsole removing dust or other material from the surface. Either type of impression

SoleSearcher: FBI Questioned Footwear Examination

SoleSearcher is an innovative investigative tool used by the Questioned Documents Unit (QDU) of the FBI to help determine the brand name or manufacturer of footwear from questioned footwear impressions left behind at crime scenes. Hundreds of these database searches are requested annually by law enforcement agencies in cases that range from burglary to terrorism. The database contains more than 11,000 different shoe outsole designs from more than 370 different footwear manufacturers from around the world. New outsole images are added to the database daily. With so many different manufacturers and a majority of shoes being manufactured overseas, keeping up with all of these shoes is a monumental task.

SoleSearcher was designed using different design icons commonly found on the outsole of shoes. These different design icons are placed on a blank shoe canvas to resemble the questioned footwear impressions. The image is then searched through the database using the specific criteria. If the database produces a shoe outsole that is consistent with the questioned impression, a printout of this shoe outsole is sent back to the contributor. If the contributor has or develops a suspect or a person of interest, the contributor can use this print during a search to obtain the shoes needed to conduct further footwear examinations.

The FBI Laboratory's first footwear reference collection dates back to around 1935. This collection consisted of a series of photographs of shoe soles and heels. These photographs were placed in file cabinets according to manufacturer. The reference-file searches were all conducted by hand. Back in the 1930s, this collection was considered complete because all shoes at that time were manufactured in the United States. The shoes were primarily dress shoes and perhaps a small number of casual shoes. Specialty athletic shoes were not introduced until the 1950s and 1960s.

The first computerized footwear database was introduced in the mid-1980s and was housed on a mainframe computer system. In the early 1990s, the footwear database was placed on a personal computer. The manufacturer would send in catalogs, and these outsole images would be scanned into the database. At that time, the database had approximately 2,000 outsole images. Today, the database contains more than 11,000 images. In the future, the Laboratory plans to make SoleSearcher available through a secure network so other law enforcement agencies may electronically contribute and search footwear images in the database.

Source: Courtesy of the Federal Bureau of Investigation (2006). *FBI Laboratory 2006 Report*. Quantico, Virginia: FBI Laboratory Publication. Retrieved September 16, 2009, from http://www.fbi.gov/hq/lab/lab2006/labannual2006.pdf.

has the potential of being suitable for comparison and identification. Impressions made in dust, particularly those made on hard, smooth surfaces, are the most likely to yield a conclusive identification. Investigators should be alert to the possibility of such impressions when broken glass is found in an area where the perpetrator could have stepped. Another potential source of such impressions is plastic windows in out-of-the-way areas that are rarely cleaned (such as in gas stations or warehouses). In some cases, the window will be forced out with the foot, or once forced will be stepped on, removing dust from the surface and creating a negative record of the outsole.

Whenever possible, submit the original item on which the impression appears. Lifts made in the field and improperly taken photographs rarely show the fine detail necessary to make a conclusive identification.

Recovery of Surface or Dust Impressions

(1) Photograph the entire area to show walking pattern or relation of impressions to surrounding objects. This means being cognizant of taking proper overall and midrange photographs. (2) Take a close-up photograph without a scale and then take a proper scaled photograph of each footwear impression. (3) Prepare a proper sketch showing the position of the impression(s) in relation to fixed objects at the scene, utilizing a proper mapping method.

Recovery of Objects and/or Impressions

(1) Recover and preserve the object bearing the impression. (2) If this is impossible, the impressions should be lifted by using an electrostatic lifter, a transparent footwear residue lifter, or a rubber footwear lifter. (3) The object bearing the impression or the lift should then be submitted to the forensics laboratory, along with copies of diagrams, photographs, etc.

■ Tire Impressions

Just as is the case with footwear impressions, tire tracks are often forgotten or overlooked at crime scenes. This may be due to a lack of knowledge or experience with regards to locating, documenting, and collecting such evidence. However, tire tracks are crucial evidence in proving that a suspect vehicle was present at the crime scene.

Tire impressions reflect the tread design and dimensional features of the individual tires on a vehicle. Tire tread impressions can be compared directly with the tread design and dimension of the tires for a suspect vehicle. If the surface retains sufficient detail, this comparison can result in a positive identification.

Tire tracks are the relation dimensions between two or more tires of a vehicle. By measuring the dimensions of tire tracks at the crime scene, it may be possible to determine or approximate the track width, wheelbase, or turning diameter of the vehicle. Tire tracks can be used to profile the type or size of vehicle used and for other purposes that can help to include or exclude a suspect vehicle.

Tire impressions sometimes permit identification of type and make of the tire. In some cases, a particular tire may be conclusively identified as the maker of an impression.

Recovery of Tire Evidence

Tire tracks should be documented through photographs, casting, and measurements (**Figure 12.6**), being certain to take overall, midrange, and close-up photographs. This will show both the orientation of tracks and track detail. Overlap close-up photos and always use a scale so that photos can be blown up 1:1 and spliced to be used for comparison purposes. It is important that the scale be placed on the same plane as the impression. Casting a long tire impression is important and challenging. Any cast of 4 feet or less in length should be cast in its entirety. If the impression is longer than 4 feet, the impression should be cast in manageable sections. Identify each cast with an impression number placard. Be sure to photograph the scene after the casts have been poured, for documentation purposes.

Tool Marks

For the purposes of this chapter, a tool is any instrument or object capable of making a mark on another object. Tool mark identification techniques may be applied to many types of investigations—knife

Figure 12.6 Scaled Photograph of Tire Impression.

marks on bone; fractured knife blades; homemade explosive devices; crimp marks on detonators; cut marks on wire; fractured radio antennas)—as well as burglary. A close examination of a tool mark may reveal the type of tool, contour of the cutting edge, blade width and color of the tool, or the presence of trace material. Tool marks may be found at points of entry and exit at victimized premises and upon objects that have been attacked.

Recovery of Tool Marks

Whenever possible, tool marks should be kept in their original condition. This may be done by recovering the whole or part of the object upon which the marks appear. In recovering the mark, it is important that it be protected against dirt, moisture, and scratching during transport. Tissue or other soft paper should be placed over the tool mark in packaging. Casting or other methods of taking impressions of tool marks should be used only as a last resort.

- Always submit an object containing the tool mark to forensic laboratory.
- If not practical to submit the object, remove the section of material containing the tool mark and submit it to the forensic laboratory.
- As a last resort, make a cast of the tool mark.
- Mark, protect, and individually package item(s) containing tool mark(s) and submit them to the forensic laboratory.

Casting of Tool Marks

In the event that, as a last resort, casting the tool mark is necessary, silicone casting materials have been found satisfactory. They are available through scientific and law enforcement supply houses. Directions for their use are contained in each kit. The crime scene investigator is warned not to use plasticine, plaster of paris, home repair patch plaster, and related materials, as they have a tendency to shrink and not accurately preserve the impression.

Recovery of Tools

Recover all suspect tools and inventory, observing the following precautions, and submit them to the forensic laboratory for examination and comparison with tool marks.

- Never place a suspect tool in contact with a questioned tool mark or cast.
- Package each tool individually to protect the tool for possible fingerprint, and/or trace material examination.

Bitemark Evidence

Bite marks may occur on the skin of victims of rape or sexual murder, or on a criminal. At times bite marks can be so characteristic that they make the definite identification of a suspect possible. The relative positions of the teeth, their widths, and the distance between them, together with ridges on the edges of the teeth and grooves on the backs or fronts, vary among different individuals and may be clearly evident in the bite mark (**Figure 12.7**).

Bite marks can involve both compression and striated evidence, dependent upon the action occurring at the scene of the crime. These impressions are an outstanding source of physical evidence that can be found on skin, furniture, foodstuffs, chewing gum, and many other objects located at the scene of a crime. There are many criminal incidents that include victims defending themselves by biting their attacker. Sometimes an attacker will bite his or her victim as part of the assault process. As odd as it may seem, there are many times when a criminal will eat at a crime scene and leave behind evidence that may have bite marks present on or in them. All of these instances can be analyzed and the identified impressions documented and compared to a suspect of victim dentition in an effort to corroborate or refute testimonial evidence, and to help place the individual at the scene of the crime.

Figure 12.7 Model of a Dental Impression.

Processing Bitemark Evidence

The potential evidence of a bite mark is not simply the pattern itself but also the saliva and thus the DNA that could be present with the impression. For this reason, the area should be processed as follows:

- Photograph, with and without a scale for reference, utilizing oblique lighting to best distinguish the characteristics of the mark.
- After this has occurred, the area should be swabbed (see Chapter 10) to collect potential DNA evidence.

RIPPED FROM THE HEADLINES

Forensic Science to Gain Another Forensic Database?

On May 15, 2008 the Associated Press reported that researchers at Marquette University, in Milwaukee, Wisconsin, had developed a first-of-its kind computer program that could measure bite characteristics. They were hopeful that their work could lead to a database of bite characteristics that could narrow down suspects and lend more scientific weight to bite-mark testimony.

Built around the assumption that every person's teeth are unique, forensic dentistry has used bite impressions to identify criminals for 40 years.

But critics say human skin changes and distorts imprints until they are nearly unrecognizable. As a result, courtroom experts end up offering competing opinions. Since 2000, at least seven people in five states who were convicted largely on bite-mark identification have been exonerated, according to the Innocence Project.

Determined to prove that bite analysis can be done scientifically, Dr. L. Thomas Johnson and his team won about $110,000 in grants from the Midwest Forensic Resources Center at Iowa State University and collected 419 bite impressions from Wisconsin soldier volunteers.

They built a computer program to catalog characteristics, including tooth widths, missing teeth, and spaces between teeth. The program then calculated how frequently—or infrequently—each characteristic appeared.

Johnson hopes to collect more impressions from dental schools across the country to expand the database into something close to law enforcement's DNA databanks. With enough samples, the software could help forensic dentists answer questions in court about how rarely a dental characteristic appears in the American population. "That would help exclude or include defendants as perpetrators," Johnson said.

He acknowledged that the team's software will probably never turn bite-mark analysis into a surefire identifier like DNA and that he would need tens of thousands of samples before his work would stand up in court.

Source: Information retrieved from Richmond, T. (2008, May 15). Scientists are building database of bite marks. Associated Press Online. Retrieved September 16, 2009, from http://www.thefreelibrary.com/Scientists+are+building+database+of+bite+marks-a01611533435.

- If the mark is an evident one, not one that requires "maturing" of the wound, then it is suggested that the crime scene investigator lay a clear piece of acetate over the wound and carefully outline the injury, using a permanent marker on the acetate. This could prove useful for a forensic odontologist to conduct an eventual comparison.
- If the individual is deceased, the medical examiner may elect to excise the portion of the skin baring the impression evidence and preserve it.
- If the area is to be cast, the crime scene investigator should make use of silicon casting material to cast the area and preserve the cast for subsequent analysis.
- Some injuries, if fresh and made to a live person, may require them to "mature" and this will require the crime scene investigator to revisit the injury to take photographs at a later time to provide the best visual evidence.

When crime scene personnel confronted with bite-mark evidence do not know the correct way to proceed with the collection efforts, it is

CASE IN POINT

Bite Marks Bite Back

Perhaps the most famous incident where bitemark evidence was found to have such strong forensic value that it was used to convict was in the case of the notorious serial killer, Ted Bundy. From 1973 to 1978 Bundy was responsible for an undetermined number of murders across the United States. The case that provided the forensic link was when he was finally tied to the murder of Lisa Levy through bite marks that were left on her buttocks during the assault and subsequent murder. An astute officer at the scene of the crime had the forethought to place a ruler next to the marks discovered upon Levy's body and were then photographed.

Due to a variety of other forensic evidence and investigative leads, Bundy was eventually arrested. Following Bundy's arrest, police obtained a search warrant that allowed them to take a dental impression from Bundy and photograph his front upper and lower teeth and gum areas. During the subsequent trial, the prosecution produced the photographic evidence of the bite marks and the evidence that had been collected of Bundy's teeth. They demonstrated the relationship between the two by placing an acetate overlay of the bite mark on the photo of the teeth. When this was done, it left no reasonable doubt in juror's minds that it was Ted Bundy's highly irregular teeth pattern that had resulted in the impressions discovered on Levy's body. Based upon this evidence, Ted Bundy was convicted and sentenced to die in the electric chair, which he did on January 24, 1989.

Source: Information retrieved from http://www.crimeandclues.com/bite_marks.htm.

important that they consult medical personnel who are knowledgeable on such matters. Many medical personnel have additional training in the recognition of bite mark evidence because they are commonly involved in tending to victims of child and spousal abuse. If this evidence is not properly recognized or processed, critical information may be lost or destroyed.

Ligature/Binding Impressions

Oftentimes, the item that is used to bind, incapacitate, or kill someone will leave an impression within or upon the skin of the victim. These items are referred to as **ligatures**. Ropes, phone cords, and other materials that sometimes are used in strangulation cases will leave behind patterned impression evidence that later can be matched to the item that was used. Depending upon the action that occurred during the criminal event, the ligature or binding impressions can be either striated or compression evidence (**Figure 12.8**).

Processing Ligature Impressions

All ligature or binding impressions should be photographed, typically with the use of oblique lighting in an effort to accentuate the impression through the use of shadows. The area then can be cast making use of a forensic casting silicone material. This should create a scaled

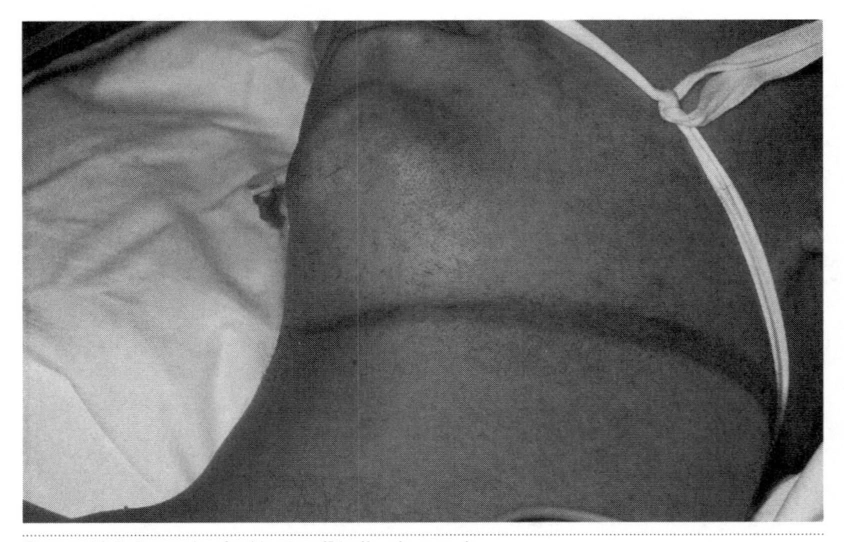

Figure 12.8 Example of a Ligature/Binding Impression.

replica of the impression that later can be compared to any items believed to have been responsible for the mark.

If the ligature or binding material is still present upon the body of a victim, it is important that the crime scene investigator observe a few key guidelines. Firstly, a body is not to be touched, moved, or otherwise tampered with without the permission or guidance of the coroner or medical examiner staff (as appropriate). Typically, such items are not removed from the body of a deceased individual until autopsy. If, however, the removal of the item is warranted, and permission is gained, then the ligature should be processed as follows.

The ligature should be photographed in place, prior to removal, both with and without a scale for reference. Next, the item should be carefully removed. If the item is one that includes a knot (e.g., necktie, rope, etc.), the knot should not be untied as this itself is evidence that can give information as to the skill and education of the individual who tied it (i.e., military, Boy Scout, etc.). Instead, the item should be severed at an area not near the knot, which allows the item to be carefully removed. The two fresh ends should then be labeled to differentiate them from any other loose ends of the same material. Some instructors teach that a string should be tied to each fresh end, thus bridging the gap between the new ends, after removal. However, it is the experience of the author that the least amount of touching, damage, and otherwise manipulation of the item of evidence will ensure the preservation of the greatest amount of evidence associated with the event. The item should then be packaged as appropriate, to minimize movement and dislodgement of trace evidence.

Fabric Impressions

Sometimes the movement of an individual at a crime scene can result in fabric impression evidence being left behind. The person may lean up against an object present at the scene of a crime and leave behind the pattern of his or her clothing. If a suspect or victim leans against wet paint, or a dust covered area, sometimes this will result in the pattern of the clothing being left behind, waiting for an astute individual to recognize the evidentiary potential. While clothing itself is not typically individualistic, the way an item of clothing touches or impresses upon another can be individualistic, similar to a jigsaw puzzle piece. Often, if properly documented and collected, the impression can be matched up to an object responsible for leaving the impression. In addition to matching the weave and pattern, the object also may be found to have relating cross-transfer material such as the paint or dust that resulted in the impression.

Processing Fabric Impressions

As with other impression evidence, proper photographs are paramount. The crime scene investigator also should collect a comparison sample of the dust, paint, or other material in which the impression was left so that it might later be compared against trace materials found on the clothing that left the impression (see Chapter 9 for more information about trace evidence). It may prove useful to cast the impression with the use of silicon casting material.

Chapter Summary

"Commit a crime and the earth is made of glass . . . you cannot wipe or cut the foot-track . . . so as to leave no inlet or clue, some damning circumstance always transpires." (Ralph Waldo Emerson, 1841). However, as the information presented within this chapter reveals, and as stated by Professor Locard over 100 years ago, "Only human failure to find it, study and understand it can diminish its value."(Kirk, 1974, p. 2).

Review Questions

1. What are the two types of impression markings?
2. What are the two major reasons why footwear impression evidence is overlooked?
3. List three benefits to the investigator casting footwear impressions.
4. What is SICAR and how does it work?
5. Casts of long tire impressions should be kept to what maximum length?
6. What information can be deduced by measuring the dimensions of tire tracks at the crime scene?
7. Why is bite-mark evidence sometimes criticized?
8. How was bite-mark evidence used in the Ted Bundy case?
9. When presented with a knot on a ligature or binding, how should an investigator proceed with evidence collection and why?
10. A ______________ impression may result from a suspect leaning against a surface covered in wet paint or dust.

References

Alexandre, G. (1996). Police Cantonale Neuchateloise, Service d'Identification Judiciaire. Computerized classification of the shoeprints of burglars' soles. *Forensic Science International, 82,* 59–65.

Ashley, W. (1996). Crime Scene Section, Victoria Science Centre. What shoe was that? The use of computerized image database to assist in identification. *Forensic Science International, 82,* 7–20.

Bodziak, W. J. (2000). *Footwear impression evidence (*2nd ed.). Boca Raton, FL: CRC Press.

Emerson, R. W. (1841). *Essays, First Series.* Compensation. Retrieved September 16, 2009, from http://www.vcu.edu/engweb/transcendentalism/authors/emerson/essays/compensation.html

Federal Bureau of Investigation (2006). *FBI Laboratory 2006 Report.* Quantico, Virginia: FBI Laboratory Publication. Retrieved September 16, 2009 from http://www.fbi.gov/hq/lab/lab2006/labannual2006.pdf.

Girod, A. (1997, April). Presentation at the European Meeting for Shoeprint/Toolmark Examiners. The Netherlands.

Hamm, E. D. (1989). Track identification: an historical overview. *Journal of Forensic Identification, 39,* 333–339.

Hilderbrand, D. S. (1999). *Footwear, the missed evidence.* Temecula, CA: Staggs Publishing.

Kirk, P. E. (1974). *Crime Investigation* (2nd ed.). New York: John Wiley & Sons.

Mankevich, A. (2002, January). *SICAR East Coast Network.* Proposal presented to the U.S. Department of Justice, Washington D.C.

Mikkonen, S., Suominen V., & Heinonen P. (1996). Use of footwear impressions in crime scene investigations assisted by computerized footwear collection system. *Forensic Science International, 82,* 67–79.

Saferstein, R. (2001). *Criminalistics, An Introduction to Forensic Science* (7th ed.). Upper Saddle River, NJ: Prentice Hall.

Firearms and Ballistic Evidence

Untutored courage is useless in the face of educated bullets.

General George S. Patton, Jr. (1885–1945)

▸▸ LEARNING OBJECTIVES

- Differentiate between ballistics and firearms identification.
- Know what information can be assembled and its potential forensic value resulting from proper processing of a firearms-related crime scene.
- Differentiate between smoothbore and rifled firearms.
- Know how the firing mechanisms of modern firearms and the components of modern ammunition allow for the interpretation and identification of firearm scenes and associated firearms.
- Understand the proper methods associated with documentation, collection, and preservation of firearms-related evidence and why they are used.

■ Introduction to Firearms

For the purposes of this text, a **firearm** is any device that expels a projectile or projectiles as a result of an explosive or propellant charge. Crime scene personnel and firearms examiners typically will come into contact with three types of firearms: handguns, rifles, and shotguns. However, it is useful to know that within each category there are several subcategories. This is important because each category and subcategory of firearm operates differently, and each may require different concerns as far as documentation and preservation methods are concerned.

▸▸ KEY TERMS

Ballistics
Bore
Breech Lock
Caliber
Comparison Microscope
Exterior Ballistics
Extractor or Ejector
Firearm
Firing Pin
Gauge
Grooves
Handgun
Integrated Ballistic Identification System (IBIS)
Interior Ballistics
Lands
National Integrated Ballistic Information Network (NIBIN)
Primer
Rifle
Rifled Firearm
Rifling
Smoothbore Firearm
Stellate Defect
Terminal Ballistics
Transitional Ballistics

Handguns

A **handgun** is a weapon designed to be held in and fired with one hand. However, anyone familiar with firearms usage will agree that the most accurate way in which to fire a handgun is through the use of two hands, as that position creates a stable shooting platform. There are two primary subcategories of handguns: pistols (semi-automatic and automatic) and revolvers.

Pistols

Most nonrevolver types of handguns are of the semi-automatic variety. A semi-automatic firearm will fire each time that the trigger is depressed and subsequently released for as long as there is ammunition remaining within the firearm. These firearms typically include a spring loaded magazine, which holds the cartridges and feeds them into the awaiting chamber after the previously spent shell casing has been extracted and ejected as a result of the slide action of the semi-automatic handgun. In some extremely rare cases, a pistol may be of the automatic variety. In this case, the weapon will continue to fire (unless interrupted by a malfunction) as long as the trigger is depressed and there is ammunition within the weapon. In either case (semi-automatic or automatic pistols), expended shell casings will be expelled from the firearm and will remain at the scene unless retrieved by the shooter (**Figure 13.1**).

Revolvers

A revolver is a type of handgun that incorporates a revolving cylinder as the method of containing ammunition and of cycling the ammunition

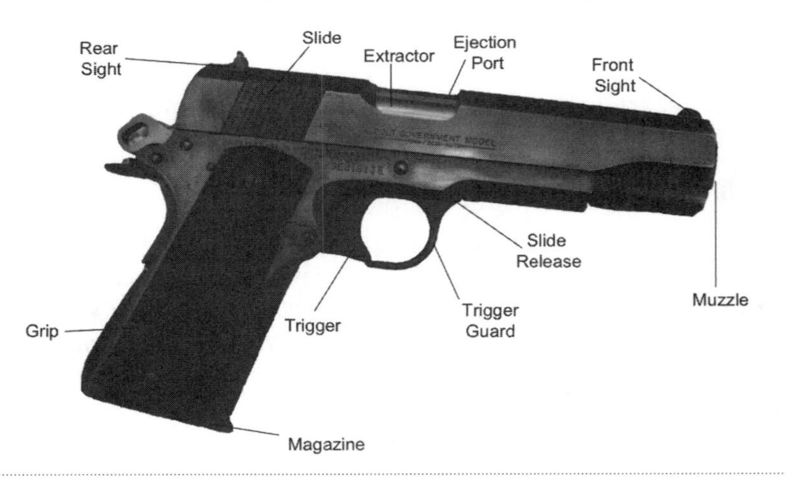

Figure 13.1 Characteristics of Modern Firearms: Handgun.

In this example, a Colt 10mm, semi-automatic pistol is shown.

into battery. There are no interchangeable magazines and shell casings of expended ammunition are contained within the cylinder rather than being expelled as in a semi-automatic pistol. At a crime scene involving the use of a firearm, therefore, failure to find expended shell casings might mean that the perpetrator made use of a revolver or else he or she picked up the shell casings.

There are two types of revolvers: single action or double action. A user of a single action revolver must manually cock it each time before firing. The user of a double action revolver need only to pull the trigger, which in turn cocks the firearm, and then subsequently fires it.

Rifles

A **rifle** is a weapon designed to be held in two hands when being fired from the shoulder. As with handguns there are various types. Some rifles are single shot, slide action types that will fire one round and then through manipulation of a bolt will extract the spent shell casing and either load another round or the user will have to manually load another. Some are semi-automatic versions, which operate similarly to semi-automatic pistols, firing one round for each depression of the trigger as long as there is ammunition available. Automatic rifles will continue to fire as long as the trigger is depressed and ammunition is available (**Figure 13.2**).

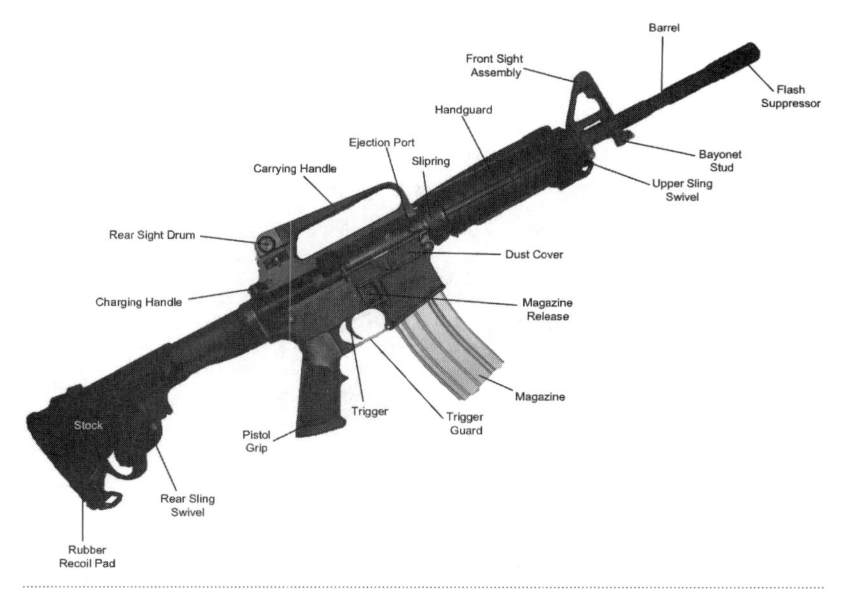

Figure 13.2 Characteristics of Modern Firearms: Rifle.

In this example, a Colt AR-15, semi-automatic rifle is shown.

Rifled Firearms

All of the aforementioned types of firearms are referred to as **rifled firearms**. A firearm is referred to as either being a smoothbore or rifled depending upon whether or not there is rifling present within the barrel (**Figures 13.3** and **13.4**). Rifled firearms contain **rifling** within the **bore** (interior of the firearm barrel). Rifling are the series of spiral grooves that are formed in the bore, which are designed to impart spin upon a projectile as it passes through the barrel to improve its accuracy. The rifling is made up of two components: grooves and lands. The **grooves** are the low-lying portions between the lands within a rifled firearm bore. The **lands**

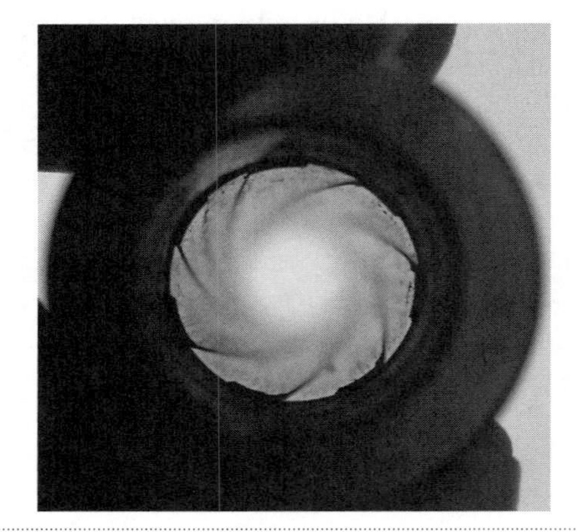

Figure 13.3 A Rifled Barrel, Showing Rifling Present within the Barrel.

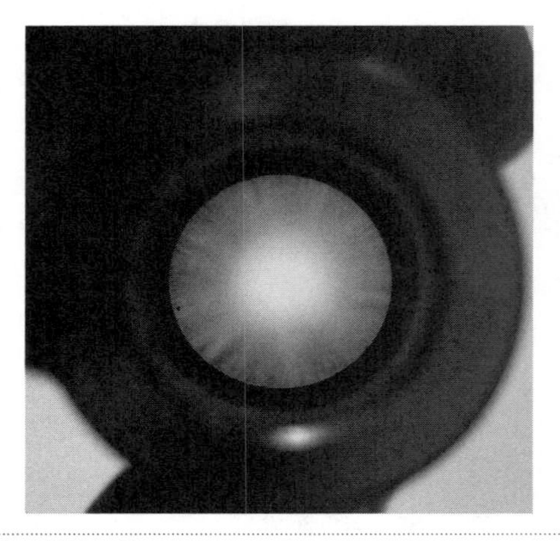

Figure 13.4 A Smoothbore Weapon.

Note the absence of rifling within the barrel.

are the raised portion between the grooves within a rifled firearm bore (**Figure 13.5**). During the manufacturing process of a firearm, minute imperfections occur within the barrel that, when combined with the lands and grooves, will impart characteristics upon a bullet passing through the barrel known as striations, a type of impression evidence discussed in Chapter 12. These are often unique to a specific barrel, much like a fingerprint is to a finger. The rifling, however, is not intended to be an identifiable "fingerprint." Instead, its design provides ballistics properties for the bullet as it exits the barrel of the firearm. Much like throwing a football, imparting spin upon the projectile gives it stability and provides for a more accurate and predictable trajectory (**Figure 13.6**).

Rifled weapons are classified by their **caliber**, which is the diameter of the bore of a rifled firearm, measured between opposing lands. The caliber is usually expressed in hundredths of an inch or millimeters, for example, .22 caliber and 9 mm.

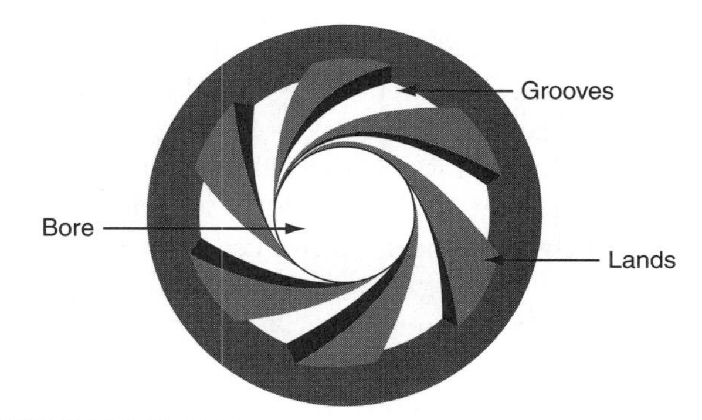

Figure 13.5 Bore Characteristics of a Rifled Firearm.

Source: Ellie Bruchez, University of Wisconsin–Platteville.

Figure 13.6 Photograph of a Bullet.

Note rifling characteristics and striations.

Smoothbore Firearms

As discussed earlier, if there is no rifling present within the firearm bore, then the firearm is referred to as being a **smoothbore firearm**. The most common example of a smoothbore firearm is a shotgun. A shotgun is a firearm that, as with a rifle, is designed to be held in two hands and fired from the shoulder. They are most commonly used to fire pellet loads rather than single projectiles (slugs). While some shotguns are able to interchange and have rifled barrels for firing slugs, the shotgun was originally designed to fire multiple projectiles (shot) that are typically fired through a smoothbore barrel.

There are other types of smoothbore weapons that typically include antique black-powder type weapons. However, because not many crime scenes involve such firearms, they are not concentrated on within the confines of this text.

Just as rifled weapons are referred to by their caliber, smoothbore weapons are referred to by their **gauge**. Gauge is the number of spherical lead balls that have the diameter of the interior of the barrel of the firearm that add up to weigh one pound. For example, 12 lead balls having the diameter of the interior of a 12-gauge shotgun barrel (roughly .75 inches) would weigh approximately 1 pound. An exception to this rule is the .410 gauge, which has an actual bore diameter is .410 inches, which is not meant to confuse the reader, but rather to recognize that referring to "gauge" is not an absolute measure with regards to smoothbore weapons.

Firing Components of Modern Firearms

There are several components within a modern firearm that significantly impact the firing of a round, which result in specific characteristics being passed along to shell casings placed within the chamber. The **breech lock** is a component that supports the base of a cartridge within the chamber. The face of this component, called the breech face, will often leave identifiable marks upon the base of the shell casing. The **firing pin** is the component that strikes the base of the cartridge and causes the initial incendiary event leading to expulsion of the bullet from the barrel. This often leaves marks upon a cartridge casing that are identifiable to a particular firearm or brand of firearm (**Figures 13.7** and **13.8**). An **extractor** or **ejector** also may be found within a firearm (nonrevolver) and is responsible for ejecting the spent shell casing after it has been fired. This hook-like object is likely to leave a compression or striated mark within the ejector groove at the base of a spent shell casing that is potentially identifiable to a particular firearm. Note that revolvers do not

Figure 13.7 Photograph of the Base of a .45 Caliber Shell Casing.
Note firing pin mark in center of primer.

eject spent shell casings and, therefore, revolver ammunition does not contain an ejector groove at its base **(Figure 13.9)**.

Characteristics and Components of Modern Ammunition

A firearm can be a centerfire or rimfire weapon based upon the location of the primer. Centerfire or rimfire refers to the location the firing pin will strike the **primer** on the cartridge to detonate the primer. The primer is a metal cup located within the center of the base of a cartridge that contains a small amount of incendiary compound, which when crushed by the firing pin sets off the initial incendiary event. The primer's explosion sets off the secondary incendiary event: the subsequent lighting of

Figure 13.8 Photograph of the Base of a 9mm Shell Casing.
Note distinctive Glock firing pin mark in center of primer.

Figure 13.9 Differences Between Semi-Automatic and Revolver Shell Casings.

Note extractor groove.

the propellant as a result of the primer flash reaching the propellant through a small hole between the anvil that separates the primer from the smokeless powder propellant. In rimfire weapons (e.g., a .22 caliber), the entire base of the cartridge serves as a primer.

A modern firearm round or cartridge consists of a shell casing, a bullet, propellant, and a primer (**Figure 13.10**). Each of these has its own potential forensic significance, which will be discussed later in the chapter (**Figure 13.11**).

Virtually every portion of contemporary ammunition has potential evidentiary value:

- Propellant
- Elemental makeup of the projectile and primer mixture
- Design characteristics of the bullet
- Wads from shotgun shells (James & Nordby, 2003)

A basic knowledge of these components and their potential evidentiary value is of critical importance for the successful investigation of a shooting incident.

Ballistics

The science of firearms identification is often incorrectly termed "ballistics." In actuality, **ballistics** is the study of a projectile (most likely from a firearm) in motion. There are four distinct areas of ballistics:

1. **Interior ballistics** is the study relating to the transition of chemical energy to kinetic energy within the barrel of a firearm and the motion of the projectile(s) as it moves through the barrel.

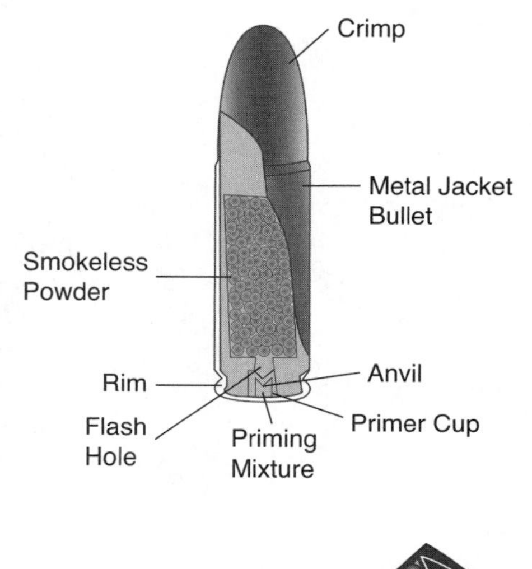

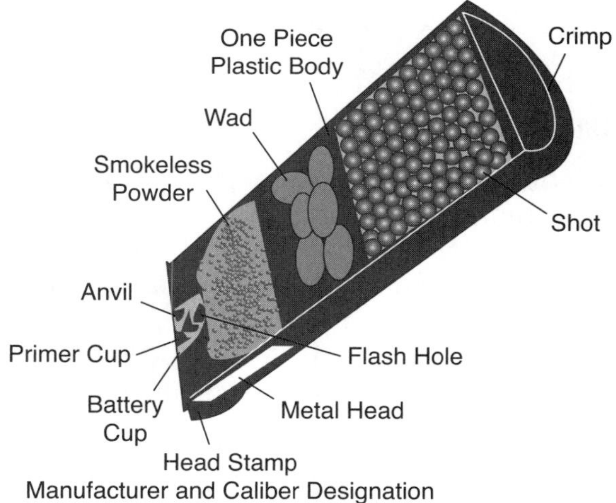

Figure 13.10 Characteristics of Modern Ammunition.

Source: Erica Lawler, University of Wisconsin–Platteville.

2. **Transitional ballistics** is concerned with the period of time in which the projectile(s) transitions from its movement through the firearm barrel to its flight through the air upon exit from the barrel.
3. **Exterior ballistics** is the study of the flight of a projectile beginning at the muzzle end the barrel of a firearm, and terminating at the target.
4. **Terminal ballistics** deals with the resulting impact and interaction between the projectile and the target matter.

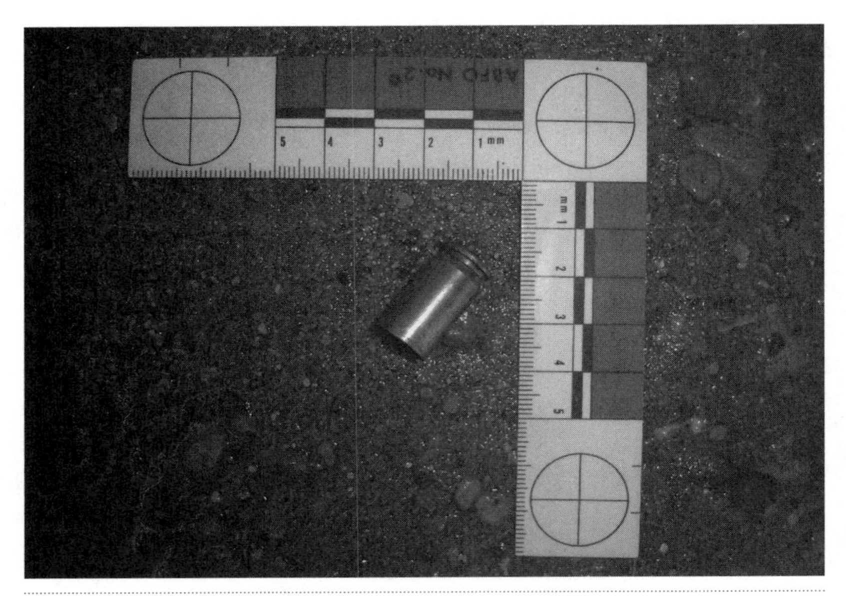

Figure 13.11 Scaled Photograph of a Shell Casing Located at a Crime Scene.

Firearms Identification

Firearms identification, unlike ballistics, is the discipline of forensic science that is primarily concerned with determining if a bullet, cartridge case, or other ammunition component was fired by a particular firearm. This determination is made by the science of examination and comparison of firearms and ammunition. This identification process typically hinges upon analysis and identification of firearms-related markings. These can include, but are not limited to:

- Striations, land/groove markings on a bullet
- Cartridge cases will be marked by a number of firearm components:
  - Firing pin marks (rimfire or centerfire)
  - Breechlock marks on primer cap
  - Extractor/ejector marks on base of cartridge in self-loading or automatic weapons
  - In weapons with magazines, the lips of the magazine may mark cartridges also. (Saferstein, 2007)

As a general rule, shotgun barrels do not leave identifiable markings on shot, wads, or shot columns.

Collection of Firearms-Related Evidence

Bullets may be difficult to recover at the crime scenes (**Figure 13.12**). Only rubber-coated or heavily taped tools should be used either to probe for bullets or to extract them, so as not to cause damage to, or impart striation evidence upon the recovered evidence. Generally, it is best for the investigator to remove the section of the building structure that contains the bullet so that the forensic firearms examiner can carefully remove the bullet in the laboratory. Care must be taken not to dislodge any trace evidence that may be on the bullet surface. It is recommended to place the bullet or cartridge in a sealed pillbox or plastic vial. The container and its seal are then marked for identification (ID). The serial numbers of all firearms seized should be recorded by investigators. For safety, loaded weapons should be unloaded before they are transported to the firearms laboratory.

Laboratory Examinations

Bullets should be examined for the presence of trace evidence as well as patterned markings. Bullets may pick up textile fibers, traces of paint, or bits of concrete and brick from intermediate targets. The goal of the initial exam is to determine general rifling characteristics of the firearm that fired the bullet.

Figure 13.12 Photograph of a Bullet That Was Removed and Cleaned after Striking a Target.

Determining Rifling Characteristics

The general rifling characteristics to be determined are:

- Caliber
- Number of lands and grooves
- Direction of twist of the rifling
- Degree of twist of the rifling
- Widths of lands and grooves (James & Nordby, 2003)

Rifled firearms have either right twist (Smith and Wesson type) rifling or left twist (Colt type). The degree of twist of the rifling and the widths of the land and grooves can be determined by microscopic measurements of the rifling marks on the bullet.

Identifying Class Characteristics

Class characteristics of firearms can be determined from expended cartridges. The significant class characteristics are:

- Caliber
- Shape of firing chamber
- Location of the firing pin
- Size and shape of the firing pin
- Size of extractors and ejectors (if any)

Test Firing for Comparison

Once it has been determined that the class characteristics of the firearm and those of the fired bullets or cartridges are consistent, the weapon must be test fired to obtain bullets and cartridges for comparative microscopical examination. To obtain test-fired bullets, the firearm is fired into a bullet trap. Most firearms laboratories use some type of water trap to catch fired bullets (**Figures 13.13** and **13.14**).

Microscopic Comparison

There are a number of ways in which an examiner will conduct analyses of ballistic evidence, and a variety of instruments which may be used (Figure 13.5). However, typically bullets are mounted for microscopical comparison on a comparison microscope. A **comparison microscope** consists of two compound microscopes connected by an optical bridge. It allows two specimens to be viewed side-by-side (**Figure 13.15**).

Serial Number Restoration

A criminal may attempt to render a firearm untraceable by filing or grinding away the serial number. However, it may be possible for a firearms examiner to restore, recover, or visualize the serial number. Magnaflux, chemical etching, electrochemical etching, and ultrasonic cavitation are several methods commonly used (Saferstein, 2007).

Figure 13.13 Portable Ballistic Trap for the Collection of Bullet Evidence.

Figure 13.14 Port View of a Portable Ballistic Trap for the Collection of Bullet Evidence.

Figure 13.15 A Firearms Examiner Uses a Stereomicroscope to Visualize Ballistic Impression Evidence on a Cartridge Casing.

■ Range-of-Fire Estimation from Powder and Pellet Patterns

In the reconstruction of shooting incidents the range and direction of fire are of paramount importance. Gunshot residue (GSR) is typically used to measure these patterns. Gunshot residue consists of particles from the gun barrel, particles from the bullet surface, particles originating from the propellant, and particles originating from the primer. This residue is projected in a roughly conical cloud in the direction of the target (**Figure 13.16**).

Firearms examiners also can estimate ranges of fire from shotgun pellet patterns. A widely used rule-of-thumb is that the pellets spread one inch for each yard down range. The sizes of shotgun pellet patterns are affected by the choke of the shotgun barrel. Sawing off the barrel of a shotgun may increase the size of the pellet patterns it will fire.

■ Visualizing Gunshot Residue

Gunshot residue is a type of trace evidence, and should be recovered in the same manner as discussed in Chapter 9 (**Figure 13.17**).Firearms

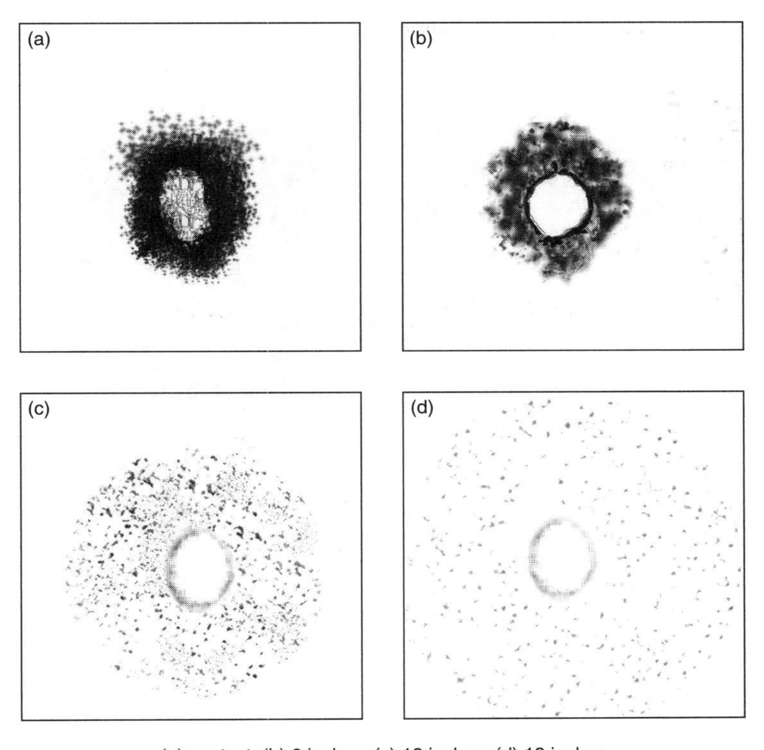

Figure 13.16 Determining Proximity of a Firearm at the Time of Discharge.

Source: Ellie Bruchez, University of Wisconsin–Platteville.

Range-of-Fire Estimation

Chapter 17 covers gunshot wound (GSW) related injuries more in depth; however, forensic pathologists usually place the range from which a GSW was inflicted into one of several categories:

- **Distant:** No detectable GSR reaches the skin or clothing of the victim. GSWs consist of a circular or elliptical defect in the skin, which is surrounded by a "marginal abrasion ring" where the skin has been stretched and torn by the bullet entry.
- **Close-Range/Intermediate Range:** GSR reaches the skin/clothing of victim and stippling/tattooing is present.
- **Near-Contact/Contact:** Often produces a **stellate defect**, which is an irregular, blown-out entrance wound. This type of wound is caused by the propellant gases separating the soft tissue from the bone and creating a temporary pocket of hot gas between the bone and the muzzle of the weapon. Blood and other tissue may be blown back into the muzzle of the weapon and onto the hand and forearm of the shooter.

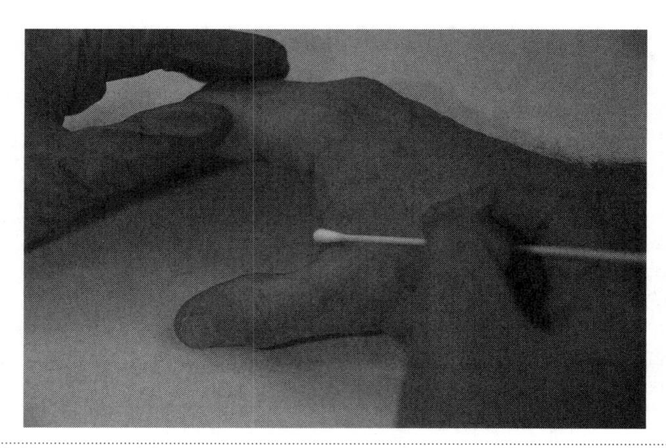

Figure 13.17 Example of Collecting GSR Evidence from the Hand of a Suspected Shooter.

examiners may need to use special techniques to visualize GSR patterns on dark or bloodstained clothing.

- Advanced light source (ALS)
- Chemical methods involve testing for and visualization of nitrates present.
  - The Griess Test: The chemical reagent reacts with inorganic nitrates in GSR to form an azo dye.
  - Maiti Test: The chemical reagent reacts with GSR and appears as blue specks on a yellow background (Saferstein, 2007).

National Integrated Ballistic Information Network

A networked computer database of fired cartridge casing and bullet images used by crime laboratories is the **National Integrated Ballistic Information Network (NIBIN)**. NIBIN was developed to link firearms evidence and to solve open cases by allowing firearms examiners to compare evidence with fired bullets, cartridge casings, shotgun casings, and firearms recovered in other jurisdictions. The heart of NIBIN is the **Integrated Ballistic Identification System (IBIS)** comprising a microscope and a computer unit that captures an image digitally. By means of a microscope attached to the system, images of bullets and cartridge casings are electronically scanned and stored for later retrieval and comparison with other case images. The system has the ability to compare the new images rapidly with images in regional and national databases. Just as with AFIS, an examiner visually compares the images to determine if there is a "hit." The system is maintained by the Bureau of Alcohol, Tobacco, Firearms and Explosives (ATF), where 236 sites are electronically joined to 16 multistate regions (ATF, 2009).

VIEW FROM AN EXPERT

Firearms and Toolmark Examination

Training, education, and desire played a crucial role in my professional development as a forensic scientist in the Firearm and Toolmark Section at the Northeastern Illinois Regional Crime Laboratory (NIRCL) in Vernon Hills, Illinois. My formal education included a Bachelor's of Science degree from the University of Wisconsin–Platteville in Chemistry-Criminalistics with a DNA emphasis. While working on my degree I had the opportunity, though my school, to work as an intern at NIRCL.

Soon, I found myself in the laboratory environment, astonished by the quantity and quality of casework the forensic scientists generated on a daily basis at NIRCL. I was requested to assist in the Firearm and Toolmark Section where I immediately was captivated by the Firearm and Toolmark discipline. After my eight week internship was complete, I was offered a training position in the Firearm and Toolmark Section, which with fervor, I accepted. As I finished my degree, I eagerly anticipated my return to the laboratory to begin my Firearm and Toolmark training.

Upon my return to NIRCL, I began my training under a former Association of Firearm and Tool Mark Examiners (AFTE) President. Quickly I discovered the discipline of firearm and tool mark identification was very much an apprenticeship; typically consisting of two years of training, followed by a year of supervised casework. As a trainee, I studied numerous readings, literature, and court rulings accompanied with written and verbal tests, practical examinations, courtroom demeanor training, and supervised casework. Although the training was long and tedious, I found it to be absolutely necessary. I feel the extended instruction was crucial in order to cement the underlying basis of the science: the reproducibility of class and individual characteristics.

Throughout my training I was afforded opportunities to tour several firearm factories, ammunition factories and tool factories; all of which expanded my appreciation for the discipline. The dedication continued as I became a certified armorer for M-16 rifles, AR-10 rifles, and Glock™ pistols. I also had the honor to partake in a television appearance titled, "Stop the Violence: Conflicts and Resolutions."

My duties as a Firearm and Toolmark Examiner include the examination of firearms and tool marks, ammunition components, comparative microscopy, serial number restoration, range proximity of gunshot residue patterns, shot pellet pattern testing, trigger pull analysis, functionality/alteration testing and examination of evidence for physical fit/fracture matches. Other duties include keeping maintenance logs of instruments, calibration and balance checks, reference collection firearm inventory, and chemical inventory. I also make entries into the Integrated Ballistics Identification System (IBIS), analyze correlations, and provide statistics on IBIS for the Bureau of Alcohol, Tobacco, Firearms and Explosives (ATF).

Throughout my time as a firearm and toolmark examiner, I have come across several firearms and tools that I had not seen previously nor could initially identify. However, firearm and toolmark examiners have many resources at hand to aid in identifying a rare firearm or tool. Some of these resources include: colleagues,

(continues)

literature, forensic journals, and AFTE. AFTE is an international professional organization of firearm and toolmark examiners and technical advisors who, through a collaborative effort, hold meetings, publish a scientific journal, and partake in forums to share the knowledge and findings of members across the globe. Due to networking, examiners in this scientific community have an increased ability to assist in many technical issues. Through the networking efforts of these individuals, answers are literally a phone call or email away.

From first stepping into the laboratory environment, to working cases, my passion and dedication to the science of firearm and toolmark identification has only increased. I thoroughly enjoy learning about the rich history of firearm and toolmark identification. I aspire to continue the endeavors of examiners before me. I can attest to the quality and expertise of those before me and those who continue to work diligently in Firearm and Toolmark discipline. This is truly a career and profession, not just a job. Through many diverse circumstances I became involved in an intriguing career, one of which only a short time ago I had no knowledge.

Gary M. Lind, B.S.
Forensic Scientist
Firearm & Toolmark ID Section,
Northeastern Illinois Regional Crime Laboratory
A Nationally Accredited Laboratory (ASCLD/LAB)

Collection and Preservation of Firearms Evidence

The following information is intended to assist the investigator in the recognition, evaluation, marking, packaging, and transmittal of firearms exhibits and related items. All exhibits should be properly inventoried. Record the description of the item, source, case number, item number, initials of person collecting, and the date and time collected. Sketch the area of recovery, indicating relative positions in feet and inches between exhibits and fixed objects, and supplement with photographs.

Firearms and fired ammunition can be delivered to the forensic laboratory either in person or via parcel post, certified mail, or United Parcel Service (UPS). Loaded ammunition must be delivered in person. United States postal regulations prohibit shipment through the mail. All firearms shipped to the forensic laboratory must be unloaded and marked on the outside package. Firearms should be securely placed within a container (typically cardboard), with the barrel end of the firearm noted on the outside of the package. Inside, the firearm should be secured in such a manner that will not allow the firearm to shift or damage the firearm (**Figures 13.18** and **13.19**).

Figure 13.18 Example of Packaging a Firearm for Evidence Submittal.

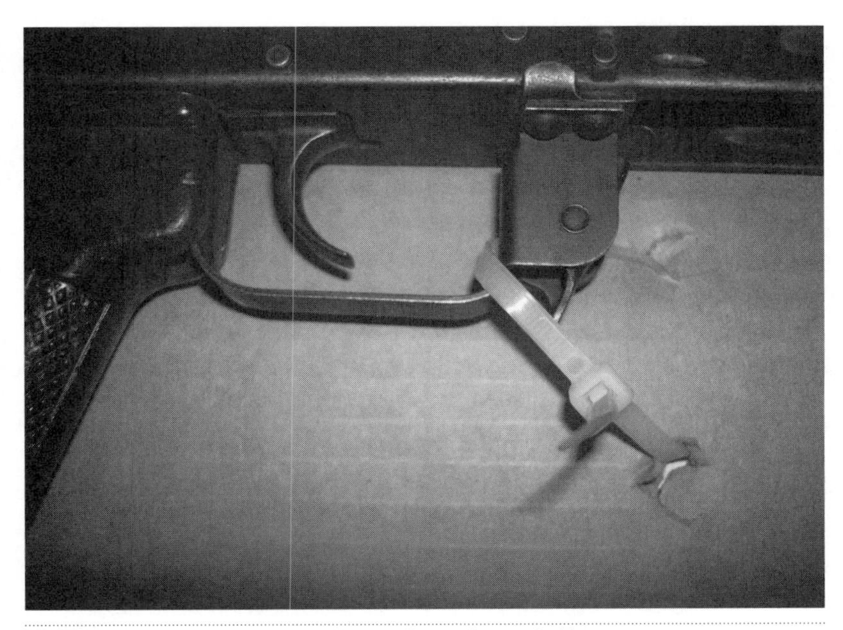

Figure 13.19 Close-Up Photograph Showing How to Secure the Firearm.

Firearms or other metal objects recovered from water (liquid) should immediately be placed in a container of the same liquid, completely submerged. When in a liquid, the oxidation process is considerably retarded, and during the period necessary to transport the firearm to the crime laboratory only a small amount of change will take place.

Marking of Firearms

Use extreme care in marking recovered firearms for purposes of identification. Attach a reinforced identification tag to the trigger guard. Mark the tag with appropriate identifying data, including the serial number and description of the firearm, source, case number, item number, initials of person collecting, and the date and time collected.

Many texts and training materials suggest that the person responsible for recovering a firearm should etch/scribe in his or her name/initials and case number on the body of the firearm. This is extremely archaic information and should never occur. The firearm should not be altered in any manner during the collection and preservation phase.

Marking of Ammunition and Components

All firearms and firearm-related items should be handled with the belief they will be fingerprinted. Therefore, only the packaging should be marked. In this manner the possibility of damage, loss, or contamination of trace evidence and destruction of possible fingerprints is greatly diminished. These items should never be packaged in cotton or sealed in plastic. All packages should be properly sealed, with initials of collector over the seal, and marked with accompanying information such as the description of the item, source, case number, item number, initials of person collecting, and the date and time collected. In situations where through-and-through penetration of the victim's body has occurred and the bullet is found on the floor, in walls, or other location, bullets or bullet fragments should not be touched with bare fingers. A small piece of clean white paper should be slipped under the bullet, then folded and placed in a rigid container, and finally sealed and identified. This procedure will minimize the possibility that the recovering officer will contaminate traces of blood that may be present on the bullet. The aforementioned recommendations also apply to shotgun pellets and wads.

Many texts suggest that collected ammunition, bullets, or ballistic components should be marked/etched with the collector's initials; however, it is the author's belief and the policy of the majority of ballistic and forensic experts that firearms evidence is no different than any other type of physical evidence. When properly packaged and a proper chain of custody is adhered to, then there is no reason to mark and otherwise contaminate or change any item of evidence.

■ Bullet Path Reconstruction

Defining a bullet's path at a shooting scene is a useful element of crime scene reconstruction. A shooter's position and final bullet location can both be defined by determining the path of a bullet or bullets through a sequence of materials. Such reconstructions are most accurate when a bullet has created both a bullet hole and a subsequent impact site or two or more bullet holes appear in successive planes of material, for example, in sheet rock on both sides of an interior wall. Inserting rods through the bullet holes (or from bullet hole to impact site) will define a bullet path that can direct the investigator to the shooter's position or to the bullet's likely location. Rods should not be inserted in any bullet hole until documentation and examination of the bullet hole has been completed. Over short distances, string can be attached to the rods to project the bullet path. This technique is especially useful in reconstructing shootings involving vehicles due to their double-panel construction. However, because the projected bullet path increases in distance from the bullet hole, greater imprecision will be introduced into the reconstruction. For bullet path reconstructions over long distances, a combination of spacer cones, rods, and lasers will offer much better precision, especially if meaningful diagramming of the reconstruction is desired (**Figure 13.20**).

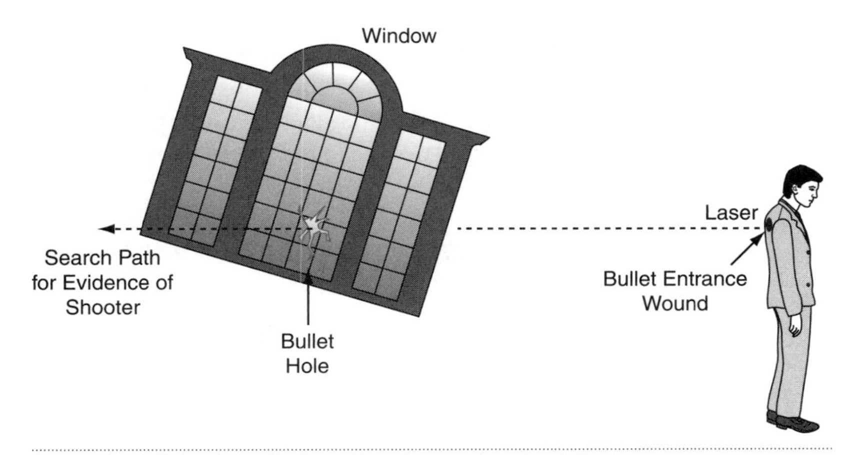

Figure 13.20 Determining Discharge Trajectory and Location of Shooter.

Source: Ellie Bruchez, University of Wisconsin–Platteville.

Unless a bullet passes through a significant thickness of material, a single bullet hole will usually not allow useful reconstruction of the bullet path. However, bullet direction can be determined from through-and-through bullet holes in many materials. As discussed with regards to the examination of glass evidence in Chapter 9, the passage of a bullet through metal will create an indentation on the metal surface facing the bullet origin and metal stretch on the surface in the direction away from bullet origin, clearly defining the direction of the bullet through the metal. Bullets that pass through auto glass, skull, and some plastics will create a crater on the side of the material away from the bullet origin. In other words, the crater opens up in the direction of bullet travel. Even a portion of a bullet hole in a destructively fractured skull can define the direction of the bullet and subsequently establish exit and entrance. The combination of glass cratering and radial glass fracture in a window can even define the sequence of shots through the window, particularly when working with vehicle shootings.

Chapter Summary

Gunshot wounds are the most common homicidal injury within the United States as this country has more firearms per capita than any other nation on earth. As such, the documentation, collection, and subsequent analysis of firearms-related evidence is often of paramount importance in an investigation involving a firearm. It is imperative that a crime scene investigator know the proper methods to collect firearms evidence that will result in the best evidentiary value. In order to achieve this end, the crime scene investigator must know the precise potential forensic value of the various components. Sometimes firearms evidence will yield only class characteristic information, but often, when properly documented and collected, firearms evidence can result in individual characteristics that are capable of being traced back to a single source.

Review Questions

1. A ______________ is any device that fires a projectile or projectiles as a result of an explosive or propellant charge.
2. ______________ is the discipline of forensic science with the primary concern of determining if a bullet, cartridge case, or other ammunition component was fired by a particular firearm.

3. ______________ is the study of a projectile (most likely from a firearm) in motion.
4. Sawing off the barrel of a shotgun may ______________ the size of the pellet patterns it will fire.
5. The diameter of the bore of a rifled firearm is known as its ______________.
6. ______________ is the number of spherical lead balls having the diameter of the interior of the barrel that weigh one pound.
7. Rifled firearms either have a twist to the ______________ like a Colt, or to the ______________ like a Smith and Wesson.
8. Which chemical test for the presence of GSR forms blue specks on a yellow background?
9. Explain what NIBIN is and how it is used in regards to ballistics evidence.
10. The heart of NIBIN is ______________, comprised of a microscope and a computer unit that captures an image digitally.

Case Studies

Research the names John Allen Muhammad and Lee Boyne Malvo as they relate to a series of firearm-related cases. Summarize how the documentation, collection, preservation, and subsequent analysis of firearms-related evidence resulted in their convictions.

References

Bureau of Alcohol, Tobacco, Firearms and Explosives. (2007, July 11). *Illinois State Police Mark 1000th 'Hit' Using ATF's Ballistic Imaging.* Retrieved September 16, 2009, from http://www2.prnewswire.com/cgi-bin/stories.pl?ACCT=ind_focus.story&STORY=/www/story/07-11-2007/004623809&EDATE=WED+Jul+11+2007,+01:09+PM

James, S. H., & Nordby, J. J. (2003). *Forensic Science.* Boca Raton, FL: CRC Press.

Patton, G. S. (1885–1945). Retrieved from http://permanent.access.gpo.gov/lps1786/vign7.html

Saferstein, R. (2007). *Criminalistics: An introduction to forensic science* (9th ed.). Upper Saddle River, NJ: Pearson Prentice Hall.

Arson and Explosive Evidence

Joseph LeFevre

Never attribute to malice that which can be adequately explained by stupidity.

Hanlon's Razor

LEARNING OBJECTIVES

- Understanding how and why fire is used to commit crimes.
- Know the proper methods associated with fire and explosives-related evidence documentation, collection, and preservation.
- Name the components needed to have fire, and explain how they can assist the investigator with crime-scene processing.
- Define and differentiate between cause and origin analyses.
- Understanding the forensic value of fire-related evidence.
- Differentiate between low and high explosives and the evidence associated with each.
- Understand the forensic value of explosives-related evidence.

KEY TERMS

Accelerant
Area of Origin
Arson
Combustible
Conduction
Convection
Direct Flame Contact
Fire Loads
Flame Point
Flammable
Flammable Limit
Flashover
Flashpoint
Fuel Load
High Explosive
Low Explosive
Plant
Point of Origin
Puddling
Pyrolysis
Radiation
Secondary Device
Spoliation
Trailers
Vapor Density

Fire Scenes

All crime scenes require some form of teamwork. At the scene of a fire, teamwork becomes an absolute. The investigation of a fire can involve many people representing different agencies and interests. The emergency services (police, fire, and medical rescue squad) typically come to mind when most people think about a fire scene. Others also will be involved, such as insurance representatives, gas or electrical utility company workers, state and federal law enforcement, and, in the case of a death, the medical examiner's office. All successful fire investigators are willing to be team players.

Typically, firefighters are not trained in how to interview suspects and have only limited experience with chain of custody for evidence. Likewise, the typical police investigator is not trained in fire suppression techniques nor do they have experience recognizing some types of physical evidence such as smoke color. Therefore, it is imperative that crime scene investigators familiarize themselves with the skill sets that will enable them to conduct a successful complete fire investigation.

Fire as a Crime

Most states have a statute covering the crime of **arson**. In general terms arson can be defined as "the willful and malicious burning of a person's property" (DeHaan, 2006, p. 605). Every fire scene must be approached by an investigator as being arson until proven to be otherwise.

From a CSI standpoint, the crime scene investigator does not need to prove the property actually destroyed was the target of the arson. This means that a specific intent is not important, simply that there was intent to start a fire. The reasons for a person to commit arson are varied and can be classified as profit, vandalism, excitement, revenge, concealment of a different crime, or terror-related goals (DeHaan, 2006). Understanding these motives can help a crime scene investigator in the search for evidence.

Profit

An arsonist may start a fire for direct monetary gain or some indirect profit. Commercial or industrial business owners may have inventory they cannot sell so burning it clears out the inventory and insurance pays for the loss. The business may be failing, so burning may be an easy way to take an early retirement. A piece of land may be worth more without a building on it and, therefore, burning it saves the time and cost of legal demolition.

In residential settings, property owners facing a foreclosed mortgage often will burn the home before handing it back over to the bank. In times of divorce, one spouse may destroy a home or property to ensure the other cannot take it. Homeowners also may be looking for an upgrade they cannot afford. The insurance payoff from small-contained fires can be enough to get that kitchen remodel and all new appliances.

Vandalism

This type of arson is typically associated with boredom and youth. Targets for vandalism tend to be schools, abandoned property, or woodlands. There is no pattern to these fires and targets are random places where opportunity has presented.

Excitement

Persons needing some excitement, looking for attention, and in very rare situations sexual stimulation may set fires also. The profile of an excitement fire setter has them starting off small fires in trashcans, then working up to larger items, such as dumpsters, until they work themselves up to the large and risky, buildings. Some people like the excitement of starting the fire, some like the lights and sirens of responding units, and some have a hero complex.

Persons with a hero complex can be dangerous. Examples of hero complex, or excitement-seeking arsonistic individuals, include: The police officers looking to get glory for putting out the fire before the firefighters show up, a bored firefighter looking to ensure "action" during his or her shift, the volunteer firefighter wanting to drum up headlines and public support for the volunteer firehouse, or even a private security guard looking to ensure his or her continued employment.

Revenge

Fires may be set as retribution for a real or imagined insult to an individual. The target(s) for revenge may be an individual or a group that the arsonist feels committed a wrong against them. Persons also may seek revenge against an institution such as a bank, church, or government office that has caused them pain.

Crime Concealments

Fire is the great destroyer. After committing a crime in a location, a suspect fearing physical evidence identifying them can attempt to destroy the evidence by fire. While true a fire can obscure and render much of the physical evidence discussed in this text difficult to use, most evidence will actually survive a fire.

In the case of murder, the arson may be set up to make the death look accidental. The arsonist may use fire after a murder also in an attempt to make identification of the victim impossible. Fortunately for an investigator, it takes significant long-term exposure to combustion for the body to no longer hold evidence as to the manner of death or personal identifications.

Arson is not limited to covering up violent crimes. Criminals may seek to cover up their thefts, burglaries, employee embezzlements, or other acts of fraud. The arson aspect of these crimes is typically an afterthought.

Terrorism

Extremists groups or social protestors have long used fire as a weapon to further their causes. Unlike military explosives and firearms, fire is not a restricted sale item. The simple combination of fire with a flammable liquid can make a fire explosive (i.e., the Molotov cocktail). The ease of using fire does not necessitate that a group be highly organized to use it in terror-like attacks. Grassroots environmental groups have been responsible for the burning of housing developments encroaching on wildlife habitats. Other groups have used fire against abortion clinics, drug testing labs, university research labs, and even business executives' offices.

What Is Fire?

The following information was adapted from the welcome packet for students attending the Federal Government's National Fire Academy (NFA), entitled, "What Is Fire?" (NFA, 2007).

Those who are responsible for the investigation of fire scenes must have a working knowledge of the behavior of fire because they often are required to interpret the aftermath of a fire, and frequently are required to use both technical and/or general explanations of fire behavior in legal proceedings.

A knowledgeable understanding of the behavior of fire helps to demonstrate the investigator's credibility in the area of fire cause determination. This understanding, coupled with the ability to explain the behavior of fire, will add credibility to court testimony and to the investigator's opinion into the origin and cause of the fire.

One of the more popular tactics used by arson defense attorneys is to attack the investigator's credibility with regard to fire behavior. Therefore, a solid, basic understanding of the behavior of fire is the foundation from which any fire cause investigation will be developed, and through which the investigator's credibility as an expert witness will be determined.

Fire is a complex chemical process, and fire investigators must understand the basic chemistry and physics involved, which will enable them to formulate their opinions based on these sound scientific principles. If the investigator is not able to explain the technical aspects of fire behavior, they will have a tough time qualifying as an expert witness. The opposing attorneys easily can use questions about the chemistry and physics of fire to discredit a fire investigator effectively.

Elements of a Fire

Fire consists of three basic elements: fuel, oxygen, and heat. These basic components have been recognized in the science of fire protection for over 100 years. The diffusion flame process is defined as "a rapid self-sustaining oxidation process accompanied by the evolution of heat and light of varying intensities" (DeHaan & Icove, 2008).

Six elements are needed in the life cycle of fire: input heat, fuel, oxygen, proportioning, mixing, and ignition continuity. All six elements are essential for both the initiation and continuation of the diffusion flame process. The first three elements—heat, fuel, and oxygen—are represented by the fire triangle.

The combustion reaction can be depicted more accurately by a four-sided solid geometric form called a tetrahedron. The four sides represent heat, fuel, oxygen, and uninhibited chain reactions **(Figure 14.1)**.

Input Heat

Solid or liquid materials do not burn. For combustion to take place, these materials must be heated sufficiently to produce vapors. It is these vapors that actually burn. The lowest temperature at which a solid or liquid material produces sufficient vapors to burn under laboratory conditions is known as the **flashpoint**. A few degrees above the flash-point is the **flame point**, the temperature at which the fuel will continue to produce sufficient vapors to sustain a continuous flame. The temperature at which the vapors will ignite is the ignition temperature, sometimes referred to as the auto ignition temperature. If the source of the heat is an open flame or spark, it is referred to as piloted ignition.

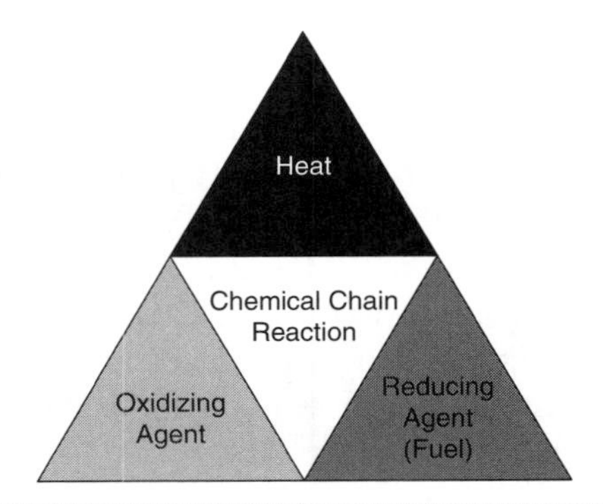

Figure 14.1 The Fire Tetrahedron.

Removing any of the four factors—fuel, heat, oxidizing agent, or chemical chain reaction—will control, suppress, or prevent a fire.

Source: Ellie Bruchez, University of Wisconsin–Platteville.

For example, gasoline has a flashpoint of –45°F (–42.8°C) and an ignition temperature of 536°F (280°C). This means that at any temperature at or above –45°F (–42.8°C), the gasoline will be producing sufficient vapors to be ignited if exposed to an open flame, spark, or any heat source of 536°F (280°C) or greater (DeHaan, 2006, p. 75).

Fuel

Initially, the fuel may be in the form of a gas, liquid, or solid at the ambient temperature. As discussed previously, liquid and solid fuels must be heated sufficiently to produce vapors.

In general terms, **combustible** means capable of burning, generally in air under normal conditions of ambient temperature and pressure, while **flammable** is defined as capable of burning with a flame. This should not be confused with the terms flammable and combustible liquids.

Flammable liquids are those that have a flashpoint below 100°F (37.8°C), such as gasoline, acetone, and ethyl alcohol. Combustible liquids are those that have a flashpoint at or above 100°F (37.8°C), such as kerosene and fuel oil.

Oxygen

The primary source of oxygen normally is the atmosphere, which contains approximately 20.8% oxygen. A concentration of at least 15% is needed for the continuation of flaming combustion, while charring or smoldering (**pyrolysis**) can occur with as little as 8%. Pyrolysis is defined as the transformation of a compound into one or more other substances by heat alone. While the atmosphere is usually the primary source of oxygen, certain chemicals, called oxidizers can be either the primary or secondary source. Examples are chlorine and ammonium nitrate.

Mixing and Proportioning

Mixing and proportioning are reactions that must be continuous in order for fire to continue to propagate. The fuel vapors and oxygen must be mixed in the correct proportions. Such mixture of fuel vapors and oxygen is said to be within the **flammable limits**. Flammable limits are expressed in the concentration (percentage) of fuel vapors in air. A mixture containing fuel vapors in an amount less than necessary for ignition to occur is too lean, while a mixture that has too high a concentration of fuel vapors is too rich.

For example, the flammable limits for propane are 2.15 to 9.6. This means that any mixture of propane and air between 2.15% and 9.6% will ignite if exposed to an open flame, spark, or other heat source equal to or greater than its ignition temperature, which is between 920°F (493.3°C) and 1,120°F (604.4°C).

Another important characteristic of gases is **vapor density**, the weight of a volume of a given gas to an equal volume of dry air, where air is given a value of 1.0. A vapor density of less than 1.0 means that the gas is lighter than air and will tend to rise in a relatively calm atmosphere, while a vapor density of more than 1.0 means that the gas is heavier than air and will tend to sink to ground/floor level.

Ignition Continuity

Ignition continuity is the thermal feedback from the fire to the fuel. Heat is transferred by conduction, convection, radiation, and direct flame contact.

- **Conduction** is the transfer of heat by direct contact (**Figure 14.2**).
- **Convection** is the transfer of heat caused by changes in density of liquids and gases. It is the most common method of heat transfer; when liquids or gases are heated, they become less dense and will expand and rise (**Figure 14.3**).
- **Radiation** is the transfer of heat by infrared radiation (heat waves; e.g., the sun), which generally is not visible to the naked eye (**Figure 14.4**).
- **Direct flame contact** is a combination of two of the basic methods of heat transfer. As hot gases from the flame rise into contact with additional fuel, the heat is transferred to the fuel by convection and radiation until the additional fuel begins to vaporize. It is the vapors that will be ignited by the flames.

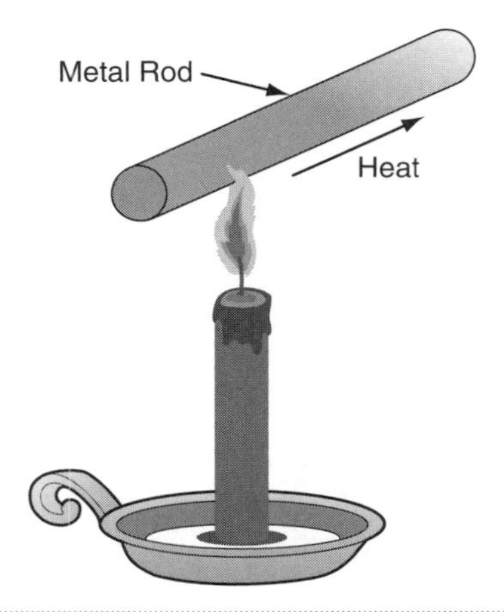

Figure 14.2 Example of Heat Conduction.

Thermal energy transfers from a heated surface to excite (heat) the molecules of a neighboring surface.

Source: Ellie Bruchez, University of Wisconsin–Platteville.

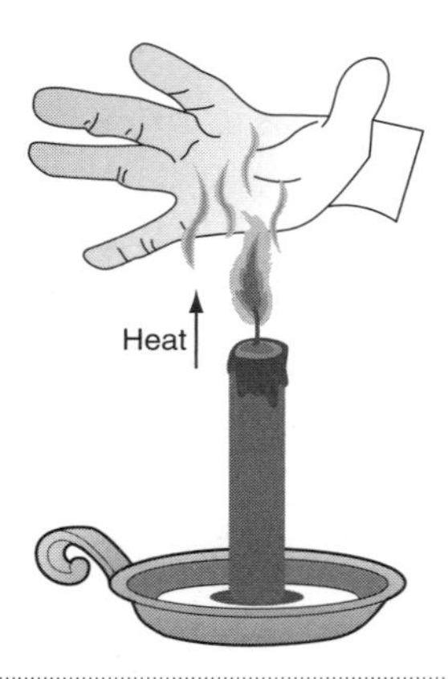

Figure 14.3 Example of Heat Convection.

Source: Ellie Bruchez, University of Wisconsin–Platteville.

The amount of heat generated is measured in British thermal units or Btu's. One Btu is the amount of heat required to raise the temperature of 1 pound of water 1°F.

■ Classification of Fires

Fires are classified by the types of materials that are burning.

- Class A fires: ordinary combustible materials such as wood, cloth, paper, rubber, and many plastics
- Class B fires: flammable/combustible liquids, greases, and gases

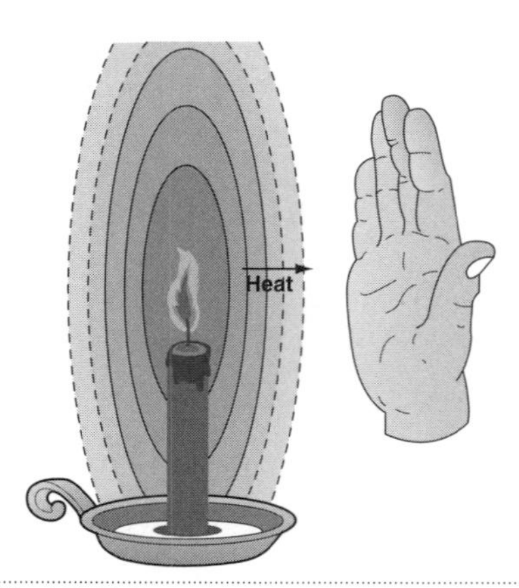

Figure 14.4 Example of Heat Radiation.

Thermal electromagnetic energy is emitted by the flame's heat, exciting the photons in the surrounding environment to affect the temperature of the hand.

Source: Ellie Bruchez, University of Wisconsin–Platteville.

- Class C fires: energized electrical equipment
- Class D fires: combustible metals such as magnesium, titanium, zirconium, sodium, and potassium

All fires produce combustion products. Combustion products fall into four categories: heat, gases, flame, and smoke. Heat is defined as a form of energy characterized by the vibration of molecules and is capable of initiating and supporting chemical changes and changes of state (solid, liquid, gas). Gases are substances that have no shape or volume of their own and will expand to take the shape and volume of the space they occupy. Fire gases include carbon monoxide, hydrogen cyanide, ammonia, hydrogen chloride, and acrolein. Flame is the luminous portion of burning gases or vapors. Smoke is the airborne particulate products of incomplete combustion, suspended in gases, vapors, or solid or liquid aerosols. Soot, black particles of carbon, is contained in smoke (**Figure 14.5**).

Stages of Fire Development

A fire in a room or a defined space generally will progress through three predictable developmental stages. The behavior of a fire in a corridor is affected by the same conditions as a room or defined space fire. The physical configuration of a corridor can cause the fire to spread rapidly, because the corridor will function as a horizontal chimney or flue. Rapid fire spread in a corridor can occur with normal materials

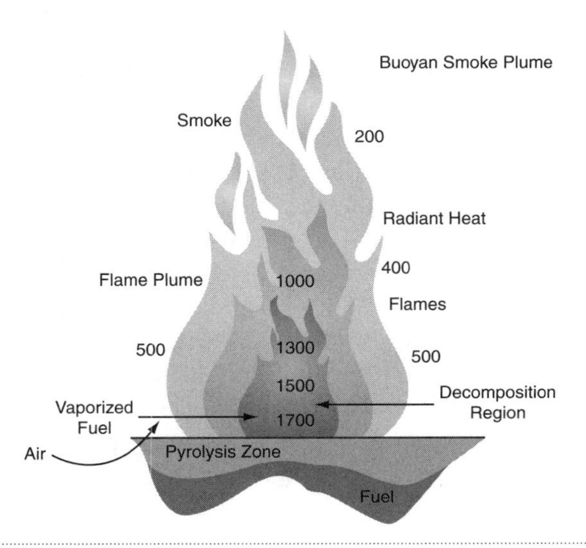

Figure 14.5 Example of Typical Flaming Combustion.

The numbers indicate Btu.

Source: Ellie Bruchez, University of Wisconsin–Platteville.

providing the **fuel load**. In order to determine the origin and cause of a fire, the investigator must be able to interpret the effects of the three stages of fire development during the examination of the fire scene.

Incipient Stage

The first stage of fire development is the incipient stage (growth). This begins at the moment of ignition, and at this time the flames are localized. At this stage the fire is fuel-regulated. That is, the fire propagation is regulated not by the available oxygen but by the configuration, mass, and geometry of the fuel itself. The oxygen content is within the normal range and normal ambient temperatures still exist. A plume of hot fire gases will begin to rise to the upper portions of the room. As convection causes the plume to rise, it will draw additional oxygen into the bottom of the flames. Fire gases such as sulfur dioxide, carbon monoxide, and others will begin to accumulate in the room. If there is any solid fuel above the flame, then both convection and direct flame contact will cause upward and outward fire spread, producing the characteristic "V" pattern charring (more on fire patterns later in this chapter).

Free-Burning Stage

Second is the free-burning stage (development), where more fuel is being consumed and the fire is intensifying. Flames have spread upward and outward from the initial point of origin by convection, conduction, and direct flame impingement. A hot, dense layer of smoke and fire gases is collecting at the upper levels of the room and is beginning to radiate heat downward. This upper layer of smoke and fire gases contains not only soot but also toxic gases such as carbon monoxide, hydrogen cyanide, hydrogen chloride, acrolein, and other gases. Unless the room of origin is sealed tightly, the smoke and fire gases will be spread throughout the building. The temperature at the ceiling level has begun to rise rapidly while the floor temperature is still relatively cool. It is still possible to survive in the room at the cooler lower level.

The fire continues to grow in intensity and the layer of soot and fire gases drops lower and lower. The soot and combustible gases continue to accumulate until one (or more) of the fuels reaches its ignition temperature. Rollover occurs when ignition of the upper layer results in fire extending across the room at the ceiling level. This rollover causes the ceiling temperature to increase at an even greater rate and also increases the heat being radiated downward into the room. Secondary fires can and do result from the heat being generated. The fire is still fuel-regulated at this time.

Flashover Stage

When the upper layer reaches a temperature of approximately 1,100°F (593.3°C) sufficient heat is generated to cause simultaneous ignition of all fuels in the room. This is called **flashover**. Once flashover has occurred, survival for more than a few seconds is impossible. Temperatures in the space will reach 2,000°F (1,093.3°C) or more at the ceiling level down to over 1,000°F (538.8°C) at the floor. At the point of flashover, the fire is still fuel-regulated; however, if the fire stays confined to the room of origin it quickly becomes oxygen-regulated. The rapid temperature rise associated with flashover generally results in windows breaking, which then produces an unlimited supply of oxygen causing the fire to transfer back to the fuel-regulated phase. As a general rule, once flashover has occurred, full involvement of the structure quickly follows.

Flashover results in intense burning of the entire room and its contents. Flashover will produce heavy floor-level burning and even can result in burning on the underside of objects in the room.

The length of time necessary for a fire to go from the incipient stage to flashover depends upon the fuel package, the room geometry, and ventilation. In the typical residential accidental fire setting, this time may be as quick as 2 to 3 minutes.

Structural loads and loading is another concept that has an outcome in the fire investigation field. There are four types of loads: (1) Dead load is the weight of the building and any equipment permanently attached or built-in. (2) Live load is any load other than a dead load. Live loads vary with intended usage. Examples are occupants, storage, and furnishings not permanently attached or built-in. Fire operations increase the live loads both in water accumulation and in fire personnel. (3) Impact loads are delivered in a short period of time. They can be more harmful when supported as dead or live loads. Examples of common impact loads are explosions, wind, and earthquakes. (4) **Fire loads** are the total number of Btu's that might evolve during a fire in the building or area under consideration and the rate at which the heat evolves. Occupancy type has a direct relationship to fire load and generally dictates the possible fire load.

■ Legal Authority to Investigate

To start any investigation of a fire, the investigator needs to ensure that they have the legal authority to be conducting the investigation. The Fourth Amendment protects citizens from unreasonable searches by agents of the government. The U.S. courts typically have granted authority to fire departments to conduct investigations into the cause and origin of fire. Many states have statutes that require an investigation of fires that cause damage over a specified dollar amount. Much like the mandatory reporting of car crashes, these fire reports must be documented by a city official and forwarded to the state.

Due to the doctrine of exigent circumstances, the fire department is always able to enter a private home or businesses for the purpose of extinguishing a fire. The Supreme Court has held that immediately following that extinguishment, the fire department can remain on scene for the time it takes to reasonably conduct an investigation into why that fire started.

Unfortunately, there are times that a fire department may not have the ability to conduct the investigation with personnel who are on-scene during fire suppression. An investigator may come sometime later. In such cases, the investigation is not viewed as being within that exigent circumstance time frame. In these cases it is possible for the fire investigator to get an inspection warrant, which is an administrative warrant issued by a judge. It differs from a search warrant in that the investigating officer does not need to show probable cause other than real property burning. Keep in mind that inspection warrants are limited in scope. An investigator can only use these to inspect the scene for what may have caused the fire. If an inspection points to any criminal activity, the investigator needs to leave the scene, secure it, and apply for a search warrant.

Typically, an investigation will be able to be conducted by simply getting the consent of the person in control of the property. Insurance policies often have clauses requiring an investigation into cause and origin before paying out on a fire claim. Property owners looking to recoup the losses from fire will need to cooperate with fire investigation personnel. It is suggested that consent be given in writing. Having written consent prevents "I said/you said" arguments later in court.

Fire Suppressions Effects on Investigations

Police officers who work the collection and preservation of evidence often make the joke of calling fire department personnel the "evidence eradication squad." Not understanding the reasoning behind why firefighters take some destructive measures when fighting a fire can look like wanton destruction. The process of fire suppression actions changes the scene from its original state. A successful investigation will require the investigator to document damage caused by fire versus the suppression efforts of firefighters. A thorough investigation will look into where items had been originally before the suppression operations were started.

For some suppression efforts, the fire department may require the use of gasoline-powered tools. Chainsaws, K12 saws, high speed gas-driven fans, or gas-engine power generators all are regularly used at the fire scene. Because all of these items are powered by gas, they bring into the fire scene an accelerant (more on that term later). An arson investigator needs to ask firefighting personnel where they had gas-powered tools in use. The investigator also has to find out if and where these tools were refueled. Gas-powered tools with a leaky fuel tank or gasoline spilled during refilling these tools could provide false positive readings for an accelerant in the crime scene.

Cause and Origin Investigations

The fundamental reason to conduct a post-fire examination of a fire scene is to figure out the cause and origin of the fire. All fires should be investigated to find out why they started. The identification of why a fire starts can help to prevent future fires regardless of if it was a criminal act, a manufacturing defect in a home appliance, or an accident.

The overall investigation will start by looking for an **area of origin**. An area of origin is the large track of space or area where the fire would have started (DeHaan, 2006), and subsequently was able to grow and develop. Working back through the property with the least destruction to property displaying the most destruction will help the investigator to find the area of origin.

With an area of origin located, the investigator then works towards finding a **point of origin**. The point of origin is the location where the fire actually started, the place of its beginning. Keep in mind that multiple points of origin to a fire can exist. When confronted with multiple points of origin, the investigator should be thinking about arson. There will be times when an investigator will not be able to locate the point of origin due to the damage done within the area of origin.

Cause Classifications

After the origin of the fire is found, the cause of the fire needs to be determined. In theory, a fire can be classified into one of four cause classifications (O'Connor & Redsicker, 1997):

1. Natural: an "act of god" such as lightning
2. Accidental: unintentionally set yet able to be explained
3. Incendiary: intentionally set, typically viewed as criminal
4. Undetermined: cause unknown or unable to be identified

There was a time when the term suspicious was also a classification for cause. According to the National Fire Protection Agency (NFPA) in their manual for fire investigations titled NFPA 921 (2004), suspicious is now a term to avoid using.

Physical Examination of the Fire Scene

As mentioned in Chapter 1, crime scenes need to be looked at by applying the scientific method. The fire scene also must be viewed logically when conducting an investigation. The initial step of the scientific method is to recognize a problem, in this case, to have a fire scene to investigate. Second, the problem needs to be defined. In order to define the problem at a fire scene, the fire needs to be classified into one of the four categories mentioned previously. To classify a fire, data must be collected about the fire. Post-fire scene documentation, documentation about firefighting efforts, witness statements, photography, physical evidence collection, and all other forms of crime scene documentation are incorporated into the data collection phase.

To achieve the fourth step of the scientific method, the investigators analyze the data looking for facts. Using their training and experience, the investigators are able to take data they have in hand and use inductive reasoning to determine what happened. This allows the investigators to form a hypothesis or multiple hypotheses about what actually took place. With one or more hypotheses in hand, the investigators next test them to see if they hold up against the facts. If the investigators are unable to prove a cause, then the fire is classified as "undetermined."

Exterior Examination

Processing of a fire scene begins as firefighting efforts are still underway. As noted earlier, documentation must be made on the firefighting techniques and equipment used. Additionally, if the investigator is on scene at the time of the fire, photography of the flame color, smoke color, and the crowd of onlookers should be taken. The fire scene investigator also needs to start the interview process.

As discussed in Chapter 1, there are two types of evidence, physical and testimonial. This text is specifically dedicated to the collection and preservation of physical evidence. During fire scene investigations, some important pieces of information such as heat, smoke, and flame colors cannot always be collected as physical evidence. The testimonial evidence provided by the first few firefighters and police officers who arrived on scene thus become very important. The list of questions suggested to ask first responders to a fire scene is provided on the next page.

After assembling the statements from first responders, the task of physical evidence recovery can begin. A good practice for the investigator is to walk around the structure and make personal observations of the exterior. Note door and window positions. Note broken glass

Questions to Ask First Responders

- Were there people or vehicles in the vicinity of the fire on your arrival?
- If so, would you call their actions or demeanor suspicious? As in
  - Persons fighting?
  - Persons overly eager to try and help or wanting to give information?
  - Those attempting to obstruct firefighting operations?
  - Anyone fleeing the scene upon seeing official personnel?
  - Persons seen at other fire scenes?
- If any of the above were observed, are there physical descriptions available?
- How much of the structure was involved in burning?
  - Is this consistent with other fires of other buildings constructed similarly?
  - Were flames and/or smoke visible? What color(s) were the fire and the smoke? (**Table 14.1**)
  - What side of the building was involved in the burning?
  - Had the fire broken out of the roof?
  - Had flames come out windows? Which windows? Note if the amount of broken glass represents the breaking force coming from inside or outside the window.
  - Were flames quietly lapping or were they in a violent roar?
- Were doors or windows open or closed?
  - Were they locked or otherwise blocked?
  - Were shades pulled or views into them otherwise blocked?
  - Were access routes to the structure blocked?
  - Did emergency personnel have to force an entry; if so who was involved and how was it done?
- Were any unusual odors noticed?
- Were hydrants, standpipes, and alarms operational?
  - Were fire hydrants or standpipes blocked, covered over, or otherwise unusable?
  - Are there smoke alarms or fire protection systems in the building but no evidence they went off?

Source: Information retrieved from O'Connor, J. & Redsicker, D. (1997). *Piratical fire and arson investigation* (2nd ed.). Boca Raton, FL: CRC Press.

around windows, asking whether the pattern is consistent with being broken from the inside or the outside of the building.

While doing this exterior walk around, the investigator should be looking at the fire damage. See from the outside where the most damage will be on the inside. Look also for all of the points where fire was able to break outside of the building.

Interior Examination

Officer or investigator safety is paramount whenever you are conducting an investigation (see Chapter 5). Remember that when entering a structure that has been burned, flames and water may have compromised the

TABLE 14.1 Types of Fuel and Corresponding Smoke and Flame Colors

Fuel	Color of Smoke	Color of Flame
Wood	Gray to Brown	Yellow to Red
Paper	Gray to Brown	Yellow to Red
Gasoline	Black	Yellow to White
Lubrication Oil	Black	Yellow to White
Turpentine	Brown to Black	Yellow to White
Cooking Oil	Brown	Yellow
Kerosene	Black	Yellow

Source: Adapted from Department of Homeland Security. (2002). *Arson Detection for the First Responder: ADFR-Student Manual*. Washington, DC: US Government Printing Office.

support structures. Always test floors or stairs before stepping on them. Wearing personal protective equipment is also of paramount importance. All persons inside of a burned structure should wear firefighter style turnout gear, including a helmet.

Burn Patterns

Once inside the structure, the investigator can follow the burn patterns to find the area of the building containing the most severe damage. Generally speaking, this area will hold the point of origin. The exception to this rule is in the case of an arsonist using a flammable material trail to be able to light a fire a safe distance from where he or she wants the fire to start (**Figures 14.6** and **14.7**).

One of the most common patterns a fire investigator can use is the "V" or conical pattern. As fire burns, it grows moving up and out. Once the fire is out the damage is often times described as looking like

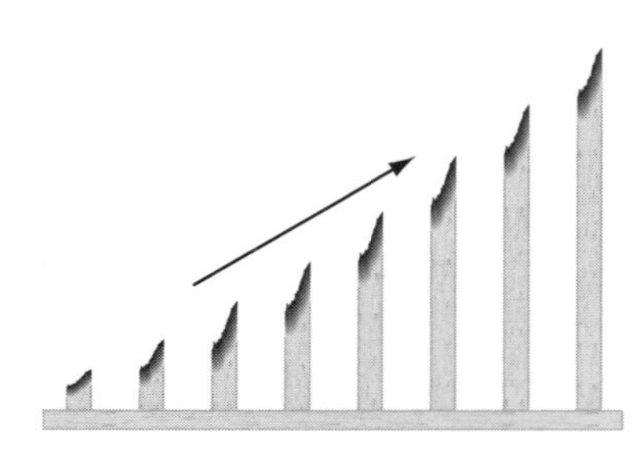

Figure 14.6 Example of Identifying Fire Travel via Sustained Damage to Wall Studs in a Building.

Source: Ellie Bruchez, University of Wisconsin–Platteville.

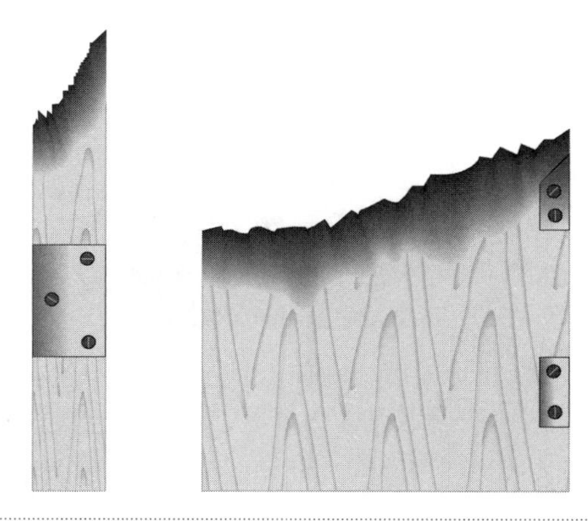

Figure 14.7 Example of Identifying Fire Travel via Sustained Damage.
Note the uneven edges and height of the burned wall studs.

Source: Ellie Bruchez, University of Wisconsin–Platteville.

the letter V. This V shape can be seen on more than simply charred drywall. Large items of furniture often exhibit this pattern. Sections of wall where the drywall or plaster are missing also show the V pattern, and burnt stud sections still show the stepping down of a V pattern.

Following the V pattern, an investigator can find the lowest point of the burn. Typically the examination of the lowest burn point will show the cause of a fire. Should this lowest point consist of the entire surface area of a floor or the joint of floor and wall over the length of a wall, suspicions about the fire should be investigated. In cases like this, a search for a liquid accelerant needs to be done.

Other Pattern Evidence

Smoke levels or the smoke horizon is also a pattern needing to be documented. Walls and other surfaces will become stained by smoke. These patterns can show airflow through the building during a blaze. When the fire involves a death, this written observation can be very important. The smoke horizon can show the likelihood of occupants finding exits to be visible or obscured. It also can show how close lethal gases got to where the occupants were located.

Along with looking at the smoke layers, the investigator needs to search for the areas protected from smoke and debris. Furniture and wall hangings will protect surfaces from some damage (**Figure 14.8**). To reconstruct the scene, an investigator should return furniture moved during salvage operations back to their locations as indicated by the protected areas. When the contents of a room are clustered together

Figure 14.8 Example of Identifying Fire Travel via Sustained Damage.

Here the location and colors of the flames and smoke should be noted while the building is burning.

Source: Ellie Bruchez, University of Wisconsin–Platteville.

without evidence of other protected areas, and firefighter interviews do not indicate that salvage took place in the room, it might be a clue for arson. The suspect may have moved all of the room's contents into one location hoping to increase the fuel load to heat up the fire.

Additional Scene Documentation

As the investigator works the scene, they should be developing an inventory of the contents of the structure. Suspicion should start to arise when an investigator does not see clothing, jewelry, personal photos, or other irreplaceable items in a residential fire. Commercial or industrial locations missing equipment, inventory, or irreplaceable documents should draw suspicion also. It is common for persons looking to commit insurance fraud to substitute the regular contents of a building with junk and empty boxes.

Investigators should ask the occupants of a structure to prepare an itemized list of property. That list can be compared to the actual inventory of the house. Claims of leather couches and big screen TVs can be compared to your documented inventory of lower value items actually found to uncover insurance fraud.

Criminal Burn Patterns

Flammable or combustible liquids are typically referred to as **accelerants.** An accelerant is "any substance used to accelerate (and sometimes direct) the spread of a fire" (O'Connor & Resicker, 1997). Given this definition, the investigator needs to be thinking about more than flammable liquids as it relates to criminal fires. Solid fuels as well as chemical powders could be used as accelerants.

A **plant**, also called a booster, is a pool of flammable liquids or pile of combustibles (newspapers, rags, etc.) that is used to heat up a fire at a select location. A plant is designed to produce great heat at an area the arsonist wishes to cause high levels of damage. A plant will not make definitive pattern marks, as the size and shape can vary dependent on the accelerant used and quantity placed.

Trailers, sometimes called streamers, are the arrangement of a combustible or flammable material (solids, liquids, or combinations of both) to ensure fire is carried from one location to another. A trailer is regularly placed from a point of exit to the area the arsonist wants to be the area of origin. This allows the arsonist the safety of setting the fire from outside of the structure while giving an appearance of the fire starting well within the heart of a structure. Trailers also are used to connect multiple plants within a structure.

A trailer will leave burn marks that create paths that are not natural to regular fire progression. The substance used for the trailer typically will create a protected area under the material. The trailer can then produce a line of unburned flooring surrounded by burnt and charred flooring materials. Trailers are typically made by pouring a flammable liquid onto the flooring, or tying combustible materials such as towels or newspaper end-to-end.

Puddling is another pattern that can be found when a flammable liquid is poured onto a floor to make a pool. When that pool is burning, only the vapors burn but not the actual liquid. A pool of liquid burns from the perimeter in towards the center. This leaves heavier burning on the outer edges than in the center part of where the pool was located.

■ Physical Evidence Collection

At some point the investigator will need to collect the physical evidence to prove her or his theory on the cause of the fire. Any items found near the point of origin that could be a source of ignition need to be collected as evidence.

Before talking about the collection of physical evidence, the fire investigator needs to be aware of the concept and issues dealing with **spoliation**. The definition of spoliation is the intentional or negligent altering of evidence. Fire investigators may locate the source of a fire as coming from an electrical appliance. An overzealous investigator may take that appliance apart to inspect the inner workings to look for a defect or criminal tampering. However, a typical investigator does not have the level of technical expertise to know how the home electronics actually operate. When a piece of equipment or home appliance is suspected as starting a fire, investigator needs to collect it whole and have a forensic engineer investigate its inner workings.

Physical Evidence

Physical evidence at a fire scene can be in solid and liquid forms. An investigator needs to be prepared to collect these items of evidence by having the proper equipment. Shovels, axes, and power tools such as saws all need to be readily available. These tools also cannot be the same ones used in fire suppression operations due to issues of contamination. Items collected are placed into a previously unused metal paint style can. With the proper tools in hand and knowledge of burn patterns, the investigator is now ready to collect physical evidence.

When confronted with chemical burn patterns, the investigator will want to collect samples for crime lab analysis. Additionally, the investigator will collect comparison samples. For example if a carpet shows signs that a flammable liquid was poured onto it, the investigator will have to collect two samples from that same carpet: the suspect sample from the area where the investigator believes a flammable liquid was poured on it and a control sample from an area that would not have such potential evidence. The reason for this is that some objects are manufactured by processes that use flammable chemicals. By collecting a control sample, the lab can determine if the presence of a flammable liquid is inherent to the carpet or if it has been added, presumably for criminal purposes.

An example of this is pine wood flooring or trim board. Turpentine is a highly flammable liquid made from the resin of pine trees. The flashpoint of turpentine is only about 35°C (95°F). All pine wood products will test positive for turpentine (DeHaan, 2006, p. 504).

The next thing for an investigator to consider when collecting physical evidence is that only one type of material goes into each paint can. For example, a modern home floor may have carpet, a carpet pad, and multiple layers of wood subflooring. Only material from one layer goes into a can. So, a suspected flammable liquid pour could require more that three individually-packaged pieces of evidence for the suspect sample and the same number for the control sample. A sample of the carpet, a sample of carpet pad, and a sample from each layer of subflooring at both locations have to be collected.

When placing samples into a metal paint can, remember to keep the sample size appropriate to the size of the can. Filling the can a little less than half full is typically considered to be appropriate. Even in smaller size paint cans this amount of material is enough for the lab to test. It also allows room within the can for vapors and outgassing to occur. Filling a can full will not allow vapors to collect in the can and could be dangerous to persons handling or opening it later.

Remember that accurately documenting a scene involves photographing evidence as discussed in Chapter 6. The suspected criminal burn patterns must be photographed just like all other physical evidence. It is recommended that investigators also photograph the evidence container at the location of the sample to be collected. A second set of midrange and close-up photographs will need to be made showing the metal can(s) used to collect both the suspect samples and control samples. The metal cans should be placed at the location the sample was taken from and, if possible, be clearly marked with the item number noted on the evidence log sheet.

For items that do not fit into paint cans, crime scene companies have developed nylon evidence collection bags. Plastic evidence bags and plastic cans need to be avoided in the collection of physical evidence at a fire scene. Plastic is a petroleum by-product. Many flammable liquids are also petroleum-based, so there is risk of contamination of samples and/or the breakdown of the evidence bag.

■ Explosive Evidence

At times an investigator will respond to the scene of an explosion. Just as all fires are not criminal in nature, not all explosions are criminal. Construction accidents, defective home heating devices, or a tanker truck hauling a violate chemical all could cause an explosion. However, there are explosions that do occur for criminal or terrorist reasons.

According to the Department of Homeland Security (2008) *Incident Response to Terrorist Bombings*, explosives can be divided into three categories: pyrotechnics, propellants (both of which are called low explosives), and high explosives. These categorizations depend, in part, on the manner of use of the materials.

Low Explosives

Low explosives typically involve pyrotechnics that create smoke, light, heat, and sound. The pyrotechnics most people are familiar with are those used for entertainment purposes, such as firework displays. Propellants are designed to burn rapidly. Black powder and smokeless powder are the propellants. Low explosives burn or detonate at a speed lower than 3,300 feet per second.

High Explosives

High explosives are designed to detonate and yield a near instantaneous release of energy. High explosives chemically detonate at a speed greater than 3,300 feet per second.

Components of an Explosive Device

An explosive device will typically have four main parts: the power supply, initiator, explosives (mentioned previously), and some kind of switch. In the case of a pipe bomb, the power supply, initiator, and switch can be the fuse. An understanding of these parts will aid an investigator when searching for evidence at the post-blast scene.

For the power supply, many bombs contain an electric initiator and, as such, require an electric power source. Most commercially available batteries that come in many shapes and sizes can reliably supply power to initiators. As an alternative, mechanical action, such as a spring under pressure, can store sufficient energy to cause the function of a nonelectric initiator.

Most explosives are sensitive to shock. The initiator provides the necessary energy to start a chain reaction with the explosive, causing it to burn or detonate. The most common types of initiators are squibs and blasting caps. An effective homemade initiator is a flash bulb with the filament exposed.

Switches fall into one of two categories: mechanical or electrical. The switch completes an electric circuit or causes a mechanical action to initiate the improvised explosive device (IED). Switches generally perform the function of safely arming the device and/or detonating it.

Explosive Investigation

There are two types of bombing incidents: pre-detonation and post-detonation. These situations constitute the categories of incident response involving energetic materials. Pre-detonation incidents include bomb threats, suspicious item incidents, and suspected suicide bombers. Post-detonation incidents include situations where explosives have been detonated.

Pre-Detonation

Bomb threats are the most common type of pre-detonation incident. Most are hoaxes. When one occurs, expect others. Bomb threat callers often communicate their threats as a means of causing evacuation of a workplace or school. For many years, bomb threats were considered "pranks." However, many agencies, including school districts, now impose significant penalties on citizens who communicate bomb threats. Statutes covering these situations have been upgraded in many jurisdictions and callers can face large fines and imprisonment.

The pre-detonation investigation often involves searching for a bomb. One of the most important people to interview at a facility is the maintenance custodian. While individual occupants of the facility may provide good information on their areas of responsibility, a maintenance custodian probably has the most information about common areas such as building perimeters, halls, stairways, lounges, restrooms, and any other part of the facility where a potential bomber could visit without being discovered or challenged by an occupant.

The initial first responder at a suspicious item incident is typically a law enforcement officer. He or she should immediately evacuate persons from around the suspected item ore areas where devices could be located. In no case should anyone other than a certified bomb technician attempt to handle or render safe any suspected explosive device. The Department of Homeland Security recommends the actions of first responders at a pre-detonation incident not be observed or recorded by personnel who do not have a legitimate requirement to document such actions. Written summaries, audio or video recordings, or police responses can be used by bombers in planning subsequent attacks.

Post-Detonation

During the initial response to a post-detonation incident, the first priority is safety of the public and first responders. Every effort must be made to avoid additional casualties. There is an understandable tendency for emergency responders to conduct rescue operations without regard to personal safety. However, dead or injured responders cannot contribute to subsequent operations and they increase the workload for other responders.

All responders should be alert for any item that could be a secondary device or any location where a secondary device can be concealed. A **secondary device** is a second bomb placed at a scene to detonate after the original explosion. Secondary devices are typically targeted at

Responses to Pre-Detonation Situations

- Do not touch an item that could contain explosive material.
- Always move people away from a suspicious item—never try to move the item away from people.
- Never use a radio, cellular telephone, or other transmitter within a minimum of 300 feet of a location where there is a suspected or actual explosive device.
- Pay close attention to appropriate evacuation distances.
- Be aware of the potential for secondary devices.

Source: Adapted from Department of Homeland Security. (2008). *Incident response to terrorist bombings*. Washington, DC: US Government Printing Office.

emergency responders and investigators at a bombing. Law enforcement officers, firefighters, medical personnel, and other emergency response personnel should continually observe their operational areas for any signs of secondary devices. If a suspected secondary device is observed, the incident commander should be notified and recall procedures implemented immediately. The search for secondary devices should include more than just operational areas; it should also include staging areas, command posts, rest and rehab areas, and triage areas.

Sometimes during rescue efforts items need to be moved. Evidence that has been moved or handled before being recorded and inventoried by an investigative team may be considered contaminated. If obvious or suspected items of evidence are destroyed or moved because of rescue requirements, that action should be reported to a representative of the investigative team. Such reports also should include a description of the item along with a detailed description of its original location.

Control of the scene is a critical step in supporting all other operations. Traffic flow must be controlled so that emergency vehicles (fire apparatus and ambulances) can be positioned to perform critical rescue functions and provide medical evacuation. Potential victims must be identified and medically evaluated before they depart the area. Witnesses and potential perpetrators must be identified for further action by law enforcement. Evidence must be protected from contamination or theft. In addition, curious citizens and untrained "volunteers" must be kept out of harm's way. It is also important to furnish a location for the media to be kept informed by the Public Information Officer. Scene control is an essential task that is usually the principal responsibility of law enforcement during the initial phases of a response.

Zones and Perimeters

Hot Zone

For bombing incidents, the hot zone includes the area that encompasses significant structural collapse or damage and extends to the point where the person would be relatively safe from blast, fragmentation, or shrapnel from secondary devices. This area can differ significantly in size and will shrink as the threat of secondary devices diminishes. The only personnel authorized in the area are first responders who have been trained and equipped to operate in the high-hazard environment and have a legitimate need to be in the area. This group includes firefighters and emergency medical personnel conducting rescue operations, bomb technicians, and specially-trained law enforcement officers.

Warm Zone
The diameter of the warm zone is determined by identifying the distance from the seat of the explosion (or ground zero) to the farthest point where evidence can be visually identified, then adding half that distance.

Cold Zone
The cold zone encompasses the area required for the command post, staging areas, temporary morgue, and a designated area for victims' families. This special area for victims' families is typically used to discourage family members from trying to approach the target to assist in rescue operations or attempt to obtain information concerning their loved ones.

Investigation

With the perimeter set and the scene stabilized as much as possible, the investigation can begin. The first part of the investigation of an explosion can be described with the "Four Rs": Recognition, Recovery, Reassembly, and Reconstruction. Recognition includes the both the blast damage and the identification of possible evidence. Recovery involves both the physical picking up of items of evidence and proper documentation of these items. With the first two of the "Rs" accomplished on scene, it may be possible for personnel back at the lab to reassemble the suspected device to reconstruct how it worked.

When looking for physical evidence it is important to remember that many materials can be used to construct an explosive device. For instance, water can be made into an explosive. Liquids contained in a sealed vessel heated to the point of boiling turn into vapor. Vapor takes up more space than a liquid. If the vessel is not able to vent off the extra vapor, eventually the vessel will rupture violently. The eruption is called a Boiling Liquid Expanding Vapor Explosion (BLEVE) (**Figure 14.9**). A typical home water heater, about 40 gallons, is more than capable of demolishing a residential home.

The search for evidence is a slow process using search patterns, as discussed in Chapter 6. Small pieces of metal, plastic pipe, wires, or cardboard could be overlooked. Some blast investigators recommend sweeping up the scene after all the evidence has been collected. Later the sweepings can be sifted if the need arises.

Evidence Collection

Much like fire debris, post-blast evidence should not be placed into paper or plastic bags. The use of nylon arson bags or empty metal cans is recommended. For some objects that are not practical to collect, such as a building, the surface needs to be swabbed with a cotton swatch to test for residue. The swabs should then be placed into a glass jar or vial.

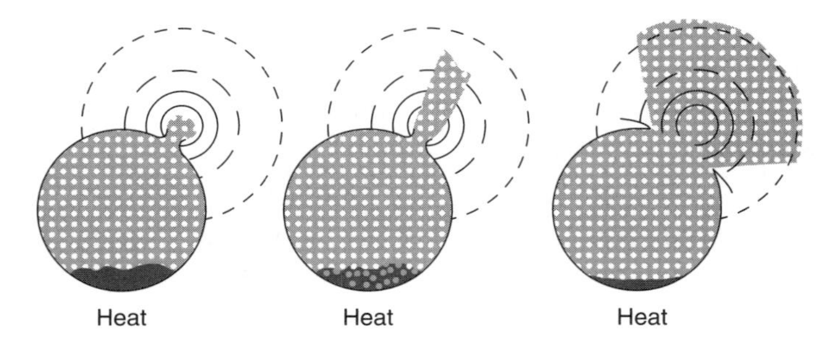

Figure 14.9 Example of a Boiling Liquid Expanding Vapor Explosion (BLEVE).

Vapor from a liquid substance in a confined space will expand when heated until it explodes from the container.

Source: Ellie Bruchez, University of Wisconsin–Platteville.

Chapter Summary

To properly investigate a fire, the investigation needs to begin during firefighting operations. First responders' observations of people in the area of the fire and what the fire looked like can help in an investigation and are of vital importance. The investigator does not need to be a firefighter; however, a basic understanding of firefighting techniques will aid investigators in the interpretation of a fire scene. Fire and explosive scenes must be approached in the same, methodical manner as all other crime scenes. The crime scene investigator should approach the fire scene or scene of an explosion as though it were a criminal event and allow the physical evidence found to point them in the direction of classifying the fire or explosion as intentional, accident, or an act of god.

Review Questions

1. What percent of arson or suspected arson cases result in convictions?
2. What are the three elements of the fire triangle?
3. ______________ is the transformation of a compound into one or more other substances by heat alone.
4. One Btu is equal to the amount of heat required to raise the temperature of______________ of water______________ degree(s) F.
5. The simultaneous ignition of all of the fuels in the room is called ______________.
6. How does an inspection warrant differ from a search warrant?

7. What is the difference between the area of origin and the point of origin?
8. Which pattern is the most commonly seen by fire investigators?
9. What speed is the dividing point between low and high explosives?
10. What is the definition of BLEVE and why should an investigator be aware of it?

References

Department of Homeland Security. (2008). *Incident response to terrorist bombings*. Washington, DC: U.S. Government Printing Office.

Department of Homeland Security. (2002). *Arson detection for the first responder: ADFR-student manual*. Washington, DC: U.S. Government Printing Office.

DeHaan, J. (2006). *Kirk's fire investigation* (6th ed.). Upper Saddle River, NJ: Brady Fire, Pearson Higher Education.

DeHaan, J. & Icove, D. (2008). *Forensic fire scene reconstruction* (2nd ed.). Upper Saddle River, NJ: Brady Fire, Prentice Hall.

Hanlon, R. J. (1977). *Murphy's law and other reasons why things go wrong*. Los Angeles, CA: Price Stern Sloan Publishers.

National Fire Academy. (2007). *Fire/arson origin and cause investigations pre course reader.* Washington, DC: U.S. Government Printing Office.

National Fire Protection Association (NFPA). (2004). *NFPA 921. Guide for fire and explosion investigations*. Quincy, MA: NFPA.

O'Connor, J. & Redsicker, D. (1997). *Piratical fire and arson investigation* (2nd ed.). Boca Raton, FL: CRC Press.

Drug Evidence

Half of the modern drugs could well be thrown out of the window, except that the birds might eat them.

Dr. Martin Henry Fischer (1879–1962)
Professor Emeritus in Physiology, University of Cincinnati

LEARNING OBJECTIVES

- Differentiate between the terms drug and controlled substance.
- Know the legislation, the Controlled Substances Act, and how it impacts drug-related evidence and investigations.
- List and differentiate between the categories of controlled substances.
- Understanding how and why to correctly document, collect, package, and preserve drug-related evidence.
- Define the forensic value and the analysis of drug-related evidence.

KEY TERMS

Controlled Substance
Controlled Substances Act (CSA)
Drug
Depressant
Hallucinogen
Narcotic
Scheduling
Stimulant

Drug-Related Evidence in Crime Scene Investigation

While illegal drug use within the United States appears to have leveled off, and in some cases even declined, it is by no means absent within society or within crime scenes. It is important for the crime scene investigator to have a good foundation as to what substances are considered to be "drugs" and how they should properly document, collect, and preserve such substances. Typically, when one talks of "drugs" one is in fact referring to "controlled substances." However, it is important to differentiate between the two, because the distinctions between the two terms relate to the investigative importance and criminal wrongdoing associated with each.

A **drug** is any chemical substance, other than food, which is intended for use in the diagnosis, treatment, cure, mitigation, or prevention of disease or symptoms. These can be dispensed and used either by prescription or over-the-counter (OTC). Possession of an OTC drug does not require a prescription and is not illegal (except in bulk as is associated with the manufacture of methamphetamine, which will be discussed later within this chapter).

TABLE 15.1 Types of Controlled Substances

Types of Controlled Substances	Examples	Legitimate Uses
Stimulants	cocaine, amphetamines, methamphetamine	Legitimate uses include increased alertness, reduced fatigue, weight control, and topical analgesic (pain killing) action.
Depressants	barbiturates, sedatives, and tranquilizers	Legitimately used to obtain release from anxiety, for the treatment of psychological problems, and as mood elevators.
Narcotics	opium, morphine, heroin, methadone, codeine, Dilaudid	Legitimately used for pain relief, antidiarrheal action, and cough suppression.
Cannabis	marijuana, hashish, cannabis plants, sinsemilla, and hashish oil (all of which are collectively referred to as marijuana)	No legitimate use according to Federal law.
Hallucinogens	lysergic acid diethylamide (LSD); phencyclidine (PCP); peyote; mescaline; psilocybin; 3,4methylene-dioxyamphetamine (MDA); methylenedioxy-methamphetamine (MDMA)	No legitimate use according to Federal law.
Anabolic steroids	nandrolene, oxandrolene, oxymetholone, and stanoxolol	Used legitimately for weight gain, for treatment of arthritis, anemia, and cancer treatments.
Inhalants	nitrous oxide, super glue, gasoline, chloroform, freon, and toluene	Legitimately used for a variety of purposes, but very few of them related to inhalation.

Controlled substances, on the other hand, are those substances (typically drugs) whose possession or use is regulated by the government (**Table 15.1**). Title 21 of the United States Code (21 USC) defines these substances (USDOJ, 1990).

The Controlled Substances Act

The **Controlled Substances Act (CSA)**, Title II and Title III of the Comprehensive Drug Abuse Prevention and Control Act of 1970 are the legal foundation of the U.S. Government's fight against the abuse of drugs and other substances. This law is actually a consolidation of numerous laws regulating the manufacture and distribution of narcotics, stimulants, depressants, hallucinogens, anabolic steroids, and chemicals that are used in the illicit production of controlled substances.

Drug Scheduling

The CSA places all substances that were in some manner regulated under existing federal law into one of five **schedules**. This placement is based upon the substance's medical use, potential for abuse, and safety or liability for psychological and/or physical dependence.

Proceedings to add, delete, or change the schedule of a drug or other substance may be initiated by the Drug Enforcement Administration (DEA), the Department of Health and Human Services (DHHS), or by petition from any interested person. The scheduling is conducted through the cooperation between DHHS and the DEA, with both the legal and scientific/medical community weighing in on the process.

Scheduling of Controlled Substances

According to the 2005 edition of the DEA publication, *Drugs of Abuse*, the five schedules into which substances may be placed are as follows:

Schedule I

- The drug or other substance has a high potential for abuse.
- The drug or other substance has no currently accepted medical use in treatment in the United States.
- There is a lack of accepted safety for use of the drug or other substance under medical supervision.
- Examples: heroin, LSD, marijuana, and MDMA.

Schedule II

- The drug or other substance has a high potential for abuse.
- The drug or other substance has a currently accepted medical use in treatment in the United States or a currently accepted medical use with severe restrictions.

(continues)

Scheduling of Controlled Substances (continued)

- Abuse of the drug or other substance may lead to severe psychological or physical dependence.
- Examples: morphine, PCP, cocaine, methadone, and methamphetamine.

Schedule III
- The drug or other substance has less potential for abuse than the drugs or other substances in schedules I and II.
- The drug or other substance has a currently accepted medical use in treatment in the United States.
- Abuse of the drug or other substance may lead to moderate or low physical dependence or high psychological dependence.
- Examples: Anabolic steroids, codeine and hydrocodone with aspirin or Tylenol®, and some barbiturates.

Schedule IV
- The drug or other substance has a low potential for abuse relative to the drugs or other substances in Schedule III.
- The drug or other substance has a currently accepted medical use in treatment in the United States.
- Abuse of the drug or other substance may lead to limited physical dependence or psychological dependence relative to the drugs or other substances in Schedule III.
- Examples: Valium® and Xanax®.

Schedule V
- The drug or other substance has a low potential for abuse relative to the drugs or other substances in Schedule IV.
- The drug or other substance has a currently accepted medical use in treatment in the United States.
- Abuse of the drug or other substances may lead to limited physical dependence or psychological dependence relative to the drugs or other substances in Schedule IV.
- Examples: cough medicines with codeine.

Introduction to Drug Classes

The CSA regulates five classes of drugs: narcotics, depressants, stimulants, hallucinogens, and anabolic steroids. Each class has distinguishing properties, and drugs within each class often produce similar effects. However, all controlled substances, regardless of class, share a number of common features. It is the purpose of this chapter to familiarize the reader with some of these shared features and to give definition to terms frequently associated with these drugs.

All controlled substances have abuse potential or are immediate precursors to substances with abuse potential. With the exception of anabolic steroids, controlled substances are abused to alter mood, thought, and feeling through their actions on the central nervous system (brain and spinal cord). Some of these drugs alleviate pain, anxiety, or depression. Some induce sleep and others energize. Though therapeutically useful, the "feel good" effects of these drugs contribute to their abuse. The extent to which a substance is reliably capable of producing euphoria increases the likelihood of that substance being abused.

In legal terms, the nonsanctioned use of substances controlled in Schedules I through V of the CSA is considered drug abuse. While legal pharmaceuticals placed under control in the CSA are prescribed and used by patients for medical treatment, the use of these same pharmaceuticals outside the scope of sound medical practice is also drug abuse.

Individuals who abuse drugs often have a preferred drug that they use, but may substitute other drugs that produce similar effects (often these are found in the same drug class) when they have difficulty obtaining their drug of choice. Drugs within a class are often compared with each other with terms like potency and efficacy. Potency refers to the amount of a drug that must be taken to produce a certain effect, while efficacy refers to whether or not a drug is capable of producing a given effect regardless of dose. Both the strength and the ability of a substance to produce certain effects play a role in whether that drug is selected by the drug abuser.

It is important to keep in mind that the effects produced by any drug can vary significantly and are largely dependent on the dose and route of administration. Concurrent use of other drugs can enhance or block an effect, and substance abusers often take more than one drug to boost the desired effects or counter unwanted side effects. The risks associated with drug abuse cannot be accurately predicted because each user has his or her unique sensitivity to a drug. There are a number of theories that attempt to explain these differences, and it is clear that a genetic component may predispose an individual to certain toxicities or even addictive behavior.

In the sections that follow, each of the five classes of drugs is reviewed and various drugs within each class are profiled. Although marijuana is classified in the CSA as a hallucinogen, a separate section is dedicated to that topic. There are also a number of substances that are abused but not regulated under the CSA. Alcohol and tobacco, for example, are specifically exempt from control by the CSA. In addition, a whole group of substances called inhalants are commonly available and widely abused by children. Control of these substances under the

CSA would not only impede legitimate commerce, but would likely have little effect on the abuse of these substances by youngsters. An energetic campaign aimed at educating both adults and youth about inhalants is more likely to prevent their abuse. To that end, a section is dedicated to providing information on inhalants.

Narcotics

The term **narcotic**, derived from the Greek word for stupor, originally referred to a variety of substances that dulled the senses and relieved pain. Today, the term refers to opium, opium derivatives, and their semisynthetic substitutes. For the purposes of this text, the term narcotic refers to drugs that produce morphine-like effects.

Narcotics are used therapeutically to treat pain, suppress cough, alleviate diarrhea, and induce anesthesia. Narcotics are administered in a variety of ways. Some are taken orally, transdermally (skin patches), intranasally, or injected. As drugs of abuse, they are often smoked, sniffed, or injected. Drug effects depend heavily on the dose, route of administration, and previous exposure to the drug. Aside from their medical use, narcotics produce a general sense of well-being by reducing tension, anxiety, and aggression. These effects are helpful in a therapeutic setting but contribute to their abuse.

The poppy plant, *Papaver somniferum*, is the source for nonsynthetic narcotics. It was grown in the Mediterranean region as early as 5000 BC, and has since been cultivated in a number of countries throughout the world. The milky fluid that seeps from incisions in the unripe seed pod of this poppy, since ancient times, has been scraped by hand and air-dried to produce what is known as opium. A more modern method of harvesting is by the industrial poppy straw process of extracting alkaloids from the mature dried plant. The extract may be in liquid, solid, or powder form, although most poppy straw concentrate available commercially is a fine brownish powder. More than 500 tons of opium or equivalents in poppy straw concentrate are legally imported into the United States annually for legitimate medical use (United Nations, 2007).

Opium

There were no legal restrictions on the importation or use of opium until the early 1900s. In the United States, the unrestricted availability of opium, the influx of opium-smoking immigrants from East Asia, and the invention of the hypodermic needle contributed to the more severe variety of compulsive drug abuse seen at the turn of the twentieth century. In those days, medicines often contained opium without any warning label. Today, state, federal, and international laws govern the production and distribution of narcotic substances.

Morphine

Morphine is the principal constituent of opium and ranges in concentration from 4% to 21%. Commercial opium is standardized to contain 10% morphine. In the United States, a small percentage of the morphine obtained from opium is used directly (about 20 tons); the remaining is converted to codeine and other derivatives (about 110 tons). Morphine is one of the most effective drugs known for the relief of severe pain and remains the standard against which new analgesics are measured. Like most narcotics, the use of morphine has increased significantly in recent years. Since 1998, there has been about a twofold increase in the use of morphine products in the United States (United Nations, 2007).

Traditionally, morphine was almost exclusively used by injection. Today, morphine is marketed in a variety of forms, including oral solutions, immediate and sustained-release tablets and capsules, suppositories, and injectable preparations.

Codeine

Codeine is the most widely used, naturally occurring narcotic in medical treatment in the world. However, most codeine used in the United States is produced from morphine. Codeine is medically prescribed for the relief of moderate pain and cough suppression.

Heroin

First synthesized from morphine in 1874, heroin was not extensively used in medicine until the early 1900s. Commercial production of the new pain remedy was first started in 1898 (United Nations, 2007). It initially received widespread acceptance from the medical profession, and physicians remained unaware of its addiction potential for years. The first comprehensive control of heroin occurred with the Harrison Narcotic Act of 1914. Today, heroin is an illicit substance having no medical utility in the United States. It is in Schedule I of the CSA.

Four foreign source areas produce the heroin available in the United States: South America (Colombia), Mexico, Southeast Asia (principally Burma), and Southwest Asia (principally Afghanistan). However, South America and Mexico supply most of the illicit heroin marketed in the United States. South American heroin is a high-purity powder primarily distributed to metropolitan areas on the East Coast. Heroin powder may vary in color from white to dark brown because of impurities left from the manufacturing process or the presence of additives. Mexican heroin, known as black tar, is primarily available in the western United States. The color and consistency of black tar heroin result from the crude processing methods used to illicitly manufacture heroin in Mexico. Black tar heroin may be sticky like roofing tar or hard like coal, and its color may vary from dark brown to black.

In the past, heroin in the United States was almost always injected, because this is the most practical and efficient way to administer low-purity heroin (**Figure 15.1**). However, the recent availability of higher purity heroin at relatively low cost has meant that a larger percentage of today's users are either snorting or smoking heroin, instead of injecting it. This trend was first captured in the 1999 National Household Survey on Drug Abuse, which revealed that 60% to 70% of people who used heroin for the first time from 1996 to 1998 never injected it (SAMHSA, 1999). This trend has continued. Snorting or smoking heroin is more appealing to new users because it eliminates both the fear of acquiring syringe-borne diseases, such as HIV and hepatitis as well as eliminating the social stigma attached to intravenous heroin use. Many new users of heroin mistakenly believe that smoking or snorting heroin is a safe technique for avoiding addiction. However, both the smoking and the snorting of heroin are directly linked to high incidences of dependence and addiction.

Oxycodone

Oxycodone is synthesized from thebaine. Like morphine and hydromorphone, oxycodone is used as an analgesic. Historically, oxycodone products have been popular drugs of abuse among the narcotic abusing population. In recent years, concern has grown among federal, state, and local officials about the dramatic increase in the illicit availability and abuse of OxyContin® products. These products contain large amounts

Figure 15.1 Example of Heroin and Associated Paraphernalia.

of oxycodone (10 to 160 mg) in a formulation intended for slow release over about a 12-hour period.

Abusers have learned that this slow-release mechanism can be easily circumvented by crushing the tablet and swallowing, snorting, or injecting the drug product for a more rapid and intense high. The criminal activity associated with illicitly obtaining and distributing this drug as well as serious consequences of illicit use, including addiction and fatal overdose deaths, are of epidemic proportions in some areas of the United States.

Hydrocodone

Hydrocodone is an effective cough suppressant and analgesic. All products currently marketed in the United States are either Schedule III combination products primarily intended for pain management or Schedule V antitussive medications often marketed in liquid formulations. The Schedule III products are currently under review at the federal level to determine if an increase in regulatory control is warranted.

Hydrocodone products are the most frequently prescribed pharmaceutical opiates in the United States with over 111 million prescriptions dispensed in 2003 (United Nations, 2007). Despite their obvious utility in medical practice, hydrocodone products are among the most popular pharmaceutical drugs associated with drug diversion, trafficking, abuse, and addiction. In every geographical area in the country, the DEA has listed this drug as one of the most commonly diverted. Hydrocodone is the most frequently encountered opiate pharmaceutical in submissions of drug evidence to federal, state, and local forensic laboratories. Law enforcement has documented the diversion of millions of dosage units of hydrocodone by theft, doctor shopping, fraudulent prescriptions, bogus "call-in" prescriptions, and diversion by registrants and Internet fraud.

Hydrocodone products are associated with significant drug abuse. Hydrocodone was ranked sixth among all controlled substances in the 2006 Drug Abuse Warning Network (DAWN) emergency department (ED) data (SAMHSA, 2008). Poison control data, DAWN medical examiner (ME) data, and other ME data indicate that hydrocodone deaths are numerous, widespread, and increasing in number. In addition, the hydrocodone acetaminophen combinations (accounting for about 80% of all hydrocodone prescriptions) carry significant public health risk when taken in excess.

Fentanyl

First synthesized in Belgium in the late 1950s, fentanyl, with an analgesic potency of about 80 times that of morphine, was introduced into medical practice in the 1960s as an intravenous anesthetic. Illicit use of pharmaceutical fentanyls first appeared in the mid-1970s in the medical community and continues to be a problem in the United States. To date, over 12 different analogues of fentanyl have been produced clandestinely and identified in the U.S. drug traffic. The biological effects of the fentanyls are indistinguishable from those of heroin, with the exception that the fentanyls may be hundreds of times more potent. Fentanyls are most commonly used by intravenous administration, but like heroin, they may also be smoked or snorted.

Methadone

German scientists synthesized methadone during World War II because of a shortage of morphine. Although chemically unlike morphine or heroin, methadone produces many of the same effects. It was introduced into the United States in 1947 as an analgesic (Dolophine®). Today, methadone is primarily used for the treatment of narcotic addiction, although a growing number of prescriptions are being written for chronic pain management. It is available in oral solutions, tablets, and injectable Schedule II formulations.

Methadone's effects can last up to 24 hours, thereby permitting once-a-day oral administration in heroin detoxification and maintenance programs. High-dose methadone can block the effects of heroin, thereby discouraging the continued use of heroin by addicts in treatment. Chronic administration of methadone results in the development of tolerance and dependence. The withdrawal syndrome develops more slowly and is less severe, but more prolonged than that associated with heroin withdrawal. Ironically, methadone used to control narcotic addiction is encountered on the illicit market. Recent increases in the use of methadone for pain management have been associated with increasing numbers of overdose deaths.

Stimulants

Stimulants, sometimes referred to as uppers, reverse the effects of fatigue on both mental and physical tasks. Two commonly used stimulants are nicotine, which is found in tobacco products, and caffeine, an active ingredient in coffee, tea, some soft drinks, and many nonprescription medicines. Used in moderation, these substances tend to increase alertness. Although the use of these products has been an accepted part of U.S. culture, the recognition of their adverse effects has resulted in a proliferation of caffeine-free products and efforts to discourage cigarette smoking.

A number of stimulants, however, are under the regulatory control of the CSA. Some of these controlled substances are available by prescription for legitimate medical use in the treatment of obesity, narcolepsy, and attention deficit disorders. As drugs of abuse, stimulants are frequently taken to produce a sense of exhilaration, enhance self esteem, improve mental and physical performance, increase activity, reduce appetite, produce prolonged wakefulness, and to "get high." They are among the most potent agents of reward and reinforcement that underlie the problem of dependence.

Stimulants are diverted from legitimate channels and clandestinely manufactured exclusively for the illicit market. They are taken orally, sniffed, smoked, and injected. Smoking, snorting, or injecting stimulants produce a sudden sensation known as a rush or a flash. Abuse is often associated with a pattern of binge use: sporadically consuming large doses of stimulants over a short period of time. Heavy users may inject themselves every few hours, continuing until they have depleted their drug supply or reached a point of delirium, psychosis, and physical exhaustion. During this period of heavy use, all other interests become secondary to re-creating the initial euphoric rush. Tolerance can develop rapidly, and both physical and psychological dependence occur. Abrupt cessation, even after a brief two- or three-day binge, is commonly followed by depression, anxiety, drug craving, and extreme fatigue known as a crash.

Therapeutic levels of stimulants can produce exhilaration, extended wakefulness, and loss of appetite. These effects are greatly intensified when large doses of stimulants are taken. Physical side effects, including dizziness, tremor, headache, flushed skin, chest pain with palpitations, excessive sweating, vomiting, and abdominal cramps, may occur as a result of taking too large a dose at one time or taking large doses over an extended period of time. Psychological effects include agitation, hostility, panic, aggression, and suicidal or homicidal tendencies. Paranoia, sometimes accompanied by both auditory and visual hallucinations, may occur also. Overdose is often associated with high fever, convulsions, and cardiovascular collapse. Because accidental death is partially due to the effects of stimulants on the body's cardiovascular and temperature-regulating systems, physical exertion increases the hazards of stimulant use.

Cocaine

Cocaine, the most potent stimulant of natural origin, is extracted from the leaves of the coca plant (*Erythroxylum coca*), which is indigenous to the Andean highlands of South America. Natives in this region chew or brew coca leaves into a tea for refreshment and to relieve fatigue, similar to the customs of chewing tobacco and drinking tea or coffee.

Pure cocaine was first isolated in the 1880s and used as a local anesthetic in eye surgery. It was particularly useful in surgery of the nose and throat because of its ability to provide anesthesia as well as to constrict blood vessels and limit bleeding. Many of its therapeutic applications are now obsolete due to the development of safer drugs.

Illicit cocaine is usually distributed as a white crystalline powder or as an off-white chunky material. The powder, usually cocaine hydrochloride, is often diluted with a variety of substances, the most common being sugars such as lactose, inositol, and mannitol, and local anesthetics such as lidocaine. The adulteration increases the volume and thus multiplies profits. Cocaine hydrochloride is generally snorted or dissolved in water and injected (**Figure 15.2**).

Crack, the chunk or "rock" form of cocaine, is a ready-to-use freebase. On the illicit market, it is sold in small, inexpensive dosage units that are smoked. Smoking delivers large quantities of cocaine to the lungs, producing effects comparable to intravenous injection. Drug effects are felt almost immediately, are very intense, and are quickly over. Once introduced in the mid-1980s, crack abuse spread rapidly and made the cocaine experience available to a wide range of people and socioeconomic backgrounds. It is noteworthy that the emergence

Figure 15.2 Example of Cocaine and Related Paraphernalia.

of crack was accompanied by a dramatic increase in drug abuse problems and drug-related violence within the United States.

Cocaine is the third most commonly used illicit drug (following marijuana and psychotherapeutics) in the United States. According to the 2008 National Survey on Drug Use and Health, 1.9 million Americans, age 12 or older, have used cocaine in the past month.

Amphetamines

Amphetamine, dextroamphetamine, methamphetamine, and their various salts, are collectively referred to as amphetamines. In fact, their chemical properties and actions are so similar that even experienced users have difficulty knowing which drug they have taken (**Figure 15.3**).

Amphetamine was first marketed in the 1930s as Benzedrine® in an OTC inhaler to treat nasal congestion. By 1937, amphetamine was available by prescription in tablet form and was used in the treatment of the sleeping disorder, narcolepsy, and the behavioral syndrome called minimal brain dysfunction, which today is called attention deficit hyperactivity disorder (ADHD). During World War II, amphetamine was widely used to keep the fighting men going.

As use of amphetamines spread, so did their abuse. In the 1960s, amphetamines became a perceived remedy for helping truckers to complete their long routes without falling asleep, for weight control, for helping athletes to perform better and train longer, and for treating mild

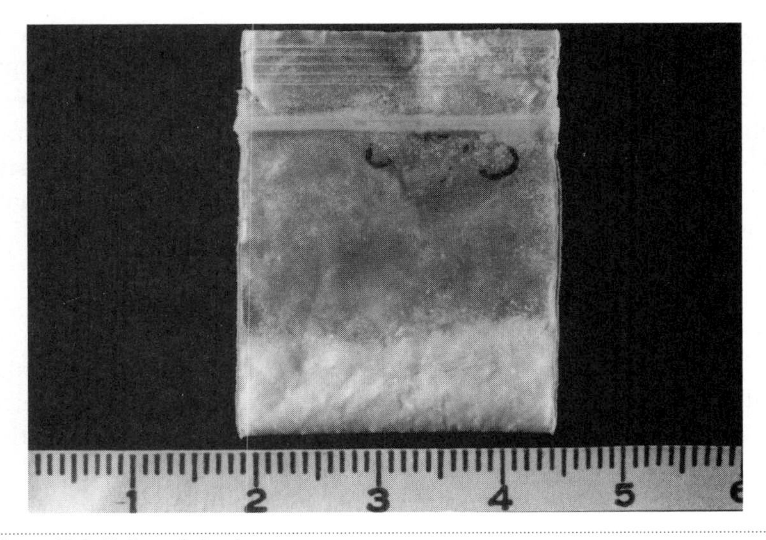

Figure 15.3 Methamphetamine.

depression. Intravenous amphetamines, primarily methamphetamine, were abused by a subculture known as speed freaks. With experience, it became evident that the dangers of abuse of these drugs outweighed most of their therapeutic uses.

Increased control measures were initiated in 1965 with amendments to the federal food and drug laws to curb the black market in amphetamines. Many pharmaceutical amphetamine products were removed from the market, including all injectable formulations, and doctors prescribed those that remained less freely. Recent increases in medical use of these drugs can be attributed to their use in the treatment of ADHD. Amphetamines are all controlled in Schedule II of the CSA.

To meet the ever-increasing black market demand for amphetamines, clandestine laboratory production has mushroomed. Today, most amphetamines distributed to the black market are produced in clandestine laboratories. Methamphetamine laboratories are, by far, the most frequently encountered clandestine laboratories in the United States. The ease of clandestine synthesis, combined with tremendous profits, has resulted in significant availability of illicit methamphetamine. Methamphetamine produced in home labs is becoming increasingly more common due to the availability and legality of the common ingredients used to manufacture the drug. One of the primary ingredients of methamphetamine is pseudoephedrine, a common OTC that has become increasingly controlled by vendors, due to this epidemic, although still not technically a controlled drug. However, due to legislation in a great many of states, possession of large amounts of pseudoephedrine tablets, or other precursor ingredients of methamphetamine, can subject the possessor to criminal charges for taking a substantial step towards, or possessing items used in, the manufacture of a controlled substance. While there has been a significant increase in "Mom and Pop" type home laboratories, there are large amounts of methamphetamine that are also illicitly smuggled into the United States from Mexico and Southeast Asia.

Amphetamines are generally taken orally or injected. However, the addition of ice, the slang name for crystallized methamphetamine hydrochloride, has promoted smoking as another mode of administration. Just as "crack" is smokable cocaine, "ice" is smokable methamphetamine. Methamphetamine in all its forms is highly addictive and toxic.

The effects of amphetamines, especially methamphetamine, are similar to cocaine, but their onset is slower and their duration is longer. In contrast to cocaine, which is quickly removed from the brain and is almost completely metabolized, methamphetamine remains in the central nervous system longer, and a larger percentage of the drug remains unchanged in the body, producing prolonged stimulant effects. Chronic abuse produces a psychosis that resembles schizophrenia and is characterized by paranoia, picking at the skin, preoccupation with one's own thoughts, and auditory and visual hallucinations. These psychotic symptoms can persist for months and even years after use of these drugs has ceased and may be related to their neurotoxic effects. Violent and erratic behavior is frequently seen among chronic abusers of amphetamines, especially methamphetamine.

Depressants

Historically, people of almost every culture have used chemical agents to induce sleep, relieve stress, and allay anxiety. While alcohol is one of the oldest and most universal agents used for these purposes, hundreds of substances have been developed that produce central nervous system depression. These drugs are called **depressants** and have also been referred to as downers, sedatives, hypnotics, minor tranquilizers, and anti-anxiety medications. Unlike most other classes of drugs of abuse, depressants are rarely produced in clandestine laboratories. Generally, legitimate pharmaceutical products are diverted to the illicit market. A notable exception to this is a relatively recent drug of abuse, gamma hydroxybutyric acid (GHB).

Barbiturates were very popular in the first half of the 20th century. In moderate amounts, these drugs produce a state of intoxication that is remarkably similar to alcohol intoxication. Symptoms include slurred speech, loss of motor coordination, and impaired judgment. Depending on the dose, frequency, and duration of use, one can rapidly develop tolerance, and physical and psychological dependence on barbiturates. With the development of tolerance, the margin of safety between the effective dose and the lethal dose becomes very narrow. That is, in order to obtain the same level of intoxication, the tolerant abuser may raise his or her dose to a level that may result in coma or death. Although many individuals have taken barbiturates therapeutically without harm, concern about the addiction potential of barbiturates and the ever-increasing number of fatalities associated with them led to the development of alternative medications. Today, less than 10% of all depressant prescriptions in the United States are for barbiturates.

Benzodiazepines were first marketed in the 1960s. Touted as much safer depressants with far less addiction potential than barbiturates, today these drugs account for about one out of every five prescriptions for controlled substances. Although benzodiazepines produce significantly less respiratory depression than barbiturates, it is now recognized that benzodiazepines share many of the undesirable side effects of the barbiturates. A number of toxic central nervous system effects are seen with chronic high-dose benzodiazepine therapy, including headaches, irritability, confusion, memory impairment, and depression. The risk of developing over-sedation, dizziness, and confusion increases substantially with higher doses of benzodiazepines. Prolonged use can lead to physical dependence even at doses recommended for medical treatment. Unlike barbiturates, large doses of benzodiazepines are rarely fatal unless combined with other drugs or alcohol. Although primary abuse of benzodiazepines is well documented, abuse of these drugs usually occurs as part of a pattern of multiple drug abuse. For example, heroin or cocaine abusers will use benzodiazepines and other depressants to augment their "high" or alter the side effects associated with over-stimulation or narcotic withdrawal.

There are marked similarities among the withdrawal symptoms seen with most drugs classified as depressants. In the mildest form, the withdrawal syndrome may produce insomnia and anxiety, usually the same symptoms that initiated the drug use. With a greater level of dependence, tremors and weakness are also present, and in its most severe form, the withdrawal syndrome can cause seizures and delirium. Unlike the withdrawal syndrome seen with most other drugs of abuse, withdrawal from depressants can be life-threatening.

Barbiturates

Barbiturates were first introduced for medical use in the early 1900s. More than 2,500 barbiturates have been synthesized, and at the height of their popularity, about 50 were marketed for human use. Today, about a dozen are in medical use. Barbiturates produce a wide spectrum of central nervous system depression, from mild sedation to coma, and have been used as sedatives, hypnotics, anesthetics, and anticonvulsants. The primary differences among many of these products are how fast they produce an effect and how long those effects last. There is a variation in scheduling (Schedule III to Schedule IV) within this group of controlled substances. Those scheduled as IV have effects that are realized in about one hour and last for about 12 hours, and are used primarily for daytime sedation and the treatment of seizure disorders. Those scheduled as III exhibit onset of action from 15 to 40 minutes, and the effects last up to six hours. These drugs are primarily used for insomnia and preoperative sedation. Veterinarians also will make use of them for anesthesia and euthanasia.

Benzodiazepines

The benzodiazepine family of depressants is used therapeutically to produce sedation, induce sleep, relieve anxiety and muscle spasms, and to prevent seizures. Of the drugs marketed in the United States that affect central nervous system function, benzodiazepines are among the most widely prescribed medications. Fifteen members of this group are presently marketed in the United States, and about 20 additional benzodiazepines are marketed in other countries. Benzodiazepines are controlled in Schedule IV of the CSA.

Benzodiazepines are classified in the CSA as depressants. Repeated use of large doses or, in some cases, daily use of therapeutic doses of benzodiazepines is associated with amnesia, hostility, irritability, and vivid or disturbing dreams as well as tolerance and physical dependence. The withdrawal syndrome is similar to that of alcohol and may require hospitalization. Abrupt cessation of benzodiazepines is not recommended and tapering-down the dose eliminates many of the unpleasant symptoms.

Given the millions of prescriptions written for benzodiazepines, relatively few individuals increase their dose on their own initiative or engage in drug-seeking behavior. Those individuals who do abuse benzodiazepines often maintain their drug supply by getting prescriptions from several doctors, forging prescriptions, or buying diverted pharmaceutical products on the illicit market. Abuse is frequently associated with adolescents and young adults who take benzodiazepines to obtain a "high." This intoxicated state results in reduced inhibition and impaired judgment. Concurrent use of alcohol or other depressant with benzodiazepines can be life threatening. Abuse of benzodiazepines is particularly high among heroin and cocaine abusers. A large percentage of people entering treatment for narcotic or cocaine addiction also report abusing benzodiazepines. Alprazolam and diazepam are the two most frequently encountered benzodiazepines on the illicit market.

Flunitrazepam

Flunitrazepam (Rohypnol®) is a benzodiazepine that is not manufactured or legally marketed in the United States, but is smuggled in by traffickers. In the mid-1990s, flunitrazepam was extensively trafficked in Florida and Texas. Known as roofies and roach, flunitrazepam gained popularity among younger individuals as a "party" drug. It has also been utilized as a "date rape" drug. In this context, flunitrazepam is placed in the alcoholic drink of an unsuspecting victim to incapacitate them and prevent resistance from sexual assault. The victim is frequently unaware of what has happened to them and often does not report the incident to authorities. A number of actions by the manufacturer of this drug and by government agencies have resulted in reducing the availability and abuse of flunitrazepam in the United States.

Gamma hydroxybutyric acid (GHB)

In recent years, gamma hydroxybutyric acid (GHB) has emerged as a significant drug of abuse throughout the United States. Abusers of this drug fall into three major groups: (1) users take GHB for its intoxicant or euphoriant effects; (2) bodybuilders who abuse GHB for its alleged utility as an anabolic agent or as a sleep aid; and (3) individuals who use GHB as a weapon for sexual assault. These categories are not mutually exclusive and an abuser may use the drug illicitly to produce several effects. GHB is frequently taken with alcohol or other drugs that heighten its effects and is often found at bars, nightclubs, rave parties, and gyms. Teenagers and young adults who frequent these establishments are the primary users. Like flunitrazepam, GHB is often referred to as a "date-rape" drug. GHB involvement in rape cases is likely to be unreported or unsubstantiated because GHB is quickly eliminated from the body, making detection in body fluids unlikely. Its fast onset of depressant effects may render the victim with little memory of the details of the attack.

GHB produces a wide range of central nervous system effects, including dose-dependent drowsiness, dizziness, nausea, amnesia, visual hallucinations, hypotension, bradycardia, severe respiratory depression, and coma. The use of alcohol in combination with GHB greatly enhances its depressant effects. Overdose frequently requires emergency room care, and many GHB-related fatalities have been reported.

GHB Usage as a Predatory Drug

GHB is a clear, odorless liquid, slightly thicker than water. Its only detection is that it has a salty taste.

It is virtually undetectable by victims, users and yes . . . even the cops!!

Its effects are not felt for two to four hours, and users commonly do not remember what happened during that time.

GHB can easily be slipped into any drink (most commonly alcoholic beverages).

GHB causes the bodies functions to slow dramatically. Breathing can decrease to six times per minute. Heart rate can slow to 25 beats per minute.

Victims commonly pass out and are unaware of what has happened to them while they were unconscious.

The abuse of GHB began to seriously escalate in the mid-1990s. For example, in 1994, there were 55 emergency department episodes involving GHB reported in the Drug Abuse Warning Network (DAWN) system. By 2006, there were 3,330 emergency room episodes. DAWN data also indicated that most users were male, less than 25 years of age, and taking the drug orally for recreational use. GHB was placed in Schedule I of the CSA in March 2000 (SAMHSA, 2006).

Cannabis

Cannabis sativa L., the cannabis plant, grows wild throughout most of the tropic and temperate regions of the world. Prior to the advent of synthetic fibers, the cannabis plant was cultivated for the tough fiber of its stem. In the United States, cannabis is legitimately grown only for scientific research.

Cannabis contains chemicals called cannabinoids that are unique to the cannabis plant. One of these, delta-9-tetrahydrocannabinol (THC), is believed to be responsible for most of the characteristic psychoactive effects of cannabis.

Cannabis products are usually smoked. Their effects are felt within minutes, reach their peak in 10 to 30 minutes, and may linger for two or three hours. The effects experienced often depend upon the experience and expectations of the individual user as well as the activity of the drug itself. Low doses tend to induce a sense of well-being and a dreamy state of relaxation, which may be accompanied by a more vivid sense of sight, smell, taste, and hearing, as well as by subtle alterations in thought formation and expression. High doses may result in image distortion, a loss of personal identity, fantasies, and hallucinations.

Three drugs that come from cannabis—marijuana, hashish, and hashish oil—are distributed on the U.S. illicit market. Having no federally accepted medical use in treatment in the United States, they remain under Schedule I of the CSA. Today, cannabis is illicitly cultivated, both indoors and out, to maximize its THC content, thereby producing the greatest possible psychoactive effect.

Marijuana

Marijuana is the most frequently encountered illicit drug worldwide (**Figure 15.4**). In the United States, according to the 2003 Monitoring the Future Study, 57% of adults aged 19 to 28 reported having used marijuana in their lifetime (Johnston, O'Malley, Bachman, 2003, p. 520). Among younger Americans, 17.5% of 8th graders and 46.1% of 12th graders had used marijuana in their lifetime (United Nations, 2007). The term marijuana, as commonly used, refers to the leaves and flowering tops of the cannabis plant that are dried to produce a tobacco-like substance. Marijuana varies significantly in its potency, depending on the source and selection of plant materials used. The form of marijuana known as sinsemilla (from Spanish, *sin semilla*: without seed), derived from the unpollinated female cannabis plant, is preferred for its high THC content. Marijuana is usually smoked in the form of loosely rolled cigarettes called joints, bongs, or hollowed out commercial cigars called blunts. Joints and blunts may be laced with a number of adulterants including phencyclidine (PCP), substantially altering the effects and toxicity of these products. Street names for marijuana include pot, grass, weed, Mary Jane, and reefer. Although marijuana grown in the United States was once considered inferior because of a low concentration of THC, advancements in plant selection and cultivation have resulted in higher THC-containing domestic marijuana (**Figures 15.5** and **15.6**).

Marijuana contains known toxins and cancer-causing chemicals. Marijuana users experience the same health problems as tobacco smokers, such as bronchitis, emphysema, and bronchial asthma. Some of the effects of marijuana use also include increased heart rate, dryness of the mouth, reddening of the eyes, impaired motor skills and concentration, and hunger with an increased desire for sweets. Extended use increases

Figure 15.4 Marijuana.

Figure 15.5 A Marijuana Grow.

Figure 15.6 Processing Evidence of Large Scale Trafficking in Marijuana.

risk to the lungs and reproductive system as well as suppression of the immune system. Occasionally, hallucinations, fantasies, and paranoia are reported. Long-term chronic marijuana use is characterized by: apathy; impairment of judgment, memory, and concentration; and loss of interest in personal appearance and pursuit of goals.

Hashish

Hashish consists of the THC-rich resinous material of the cannabis plant, which is collected, dried, and then compressed into a variety of forms, such as balls, cakes, or cookie-like sheets. Pieces are then broken off, placed in pipes, and smoked. The Middle East, North Africa, and Pakistan/Afghanistan are the main sources of hashish.

Hashish Oil

The term hash oil is used by illicit drug users and dealers, but is a misnomer in suggesting any resemblance to hashish. Hash oil is produced by extracting the cannabinoids from plant material with a solvent. The color and odor of the resulting extract will vary, depending on the type of solvent used. In terms of its psychoactive effect, a drop or two of this liquid on a cigarette is equal to a single "joint" of marijuana.

Hallucinogens

Hallucinogens are among the oldest known group of drugs used for their ability to alter human perception and mood. For centuries, many of the naturally occurring hallucinogens found in plants and fungi have been used for a variety of shamanistic practices. In more recent years, a number of synthetic hallucinogens have been produced, some of which are much more potent than their naturally occurring counterparts.

The biochemical, pharmacological, and physiological basis for hallucinogenic activity is not well understood. Even the name for this class of drugs is not ideal, because hallucinogens do not always produce hallucinations.

However, taken in non-toxic dosages, these substances produce changes in perception, thought, and mood. Physiological effects include elevated heart rate, increased blood pressure, and dilated pupils. Sensory effects include perceptual distortions that vary with dose, setting, and mood. Psychic effects include disorders of thought associated with time and space. Time may appear to stand still and forms and colors seem to change and take on new significance. This experience may be either pleasurable or extremely frightening. It needs to be stressed that the effects of hallucinogens are unpredictable each time they are used.

Weeks or even months after some hallucinogens have been taken, the user may experience flashbacks, which are fragmentary recurrences of certain aspects of the drug experience in the absence of actually taking the drug. The occurrence of a flashback is unpredictable, but is more likely to occur during times of stress and seem to occur more frequently in younger individuals. With time, these episodes diminish and become less intense.

The abuse of hallucinogens in the United States received much public attention in the 1960s and 1970s. A subsequent decline in their use in the 1980s may be attributed to real or perceived hazards associated with taking these drugs.

However, a resurgence of the use of hallucinogens is cause for concern. Fortunately, the 2008 National Survey on Drug Use and Health reports that the prevalence of past month hallucinogen use has remained relatively constant from 0.5 percent in 2002 to 0.4 percent in 2003, 2004, 2005, 2006, 2007, and 2008 (SAMHSA, 2008). Hallucinogenic mushrooms, LSD, and MDMA are popular among junior and senior high school students who use hallucinogens.

Lysergic Acid Diethylamide (LSD)

Lysergic acid diethylamide (LSD) is the most potent hallucinogen known to science as well as the most highly studied. LSD was originally synthesized in 1938 by Dr. Albert Hoffman. However, its hallucinogenic effects were unknown until 1943 when Hoffman accidentally consumed some LSD. It was later found that an oral dose of as little as 0.000025 grams (25 micrograms, which is equal in weight to a couple grains of salt) is capable of producing rich and vivid hallucinations. Because of its structural similarity to a chemical present in the brain and its similarity in effects to certain aspects of psychosis, LSD was used as a research tool to study mental illness (United Nations, 2007). LSD abuse was popularized in the 1960s by individuals like Dr. Timothy Leary, who encouraged American students to "turn on, tune in, and drop out." LSD use has varied over the years but it still remains a significant drug of abuse. In 2003, lifetime prevalence of LSD use for 8th and 12th graders was 2.1 and 5.9%, respectively.

The average effective oral dose is from 20 to 80 micrograms with the effects of higher doses lasting for 10 to 12 hours. LSD is usually sold in the form of impregnated paper (blotter acid), typically imprinted with colorful graphic designs. It has also been encountered in tablets (microdots), thin squares of gelatin (window panes), in sugar cubes and, rarely, in liquid form.

Physical reactions may include dilated pupils, lowered body temperature, nausea, "goose bumps," profuse perspiration, increased blood sugar, and rapid heart rate. During the first hour after ingestion, the user may experience visual changes with extreme changes in mood. In the hallucinatory state, the LSD user may suffer impaired depth and time perception, accompanied by distorted perception of the size and shape of objects, movements, color, sound, touch, and the user's own body image. During this period, the ability to perceive objects through the senses is distorted: a user may describe "hearing colors" and "seeing sounds." The ability to make sensible judgments and see common dangers is impaired, making the user susceptible to personal injury. After an LSD "trip," the user may suffer acute anxiety or depression for a variable period of time. Flashbacks have been reported days or even months after taking the last dose.

Psilocybin and Psilocyn

A number of Schedule I hallucinogenic substances are classified found in nature. Psilocybin and psilocyn are obtained from certain mushrooms indigenous to tropical and subtropical regions of South America, Mexico, and the United States. As pure chemicals at doses of 10 to 20 mg, these hallucinogens produce muscle relaxation, dilation of pupils, vivid visual and auditory distortions, and emotional disturbances. However, the effects produced by consuming preparations of dried or brewed mushrooms are far less predictable and largely depend on the particular mushrooms used and the age and preservation of the extract. There are many species of so-called "magic" mushrooms that contain varying amounts of these tryptamines, as well as uncertain amounts of other chemicals. As a consequence, the hallucinogenic activity as well as the extent of toxicity produced by various plant samples are often unknown.

Peyote and Mescaline

Peyote is a small, spineless cactus, *Lophophora williamsii*, whose principal active ingredient is the hallucinogen mescaline (3, 4, 5-trimethoxyphenethylamine). From earliest recorded time, peyote has been used by Natives in northern Mexico and the southwestern United States as a part of their religious rites.

The top of the cactus above ground—also referred to as the crown—consists of disc-shaped buttons that are cut from the roots and dried. These buttons are generally chewed or soaked in water to produce an intoxicating liquid. The hallucinogenic dose of mescaline is about 0.3 to 0.5 grams and lasts about 12 hours. While peyote produced rich visual hallucinations that were important to the Native American peyote users, the full spectrum of effects served as a chemically induced model of mental illness. Mescaline can be extracted from peyote or produced synthetically. Both peyote and mescaline are listed in the CSA as Schedule I hallucinogens.

Methylenedioxymethamphetamine (MDMA, Ecstasy)

3,4-Methylenedioxymethamphetamine (MDMA, Ecstasy) was first synthesized in 1912 but remained in relative obscurity for many years. In the 1980s, MDMA gained popularity as a drug of abuse resulting in its final placement in Schedule I of the CSA. Today, MDMA is extremely popular. In 2005, it was estimated that two million tablets were smuggled into the United States every week (United Nations, 2007).

MDMA produces both amphetamine-like stimulation and mild mescaline-like hallucinations. It is touted as a "feel good" drug with an undeserved reputation of safety. MDMA produces euphoria, increased energy, increased sensual arousal, and enhanced tactile sensations. However, it also produces nerve cell damage that can result in psychiatric disturbances and long-term cognitive impairments. The user will often experience increased muscle tension, tremors, blurred vision, and hyperthermia. The increased body temperature can result in organ failure and death.

MDMA is usually distributed in tablet form and taken orally at doses ranging from 50 to 200 mg. Individual tablets are often imprinted with graphic designs or commercial logos, and typically contain 80 to 100 mg of MDMA (**Figure 15.7**). After oral administration, effects are felt within 30 to 45 minutes, peak at 60 to 90 minutes, and last for 4 to 6 hours. Analysis of seized MDMA tablets indicates that about 80%

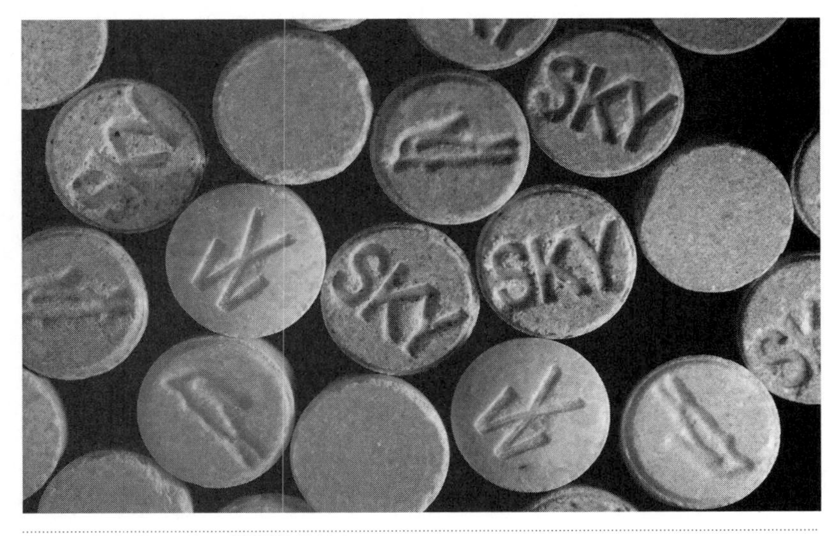

Figure 15.7 Examples of MDMA Tablets.

of all samples actually contain MDMA while the other 20% contain amphetamine, methamphetamine, or both.

Some of these differ from one another in potency, speed of onset, duration of action, and capacity to modify mood, with or without producing overt hallucinations. The drugs are usually taken orally, sometimes snorted, and rarely injected. Because they are produced in clandestine laboratories, they are seldom pure and the amount in a capsule or tablet is likely to vary considerably.

According to the National Survey on Drug Use and Health, initiation of MDMA use has increased from 1993 until 2001, when it peaked at 1.8 million new users. In 2002 the number declined to 1.1 million. Two thirds (66%) of new MDMA users in 2002 were 18 or older, and 50% were male.

Phencyclidine and Related Drugs

In the 1950s, phencyclidine (PCP) was investigated as an anesthetic but, due to the side effects of confusion and delirium, its development for human use was discontinued. It became commercially available for use as a veterinary anesthetic in the 1960s under the trade name of Sernylan® and was placed in Schedule III of the CSA. In 1978, due to considerable abuse, phencyclidine was transferred to Schedule II of the CSA and manufacturing of Sernylan® was discontinued. Today, virtually all of the phencyclidine encountered on the illicit market in the United States is produced in clandestine laboratories.

PCP is illicitly marketed under a number of other names, including angel dust, supergrass, killer weed, embalming fluid, and rocket fuel, reflecting the range of its bizarre and volatile effects. In its pure form, it is a white crystalline powder that readily dissolves in water. However, most PCP on the illicit market contains a number of contaminants as a result of makeshift manufacturing, causing the color to range from tan to brown, and the consistency from powder to a gummy mass. Although sold in tablets and capsules as well as in powder and liquid form, it is commonly applied to a leafy material, such as parsley, mint, oregano, or marijuana, and smoked.

The drug's effects are as varied as its appearance. A moderate amount of PCP often causes the user to feel detached, distant, and estranged from his surroundings. Numbness, slurred speech, and loss of coordination may be accompanied by a sense of strength and invulnerability. A blank stare, rapid and involuntary eye movements, and an exaggerated gait are among the more observable effects. Auditory hallucinations, image distortion, severe mood disorders, and amnesia may also occur. In some users, PCP may cause acute anxiety and a feeling of impending doom; in others, paranoia and violent hostility; and in some, it may produce a psychosis indistinguishable from schizophrenia. PCP use is associated with a number of risks, and many believe it to be one of the most dangerous drugs of abuse.

Ketamine

Ketamine is a rapidly-acting general anesthetic. Its pharmacological profile is essentially the same as phencyclidine. Like PCP, ketamine is referred to as a dissociative anesthetic because patients feel detached or disconnected from their pain and environment when anesthetized with this drug. Unlike most anesthetics, ketamine produces only mild respiratory depression and appears to stimulate, not depress, the cardiovascular system. In addition, ketamine has both analgesic and amnesic properties and is associated with less confusion, irrationality, and violent behavior than PCP. Use of ketamine as a general anesthetic for humans has been limited due to adverse effects including delirium and hallucinations. Today, it is primarily used in veterinary medicine, but has some utility for emergency surgery in humans.

Although ketamine has been marketed in the United States for many years, it was only recently associated with significant diversion and abuse and placed in Schedule III of the CSA in 1999. Known in the drug culture as Special K or Super K, ketamine has become a staple at dance parties or "raves." Ketamine is supplied to the illicit market by the diversion of legitimate pharmaceuticals (Ketaset®, Ketalar®). It is usually distributed as a powder obtained by removing the liquid from the pharmaceutical products. As a drug of abuse, ketamine can be administered orally, snorted, or injected. It is also sprinkled on marijuana or tobacco and smoked. After oral or intranasal administration, effects are evident in about 10 to 15 minutes and are over in about an hour.

After intravenous use, effects begin almost immediately and reach peak effects within minutes. Ketamine can act as a depressant or a psychedelic. Low doses produce vertigo, ataxia, slurred speech, slow reaction time, and euphoria. Intermediate doses produce disorganized thinking, altered body image, and a feeling of unreality with vivid visual hallucinations. High doses produce analgesia, amnesia, and coma.

Inhalants

Inhalants are a diverse group of substances that include volatile solvents, gases, and nitrites that are sniffed, snorted, huffed, or bagged to produce intoxicating effects similar to alcohol. These substances are found in common household products like glues, lighter fluid, cleaning fluids, and paint products. Inhalant abuse is the deliberate inhaling or sniffing of these substances to get high, and it is estimated that about 1,000 substances are misused in this manner. The easy accessibility, low cost, legal status, and ease of transport and concealment make inhalants one of the first substances abused by children.

For example, volatile solvents are found in a number of everyday products. Some of these products include nail polish remover, lighter fluid, gasoline, paint and paint thinner, rubber glue, waxes, and varnishes. The gas used as a propellant in canned whipped cream and in small lavender metallic containers called whippets (used to make whipped cream) is nitrous oxide or laughing gas (the same gas used by dentists for anesthesia). Butyl nitrite, sold as tape head cleaner and referred to as rush, locker room, or climax, is often sniffed or huffed to get high.

Inhalants may be sniffed directly from an open container or huffed from a rag soaked in the substance and held to the face. Alternatively, the open container or soaked rag can be placed in a bag where the vapors can concentrate before being inhaled. Some chemicals are painted on the hands or fingernails or placed on shirt sleeves or wrist bands to enable an abuser to continually inhale the fumes without being detected by a teacher or other adult.

Inhalants depress the central nervous system, producing decreased respiration and blood pressure. Users report distortion in perceptions of time and space. Many users experience headaches, nausea, slurred speech, and loss of motor coordination. Mental effects may include fear, anxiety, or depression. A rash around the nose and mouth may be seen, and the abuser may start wheezing. An odor of paint or organic solvents on clothes, skin, and breath is sometimes a sign of inhalant abuse. Other indicators of inhalant abuse include slurred speech or staggering gait, red, glassy, watery eyes, and excitability or unpredictable behavior.

For more information regarding inhalants, contact the National Inhalant Prevention Coalition by telephone (1-800-269-4237) or on the Internet (http://www.inhalants.org).

Steroids

When athletes gather, the issue of performance enhancing drugs, especially anabolic steroids, once again gains international attention. These drugs are used by high school, college, professional, and elite amateur athletes in a variety of sports (e.g., weight lifting, track and field, swimming, cycling, and others) to obtain a competitive advantage. Body builders and fitness buffs take anabolic steroids to improve their physical appearance, and individuals in occupations requiring enhanced physical strength (e.g., body guards, night club bouncers, construction workers) are also known to use these drugs.

Concerns over a growing illicit market, abuse by teenagers, and the uncertainty of possible harmful long-term effects of steroid use, led Congress in 1991 to place anabolic steroids as a class of drugs into Schedule III of the CSA. The CSA defines anabolic steroids as any drug or hormonal substance chemically and pharmacologically related to testosterone (other than estrogens, progestins, and corticosteroids) that promotes muscle growth.

Once viewed as a problem associated only with professional and elite amateur athletes, various reports indicate that anabolic steroid abuse has increased significantly among adolescents. According to the 2003 Monitoring the Future Study, 2.5% of 8th graders, 3.0% of 10th graders, and 3.5% of 12th graders reported using steroids at least once in their lifetime (United Nations, 2007).

Most illicit anabolic steroids are sold at gyms, competitions, and through mail-order operations. For the most part, these substances are smuggled into the United States from many countries. The illicit market includes various preparations intended for human and veterinary use as well as bogus and counterfeit products. The most commonly encountered anabolic steroids on the illicit market include testosterone, nandrolone, methenolone, stanozolol, and methandrostenolone.

A limited number of anabolic steroids have been approved for medical and veterinary use. The primary legitimate use of these drugs in humans is for the replacement of inadequate levels of testosterone resulting from a reduction or absence of functioning testes. Other indications include anemia and breast cancer. Experimentally, anabolic steroids have been used to treat a number of disorders including AIDS wasting, erectile dysfunction, and osteoporosis. In veterinary practice, anabolic steroids are used to promote feed efficiency and to improve weight gain, vigor, and hair coat. They are also used in veterinary practice to treat anemia and counteract tissue breakdown during illness and trauma.

When used in combination with exercise training and a high protein diet, anabolic steroids can promote increased size and strength of muscles, improve endurance, and decrease recovery time between workouts. They are taken orally or by intramuscular injection. Users concerned about drug tolerance often take steroids on a schedule called a cycle. A cycle is a period of between 6 and 14 weeks of steroid use, followed by a period of abstinence or reduction in use. Additionally, users tend to stack the drugs, using multiple drugs concurrently. Although the benefits of these practices are unsubstantiated, most users feel that cycling and stacking enhance the efficiency of the drugs and limit their side effects.

Another mode of steroid use is called pyramiding. With this method users slowly escalate steroid use (increasing the number of drugs used at one time and/or the dose and frequency of one or more steroids), reach a peak amount at mid-cycle and gradually taper the dose toward the end of the cycle. The escalation of steroid use can vary with different types of training. Body builders and weight lifters tend to escalate their dose to a much higher level than do long distance runners or swimmers.

The long-term adverse health effects of anabolic steroid use are not definitely known. There is, however, increasing concern of possible serious health problems associated with the abuse of these agents, including cardiovascular damage, cerebrovascular toxicity, and liver damage.

Physical side effects include elevated blood pressure and cholesterol levels, severe acne, premature balding, reduced sexual function, and testicular atrophy. In males, abnormal breast development can occur. In females, anabolic steroids have a masculinizing effect, resulting in more body hair, a deeper voice, smaller breasts, and fewer menstrual cycles. Several of these effects are irreversible. In adolescents, abuse of these agents may prematurely stop the lengthening of bones, resulting in stunted growth. For some individuals, the use of anabolic steroids may be associated with psychotic reactions, manic episodes, feelings of anger or hostility, aggression, and violent behavior.

Over the last few years, a number of precursors to either testosterone have been marketed as dietary supplements in the United States. New legislation has been introduced in Congress to add several steroids to the CSA and to alter the CSA requirements needed to place new steroids under control in the CSA.

Current and Future Drug Trends

According to the United Nations Office on Drugs and Crime's *World Drug Report 2007*, there are encouraging signs about drug use declining around the world. The first encouraging sign is that coca cultivation in the Andean countries continues to fall, driven by significant declines in Colombia. Global demand for cocaine has also stabilized, although the decline in the United States is offset by alarming increases in some European countries. Secondly, the production and consumption of Amphetamine Type Stimulants (ATS) has leveled off, with a clear downward trend in North America and, to a lesser degree, Europe. Thirdly, the health warnings on higher potency cannabis, delivered in past *World Drug Reports*, appear to be getting through. For the first time in years, there is not an upward trend in the global production and consumption of cannabis. Fourthly, opium production, while significant, is now highly concentrated in Afghanistan's southern provinces. Indeed, the Helmand province is on the verge of becoming the world's biggest drug supplier, with the dubious distinction of cultivating more drugs than entire countries such as Myanmar, Morocco, or even Colombia. Curing Helmand of its drug and insurgency cancer will rid the world of the most dangerous source of its most dangerous narcotic, and go a long way to bringing security to the region.

Another source of good news is that drug law enforcement has improved: almost half of all cocaine produced is now being intercepted (up from 24% in 1999) and more than a quarter of all heroin (against 15% in 1999) (United Nations, 2007).

Evolution of the World Drug Problem

The world's drug problem is being contained. In 2005–2006, the global markets for the main illicit drugs—the opiates, cocaine, cannabis, and amphetamine-type stimulants—remained largely stable (United Nations, 2007). Particularly notable is the stabilization seen in the cannabis market, which had been expanding rapidly for some time. In line with a long-term trend, the share of total drug production that is seized by law enforcement has also increased: some 42% of global cocaine production and 26% of global heroin production never made it to consumers. Of course, within this aggregated picture there remains considerable variation. Most notably, heroin production continued to expand in the conflict-ridden provinces of southern Afghanistan. While global heroin consumption does not appear to be growing, the impact of this surge in supply needs to be monitored carefully.

How Is Drug Production Changing?

Most of the world's drug markets start with the farmer. Unlike other crops, however, the cultivation of opium poppy, coca leaf, and cannabis take place under threat of eradication, and so the location and the number of hectares tilled vary substantially from year to year. Around 92% of the world's heroin comes from poppies grown in Afghanistan. Despite a massive increase in opium poppy cultivation in Afghanistan in 2006, the global area under poppy was actually 10% lower than in

2000. This decline was mainly due to sustained success in reducing cultivation in South-East Asia. Poppy cultivation in the Golden Triangle has fallen by some 80% since 2000 (United Nations, 2007).

Most of the world's cocaine comes from coca leaf cultivated in Colombia, Peru, and Bolivia. The global area under coca cultivation fell by 29% to some 156,900 hectares between 2000 and 2006, largely due to reductions of coca cultivation in Colombia. The areas under coca cultivation in Peru and Bolivia increased over this period but remained significantly below the levels reported a decade earlier (United Nations, 2007).

It is impossible to accurately estimate the location and total number of hectares under cannabis, because it is grown in at least 172 countries, often in small plots by the users themselves. There is an important distinction between the extent of drug crop cultivation and the extent of drug production, however. Crop yields can be affected by weather conditions and changes in production technology, among other things. As a result, long-term declines in cultivated area do not necessarily translate into declines in total production. Opium production in Afghanistan rose almost 50% in 2006, bringing global heroin production to a new record high of 606 metric tons (mt) in 2006, exceeding the previous high (576 mt in 1999) by 5%. Similarly, the success in the reduction of coca cultivation from 2000 to 2006 has not led to a commensurate decline in cocaine production, apparently due to improvements in coca cultivation and cocaine production technology. Cocaine production has remained largely stable over the last few years, estimated at 984 mt in 2006 (United Nations, 2007).

Amphetamine-type stimulants (ATS) are manufactured illicitly using legally-produced precursors, and thus global production can only be estimated indirectly. This production appears to be stable, however, at about 480 mt in 2005. At the same time, seizures of ATS labs and precursors declined dramatically, likely a result of improved precursor control and significant reductions in domestic production operations in key markets such as the USA.

■ Crime Scene Search

In a contraband drug investigation, the crime scene investigator is typically looking for evidence that has been hidden on the person, in a dwelling, or in a vehicle. When doing so, it is important to remember the rules of evidence and the requirements for establishing a chain of evidence. As with all aspects of evidence collection it is imperative that the crime scene investigator be aware of current laws regulating search and seizure activities (Fisher, 2004).

Collection and Preservation of Drug Evidence

All illicit substances should be photographed as they were found. All drugs should be accurately weighed. The gross weight of the package, including the drug and packaging materials, should be determined and recorded. Individual pills, tablets, packets, balloons, etc. should be counted and the number written on the outer package and within the written report.

Some departments/jurisdictions require that a presumptive field test be conducted on a substance that is believed to be a controlled substance, prior to its preservation and prior to charging the person in possession. This is typically performed through the use of a NARC ID kit or NARCO pouch, which is a portable chemical-color test for the presumptive analysis of a substance. This will then be followed up by a secondary presumptive test and confirmatory analysis at the forensic laboratory.

Forensic Analysis of Controlled Substances

Once at the forensic laboratory, the ID of controlled substances is divided into botanical and chemical examinations. Botanical exams identify physical characteristics specific to plants that are considered controlled substances. Chemical exams use wet chemical or instrumental techniques to identify specific substances that are controlled by statute. For each examination a series of tests is administered to the sample. Each test is more specific than the last. At the end of the sequence, the examiner is able to determine if there is a controlled substance in the sample and identify it.

Botanical Analysis

These are the most common examinations performed in the controlled substance section of a forensics lab. Examinations are typically associated with such substances as marijuana, peyote, mushrooms, and opium. Marijuana examinations typically exceed 50% of the forensic laboratory's caseload. As a rule, by education and training, the examiner is a chemist, not a biologist or a botanist. He/she has been trained in the ID of whether plant material is/is not marijuana, etc. Beyond that the examiner should not render an opinion as to the identity of the substance.

Marijuana Analysis

The analysis of marijuana is a two-step process. The first step establishes the plant or plan material as marijuana through its physical characteristics (microscopic exam). The second step is to establish the presence of the plant resin that contains the psychoactive components. A chemical color test is then used to confirm the presence of cannabinoids, specifically Delta 9 tetrahydrocannabinol (THC), which is the primary psychoactive compound.

This chemical color test is known as the Duquenois-Levine test. Additional confirmatory chemical and instrumental tests for marijuana resin include chromatographic exam, thin layer chromatography (TLC), and gas chromatography (GC) (Saferstein, 2007).

Chemical Analysis

The balance of the samples encountered by the controlled substance section requires the ID of specific compounds within a mixture. The composition of the samples may vary, but the procedure remains the same. Each sample requires a screening step, an extraction or sample preparation stem, and a confirmatory step. Tests can be further subdivided into wet chemical and instrumental procedures. Wet chemical procedures are used as a screening method or for sample preparation. Instrumental procedures are used for screening or as a confirmation tool.

Clandestine Drug Laboratories

Illicit controlled substances, such as those discussed previously, are often manufactured in clandestine locations, known as clandestine laboratories (**Figure 15.8**). The investigation of clandestine laboratories is one of the most challenging efforts of law enforcement. No other law enforcement activity relies on forensic science as heavily.

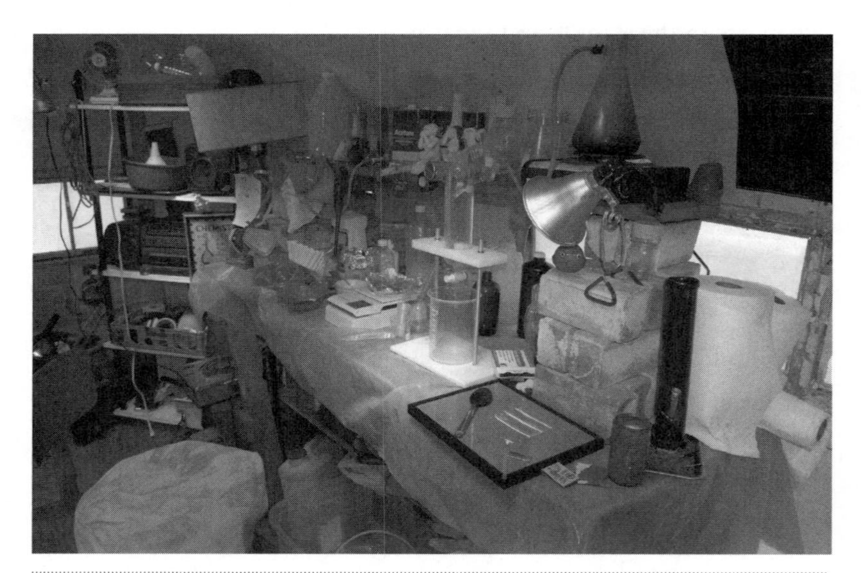

Figure 15.8 A Clandestine Methamphetamine Laboratory.

It is important in these situations that personnel take proper precautions while attempting to maximize physical evidence recovery efforts. Chapter 5 discusses the precautions that personnel must take to ensure their safety and the safety and integrity of the evidence collected within a hazardous environment.

Chapter Summary

Drug related evidence is quite prevalent at modern crime scenes. It is wise for a crime scene investigator to have a thorough understanding for the various types of drugs and controlled substances, and paraphernalia associated with each, indicators of use, and effects of their use. This information will greatly assist crime scene personnel with the correct manner in which on-scene documentation, collection, and preservation methods should progress. Because possession or use of the substance is the criminal offense, it is imperative that proper methods be followed to ensure the admissibility of evidence.

Review Questions

1. List the five classifications of drugs formed by the DEA and DHHS, and provide an example of each.
2. Explain what is meant by cycling, stacking, and pyramiding as they relate to steroid use.
3. ______________ are those substances (typically drugs) whose possession or use is regulated by the government.
4. Drug scheduling is conducted through a cooperative effort of the ______________ and the ______________.
5. All controlled substances, with the exception of ______________ are abused to alter mood, thought, and feeling through their actions on the central nervous system.
6. What does the acronym DAWN stand for and how does it affect drug evidence?
7. Which narcotic is also used in the treatment of narcotic addiction?
8. Rophies, roofies, and roach are all common names for which controlled substance?
9. What does the acronym GHB stand for and what are its effects?
10. The most frequently encountered illicit drug worldwide is ______________.
11. 3,4-Methylenedioxymethamphetamine is also known as ______________.

References

Drug Enforcement Administration (DEA). (2005). *Drugs of abuse.* Retrieved August 18, 2009, from http://www.usdoj.gov/dea/pubs/abuse/doa-p.pdf

Fischer, M. H. (1879–1962). Retrieved September 11, 2009, from http://www.quotationspage.com/quotes/Dr._Martin_Henry_Fischer/

Fisher, B. A. J. (2004). *Techniques of crime scene investigation* (7th ed.). Boca Raton, FL: CRC Press.

Johnston, L. D., O'Malley, P. M., & Bachman, J. G. (2003). *Monitoring the future national survey results on drug use, 1975–2002* (p. 520). Volume I: Secondary school students (NIH Publication No. 03-5375). Bethesda, MD: National Institute on Drug Abuse.

Saferstein, R. (2007). *Criminalistics: An introduction to forensic science* (9th ed.). Upper Saddle River, NJ: Pearson Prentice Hall.

Substance Abuse and Mental Health Services Administration. (2006). *Drug abuse warning network, 2006: National estimates of drug-related emergency department visits.* Rockville, MD: U.S. Department of Health and Human Services.

Substance Abuse and Mental Health Services Administration. (1999). *National Household Survey on Drug Abuse.* National Clearinghouse for Alcohol and Drug Information (NCADI). Retrieved September 11, 2009, from http://ncadi.samhsa.gov/govstudy/bkd376/TableofContents.aspx

United Nations. (2007). *World Drug Report 2007.* United Nations Office on Drugs and Crime, Retrieved September 11, 2009, from www.unodc.org.

U.S. Department of Health and Human Services, Office of Applied Studies. (2008). *National Survey on Drug Use and Health.* Retrieved September 10, 2009, from http://www.oas.samhsa.gov/nhsda.htm

U.S. Department of Justice. (November 29, 1990). Title 21 United States Code (USC) *Controlled Substances Act.* Retrieved September 11, 2009, from http://www.deaddiversion.usdoj.gov/21cfr/21usc/802.htm

Digital Evidence

Computers can figure out all kinds of problems, except the things in the world that just don't add up.

Isaac Asimov (1920–1992)
Author of 477 books and essays

▸▸ LEARNING OBJECTIVES

- Understand the forensic potential of digital evidence.
- Know how to properly document, collect, and preserve digital evidence.
- Understand how GPS technology can aid law enforcement with conducting investigations.

▸▸ KEY TERMS

Computer Network
Contraband
Digital Evidence
Instrumentality

■ Introduction to Digital Evidence

More and more frequently computers and other digital devices are being used as a component of criminal activity, and when such devices are found at a crime scene they should be regarded as a possible source of evidence. While the majority of agencies and departments cannot afford to train or equip personnel to specialize in the processing and analysis of digital evidence, it is important for a crime scene investigator to have a basic knowledge of the forensic potential of digital evidence, and how to proceed when confronted with such evidence. However, if at any point the investigator is unsure of the correct steps to take, a knowledgeable professional should be contacted.

Computers are being used to store records of drug transactions, money laundering, child pornography, prostitution, and many other criminal activities. The information presented in this chapter refers specifically to stand-alone digital components and computers. If the crime scene investigator suspects a network, mainframe, or some other type of system may be involved, then a computer specialist should be consulted prior to disturbing or seizing any of the components. Computer technology is constantly changing, and seizure methods may change in the future. Therefore, if the crime scene investigator has any questions, he or she should contact the forensic laboratory for technical assistance.

Criminal Use of Digital Equipment

The United States Department of Justice (USDOJ) categorizes computers that are used in a criminal manner in one of three ways: contraband, instrumentality, or "mere" evidence (USDOJ, 2008). If a digital device is found to be illegally possessed or for some legal reasons is illegal to possess, it is considered **contraband.** If a device, system, or its associated hardware played a significant role in the commission of a crime, then it is considered an item of **instrumentality**. Any system or device that is termed as "mere" evidence is oftentimes not seized, but instead the goal is the acquiring of the data that is of evidentiary value from the device, while adhering to computer forensic principles. However, before this is done, it is important to first examine the manner in which crime scene investigators and first responders should treat digital evidence.

What follows is information assembled by the USDOJ's National Institute of Justice (NIJ) in a special report entitled, *Electronic Crime Scene Investigation: A Guide for First Responders, Second Edition* (2008).

Digital Evidence

Digital evidence is information and data of value to an investigation that is stored on, received, or transmitted by an electronic device. This evidence is acquired when data or electronic devices are seized and secured for examination.

When dealing with digital evidence, general forensic and procedural principles should be applied:

- The process of collecting, securing, and transporting digital evidence should not change the evidence.
- Digital evidence should be examined only by those trained specifically for that purpose.
- Everything done during the seizure, transportation, and storage of digital evidence should be fully documented, preserved, and available for review.

First responders must use caution when they seize electronic devices. Improperly accessing data stored on electronic devices may violate Federal laws, including the Electronic Communications Privacy Act

Digital Evidence

- Is latent, like fingerprints or DNA evidence
- Crosses jurisdictional borders quickly and easily
- Is easily altered, damaged, or destroyed
- Can be time sensitive

Source: Information adapted from the National Institute of Justice Report, "Electronic Crime Investigation: A Guide for First Responders, Second Edition" (2008).

of 1986 and the Privacy Protection Act of 1980. First responders may need to obtain additional legal authority before they proceed. They should consult the prosecuting attorney for the appropriate jurisdiction to ensure that they have proper legal authority to seize the digital evidence at the scene.

In addition to the legal ramifications of improperly accessing data that are stored on a computer, first responders must understand that computer data and other digital evidence are fragile. Only properly trained personnel should attempt to examine and analyze digital evidence.

Handling Digital Evidence at the Scene

Precautions should be taken in the collection, preservation, and transportation of digital evidence. First responders may follow the steps listed below to guide their handling of digital evidence at an electronic crime scene:

- Recognize, identify, seize, and secure all digital evidence at the scene.
- Document the entire scene and the specific location of the evidence found.
- Collect, label, and preserve the digital evidence.
- Package and transport digital evidence in a secure manner.

Electronic Devices: Types, Description, and Potential Evidence

Internally attached computer hard drives, external drives, and other electronic devices at a crime scene may contain information that can be useful as evidence in a criminal investigation or prosecution. The devices themselves and the information they contain may be used as digital evidence. In this chapter, such devices are identified, along with general information about their evidential value.

Some devices require internal or external power to maintain stored information. For these devices, the power must be maintained to preserve the information stored. For additional information about maintaining power to these devices, please refer to the device manufacturer's Web site or other reliable sources of information.

Computer Systems

A computer system consists of hardware and software that process data and is likely to include:

- A case that contains circuit boards, microprocessors, hard drive, memory, and interface connections
- A monitor or video display device
- A keyboard
- A mouse
- Peripheral or externally connected drives, devices, and components

Computer systems can take many forms, such as laptops, desktops, tower computers, rack-mounted systems, minicomputers, and mainframe computers. Additional components and peripheral devices include modems, routers, printers, scanners, and docking stations. Many of these are discussed further in this chapter.

A computer system and its components can be valuable evidence in an investigation. The hardware, software, documents, photos, image files, email and attachments, databases, financial information, Internet browsing history, chat logs, buddy lists, event logs, data stored on external devices, and identifying information associated with the computer system and components are all potential evidence.

Storage Devices

Storage devices vary in size and the manner in which they store and retain data. First responders must understand that, regardless of their size or type, these devices may contain information that is valuable to an investigation or prosecution. The following storage devices may be digital evidence:

Hard Drives

Hard drives are data storage devices that consist of an external circuit board; external data and power connections; and internal magnetically-charged glass, ceramic, or metal platters that store data. First responders also may find hard drives at the scene that are not connected to or installed on a computer. These loose hard drives may still contain valuable evidence.

External Hard Drives

Hard drives can be installed in an external drive case. External hard drives increase the computer's data storage capacity and provide the user with portable data. Generally, external hard drives require a power supply and a universal serial bus (USB), FireWire, Ethernet, or wireless connection to a computer system (**Figure 16.1**).

Removable Media

Removable media are cartridges and disk-based data storage devices. They are typically used to store, archive, transfer, and transport data and other information. These devices help users share data, information, applications, and utilities among different computers and other devices (**Figure 16.2**).

Thumb Drives

Thumb drives are small, lightweight, removable data storage devices with USB connections. These devices, also referred to as flash drives, are easy to conceal and transport. They can be found as part of, or disguised as, a wristwatch, a pocket-size multi-tool such as a Swiss Army knife, a keychain fob, or any number of common and unique devices.

Memory Cards

Memory cards are small data storage devices commonly used with digital cameras, computers, mobile phones, digital music players, personal

Figure 16.1 An Example of an External Storage Device.

digital assistants (PDAs), video game consoles, and handheld and other electronic devices.

Storage devices such as hard drives, external hard drives, removable media, thumb drives, and memory cards may contain information such as email messages, Internet browsing history, Internet chat logs and buddy lists, photographs, image files, databases, financial records, and event logs that can be valuable evidence in an investigation or prosecution.

Handheld Devices

Handheld devices are portable data storage devices that provide communications, digital photography, navigation systems, entertainment, data storage, and personal information management.

Handheld devices such as mobile phones, smart phones, PDAs, digital multimedia (audio and video) devices, pagers, digital cameras, and global positioning system (GPS) receivers may contain software applications, data, and information such as documents, email messages, Internet browsing history, Internet chat logs and buddy lists, photographs, image files, databases, and financial records that are valuable evidence in an investigation or prosecution (**Figure 16.3**).

Peripheral Devices

Peripheral devices are equipment that can be connected to a computer or computer system to enhance user access and expand the computer's functions. The devices themselves and the functions they perform or facilitate are all potential evidence. Information stored on the device

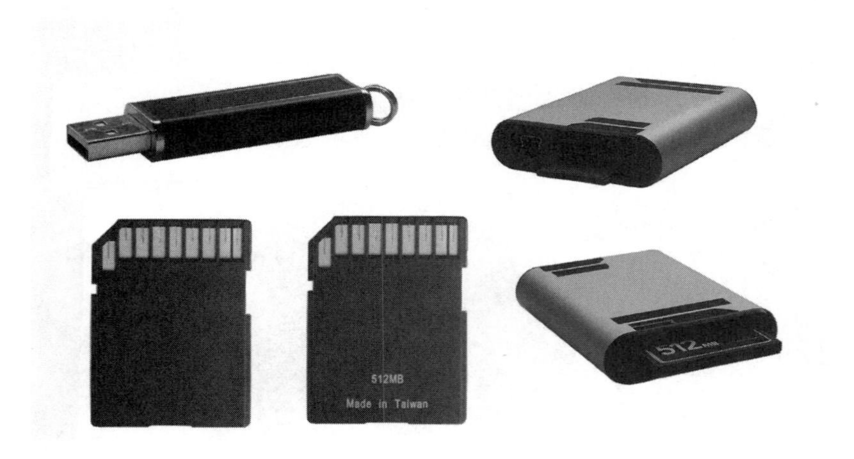

Figure 16.2 Examples of Media Cards.

regarding its use also is evidence, such as incoming and outgoing phone and fax numbers; recently scanned, faxed, or printed documents; and information about the purpose for or use of the device. In addition, these devices can be sources of fingerprints, DNA, and other identifiers.

Computer Networks

A **computer network** consists of two or more computers linked by data cables or by wireless connections that share or are capable of sharing resources and data. A computer network often includes printers, other peripheral devices, and data routing devices such as hubs, switches, and routers. The networked computers and connected devices themselves may be evidence that is useful to an investigation or prosecution. The data they contain may be valuable evidence and may include software, documents, photos, image files, email messages and attachments, databases, financial information, Internet browsing history, log files, event and chat logs, buddy lists, and data stored on external devices. The device functions, capabilities, and any identifying information associated with the computer system; components and connections, including Internet protocol (IP) and local area network (LAN) addresses associated with the computers and devices; broadcast settings; and media access card (MAC) or network interface card (NIC) addresses may all be useful as evidence.

Other Potential Sources of Digital Evidence

First responders should be aware of and consider as potential evidence other elements of the crime scene that are related to digital information, such as electronic devices, equipment, software, hardware, or

Figure 16.3 Example of a Portable Data Device.

other technology that can function independently, in conjunction with, or attached to computer systems. These items may be used to enhance the user's access of and expand the functionality of the computer system, the device itself, or other equipment.

The device or item itself, its intended or actual use, its functions or capabilities, and any settings or other information it may contain is potential evidence.

Investigative Tools and Equipment

In most cases, items or devices containing digital evidence can be collected using standard seizure tools and materials. First responders must use caution when collecting, packaging, or storing digital devices to avoid altering, damaging, or destroying the digital evidence. Avoid using any tools or materials that may produce or emit static electricity or a magnetic field as these may damage or destroy the evidence.

Should the complexity of an electronic crime scene exceed the expertise of a first responder, the first responder should request assistance from personnel with advanced equipment and training in digital evidence collection. The technical resource list at the Electronic Crime Partner Initiative (2008) Web site provides additional information for these situations.

Tools and Materials for Collecting Digital Evidence

In addition to tools for processing crime scenes in general, first responders should have the following items in their digital evidence collection toolkit:

- Cameras (photo and video)
- Cardboard boxes
- Notepads
- Gloves
- Evidence inventory logs
- Evidence tape
- Paper evidence bags
- Evidence stickers, labels, or tags
- Crime scene tape
- Antistatic bags
- Permanent markers
- Nonmagnetic tools

First responders also should have radio frequency-shielding material such as faraday isolation bags or aluminum foil to wrap cell phones, smart phones, and other mobile communication devices after they have been seized. Wrapping the phones in radio frequency-shielding material prevents the phones from receiving a call, text message, or other communications signal that may alter the evidence.

Securing and Evaluating the Scene

The first responder's primary consideration should be officer safety and the safety of everyone at the crime scene. All actions and activities carried out at the scene should be in compliance with departmental policy as well as federal, state, and local laws.

After securing the scene and all persons at the scene, the first responder should visually identify all potential evidence and ensure that the integrity of both the digital and traditional evidence is preserved. Digital evidence on computers and other electronic devices can be easily altered, deleted, or destroyed. First responders should document, photograph, and secure digital evidence as soon as possible at the scene.

When securing and evaluating the scene, the first responder should:

- Follow departmental policy for securing crime scenes.
- Immediately secure all electronic devices, including personal or portable devices.
- Ensure that no unauthorized person has access to any electronic devices at the crime scene.
- Refuse offers of help or technical assistance from any unauthorized persons.
- Remove all persons from the crime scene or the immediate area from which evidence is to be collected.
- Ensure that the condition of any electronic device is not altered.
- Leave a computer or electronic device off if it is already turned off.

Components such as keyboard, mouse, removable storage media, and other items may hold latent evidence such as fingerprints, DNA, or other physical evidence that should be preserved. First responders should take the appropriate steps to ensure that physical evidence is not compromised during documentation.

Evidence Collection

The first responder must have proper authority—such as plain view observation, consent, or a court order—to search for and collect evidence at an electronic crime scene. The first responder must be able to identify the authority under which he or she may seize evidence and should follow agency guidelines, consult a superior, or contact a prosecutor if a question of appropriate authority arises.

Digital evidence must be handled carefully to preserve the integrity of the physical device as well as the data it contains. Some digital evidence requires special collection, packaging, and transportation techniques. Data can be damaged or altered by electromagnetic fields such as those generated by static electricity, magnets, radio transmitters, and other devices. Communication devices such as mobile phones, smart phones, PDAs, and pagers should be secured and prevented from receiving or transmitting data once they are identified and collected as evidence.

Safeguarding Data

If information stored within a computer is incriminating, users can devise methods to destroy the data if an unauthorized person attempts to use the system. For this reason, it is essential that precautions are taken to safeguard the evidence when a computer is seized. When crime scene personnel discover a computer, they should immediately remove everyone from the area and not allow any further contact with the computer system, because a single keystroke can execute a program that erases information. Take a photograph of the screen to document any information that is displayed, and then immediately unplug the computer. Unplug the computer from the back of the central processing unit (CPU or main box), not from a wall receptacle. Uninterruptible power supplies (UPS) are common and can be programmed to execute destructive processes upon loss of power from a wall receptacle. Do not use the power switch, because it can be rigged to damage the hard drive or other components.

Do not execute any commands on the computer prior to pulling the plug. Commands, such as DIR and PARK can be altered to execute destructive processes. If there is a modem connected, unplug the modem cable from the wall. Leave the connections on the back of the computer undisturbed until they can be documented. Photograph the cable connections at the rear of the computer. Mark all cables with evidence tape or tags, and mark the corresponding ports on the computer in an identical manner. Photograph and document peripherals, and the marked connections on the back of the computer system prior to disconnection. This will facilitate the reassembly of the computer system and the peripherals in the office, courtroom, or at the laboratory. It is recommended that all computer hardware, software, disks, and manuals be seized. It is a reasonable and common practice to retain business records on computer media, in addition to the traditional (paper) method used to maintain and store business records and documents.

All Computer Hardware Should Be Seized

Conducting a search of a computer system, documenting the search and making evidentiary and discovery copies is a lengthy process. It is necessary to determine that no security devices are in place that could cause the destruction of evidence during the search, and in some cases, it is impossible even to conduct the search without expert technical assistance. Because it would be extremely difficult to secure the system on the premises during the entire period of the search, and computer evidence is extremely vulnerable, removal of the system from the premises will assist in retrieving the records authorized to be seized, while avoiding accidental destruction or deliberate alteration of the records. All peripherals should be seized, because some software programs are designed to work with specific hardware. Without all the components connected, it may be difficult, if not impossible, to recover and print the files of interest.

All Computer Software and Disks Should Be Seized

Specific software programs may be needed to view files stored on the system. Also, it would be impossible, without examination, to determine that disks purporting to contain standard, commercially available software have not been altered or used to store records instead. It is also possible that disks may contain information other than what is labeled, and unlabeled disks may contain files.

All Manuals and Pieces of Paper Found Near the Computer Should Be Seized

The analyst may have to refer to the manual and hand-written notes to operate the system and to recover the records authorized to be seized. Many programs can be password protected, and it is common for users to write passwords and other information in manuals and on scraps of paper.

Computers, Components, and Devices

To prevent the alteration of digital evidence during collection, first responders should first:

- Document any activity on the computer, components, or devices.
- Confirm the power state of the computer. Check for flashing lights, running fans, and other sounds that indicate the computer

or electronic device is powered on. If the power state cannot be determined from these indicators, observe the monitor to determine if it is on, off, or in sleep mode.

Computer/Digital Device "On"

If a computer is on or the power state cannot be determined, the first responder should:

- Look and listen for indications that the computer is powered on. Listen for the sound of fans running, drives spinning, or check to see if light emitting diodes (LEDs) are on.
- Check the display screen for signs that digital evidence is being destroyed. Words to look out for include "delete," "format," "remove," "copy," "move," "cut," or "wipe."
- Look for indications that the computer is being accessed from a remote computer or device.
- Look for signs of active or ongoing communications with other computers or users such as instant messaging windows or chat rooms.
- Take note of all cameras or Web cameras and determine if they are active.

Developments in technology and the convergence of communications capabilities have linked even the most conventional devices and services to each other, to computers, and to the Internet. This rapidly changing environment makes it essential for the first responder to be aware of the potential digital evidence in telephones, digital video recorders, other household appliances, and motor vehicles.

Computer/Digital Device "Off"

For desktop, tower, and minicomputers the first responder should follow these steps:

- Document, photograph, and sketch all wires, cables, and other devices connected to the computer.
- Uniquely label the power supply cord and all cables, wires, or USB drives attached to the computer as well as the corresponding connection each cord, cable, wire, or USB drive occupies on the computer.
- Photograph the uniquely labeled cords, cables, wires, and USB drives and the corresponding labeled connections.
- Remove and secure the power supply cord from the back of the computer and from the wall outlet, power strip, or battery backup device.

- Disconnect and secure all cables, wires, and USB drives from the computer and document the device or equipment connected at the opposite end.
- Place tape over the floppy disk slot, if present.
- Make sure that the CD or DVD drive trays are retracted into place; note whether these drive trays are empty, contain disks, or are unchecked; and tape the drive slot closed to prevent it from opening.
- Place tape over the power switch.
- Record the make, model, serial numbers, and any user-applied markings or identifiers.
- Record or log the computer and all its cords, cables, wires, devices, and components according to agency procedures.
- Package all evidence collected following agency procedures to prevent damage or alteration during transportation and storage.

Packaging, Transportation, and Storage of Digital Evidence

Digital evidence, and the computers and electronic devices on which it is stored, is fragile and sensitive to extreme temperatures, humidity, physical shock, static electricity, and magnetic fields.

The first responder should take precautions when documenting, photographing, packaging, transporting, and storing digital evidence to avoid altering, damaging, or destroying the data.

Packaging Procedures

All actions related to the identification, collection, packaging, transportation, and storage of digital evidence should be thoroughly documented. When packing digital evidence for transportation, the first responder should:

- Ensure that all digital evidence collected is properly documented, labeled, marked, photographed, video recorded or sketched, and inventoried before it is packaged. All connections and connected devices should be labeled for easy reconfiguration of the system later.
- Remember that digital evidence may also contain latent, trace, or biological evidence and take the appropriate steps to preserve it. Digital evidence imaging should be done before latent, trace, or biological evidence processes are conducted on the evidence.
- Pack all digital evidence in antistatic packaging. Only paper bags and envelopes, cardboard boxes, and antistatic containers

should be used for packaging digital evidence. Plastic materials should not be used when collecting digital evidence because plastic can produce or convey static electricity and allow humidity and condensation to develop, which may damage or destroy the evidence.

- Ensure that all digital evidence is packaged in a manner that will prevent it from being bent, scratched, or otherwise deformed.
- Label all containers used to package and store digital evidence clearly and properly.
- Leave cellular, mobile, or smart phone(s) in the power state (on or off) in which they were found.
- Package mobile or smart phone(s) in signal-blocking material such as faraday isolation bags, radio frequency-shielding material, or aluminum foil to prevent data messages from being sent or received by the devices. (First responders should be aware that if inappropriately packaged, or removed from shielded packaging, the device may be able to send and receive data messages if in range of a communication signal.)
- Collect all power supplies and adapters for all electronic devices seized.

Transporting the Evidence

Computers are delicate electronic instruments that are sensitive to temperature, physical shock, static electricity, and magnetic fields. When transported, care should be taken to ensure that computer components are well cushioned and protected. The original box and packaging material, if it can be located, is a good container for storing and transporting the computer. Magnetic media, such as computer hard drives, floppy disks, magnetic tapes, etc., are very sensitive to magnetic fields. When placing computer evidence in a car, remember that police radio transmitters are strong sources of magnetic fields. Do not place computer evidence in the trunk of a car if the trunk contains a police radio.

Cell Phones and Global Positioning Systems

There is yet another area of digital evidence that proves useful from an investigatory side, in a processing manner, and as a public service as well. Global Positioning Satellite (GPS) devices are available as handheld devices and are even included in most phones and personal data assistants (PDAs). Chapter 6 discussed how GPS technology can be utilized to document crime scenes; however, with the improvement of and increased access to GPS technology, law enforcement also has the resources to track individuals, corroborate information, and from a crime scene aspect, map and sketch large scenes.

Cell phone technology has become important in criminal investigations as people are relying more upon their phones for conducting business and personal activities, whether legal or illegal (**Figure 16.4**).

GPS built into cell phones allows authorities the potential to track criminals and people in need of help. Every time that a cell phone is turned on it sends a registration message, including the serial and phone numbers, to the closest cellular tower. A tower receives signals from cell phones on all sides of it. The tower then divides the area around it into three equal sectors. A GPS locator pinpoints the sector where the phone is calling. There are over 200,000 cellular towers placed across the United States.

Due to this technology, dispatchers can often deploy searchers within 100 feet of the caller's location, even if the caller is unable to ascertain his or her current position. If a caller makes use of an older phone without GPS, searchers can narrow down the caller's location by using three towers to triangulate a phone's last known spot by measuring the time that it takes signals to reach the towers. Law enforcement has three options for locating a person through his or her cell phone, dependent upon the scenario:

1. Single cell tower search
2. GPS tracking
3. Cell tower triangulation

Figure 16.4 A Type of Cellular Phone.

Most phones and digital devices allow the user to manually turn off the GPS feature; however, the device will automatically activate the GPS when a call is placed to 911.

Some ways that law enforcement currently makes use of GPS technology are as follows:

- Stolen cars that are equipped with GPS can be tracked and recovered. Instead of being involved in a police pursuit, which can create a greater risk to the public, a stolen vehicle can be located by law enforcement through the use of GPS technology. Once the stolen vehicle has reached a destination where law enforcement deems it is safe, those in the vehicle can be apprehended in a safer manner.
- Many jurisdictions have begun to use GPS bracelets on parolees as a way of monitoring their whereabouts. This technology can be used also to investigate and to enforce restraining orders through a similar method of having offenders fitted with a GPS bracelet.
- In recent times, law enforcement has made use of GPS-equipped vehicles and other items as bait in law enforcement sting operations, particularly with regards to investigating stolen vehicle crime rings, and also construction site equipment thefts.

Tracking Technology

A quick search of the Internet will yield hundreds of sites and devices that can enable someone to track a person or object through the use of GPS technology. Some of these devices are small enough to fit within purses, vehicles, or even small electronic items. Some have the purpose of legitimately tracking shipments, items, vehicles, or persons. Others, however, can be used for criminal purposes, including stalking.

The courts are somewhat split on whether or not such planting of GPS equipment is an invasion of 4th Amendment right to one's privacy. A 2009 ruling in the State of Wisconsin stated that GPS technology was simply another method of documenting a movement of a vehicle on a public roadway, rather than use personal observation, which would have otherwise been legal, and therefore a search warrant was not a requirement (Foley, 2009). During the same year, a court in the State of New York ruled that law enforcement must obtain a search warrant before such GPS tracking can occur (Chan, 2009). Hence, it is suggested to the reader to contact the district attorney of the jurisdiction in question, prior to implementing the use of GPS technology for investigatory purposes, such as tracking movement.

RIPPED FROM THE HEADLINES

Man Charged With Stalking Using GPS

The *Wisconsin State Journal* ran an article on February 15, 2008 reporting that the husband of a Madison, Wisconsin police officer had been charged with four felonies after he had secretly placed a GPS tracker in his wife's car and had used her name and password to access the police department's computer system.

The case involved the husband tracking the movements of his police wife as she was involved with an extramarital affair with another officer with whom she worked. At one point she told her lover that she believed that she was being followed by her husband and he suggested that perhaps her husband had placed a GPS tracking device within her car. The two searched her car and located such a device.

The use of digital technology to track his wife's whereabouts went deeper yet. After the stalking behavior was reported to law enforcement, police searched the husband's home computer and his computer at his place of employment. The home computer was found to contain key-tracking software that allowed the husband to identify his wife's login identity and password, and for him to then access the Madison Police department's scheduling system that contained work and vacation schedules. On his work computer, police found evidence that the husband had accessed his wife's personal emails and printed off communication between she and her law enforcement lover.

The husband was eventually found guilty of three felony counts of stalking and identity theft. As a side note, infidelity is a moral offense in the State of Wisconsin, with no criminal penalty.

Source: Information retrieved from Singletary, K. (2008, February 15). Man is charged with stalking. He's accused of putting a GPS device in the car of his estranged wife. *Wisconsin State Journal*.

There is also the possibility of utilizing Geographic Information Systems (GIS) to map criminal movement and activities. Most wireless phones have GIS technology embedded, which further adds to the ease of tracking movements. Again, it is suggested that prior to using such evidence, the district attorney be notified to see if a search warrant is a requirement or not.

Chapter Summary

We live in a digital world. Nearly all of our daily activities are captured, scheduled, or recorded on a digital device. This technology has enabled society to work more efficiently and has significantly reduced its reliance upon paper products and resources, thus proving more environmentally friendly. There are, however, drawbacks to this digital phenomenon. As with all aspects of life, criminals use this technology to further their activities, requiring that law enforcement stay abreast of the latest methods of digital evidence documentation, retrieval, and preservation. Failure to stay current on such matters will result in cases being lost. However, the very same technology that is a scourge also aids law enforcement and crime scene personnel through the production of new technology that expedites and makes more efficient the processing methodologies used while processing a modern crime scene.

Review Questions

1. If a device, system, or its associated hardware played a significant role in the commission of a crime, then it is considered an item of ______________.
2. ______________ is information and data of value to an investigation that is stored on, received, or transmitted by an electronic device.
3. A crime scene investigator needs to remember that digital evidence on computers and other electronic devices can be easily ______________, ______________, or ______________.
4. If an electronic device is discovered to be "on," what should be done before the item is unplugged?
5. Through the use of GPS technology, law enforcement is able to ______________, ______________, and ______________.
6. What are the three ways that law enforcement might use a cell phone to locate an individual?

Case Studies

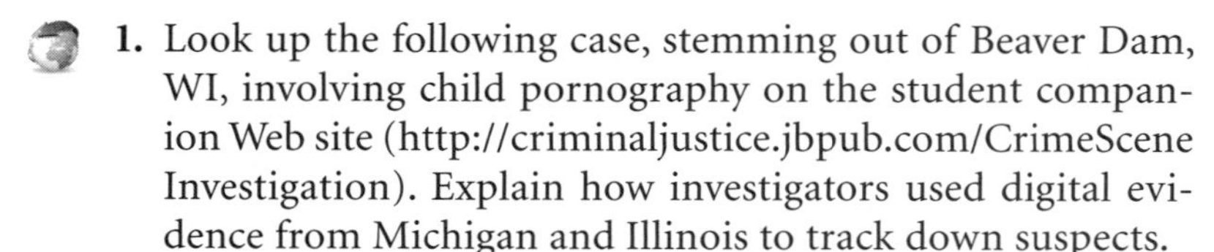

1. Look up the following case, stemming out of Beaver Dam, WI, involving child pornography on the student companion Web site (http://criminaljustice.jbpub.com/CrimeSceneInvestigation). Explain how investigators used digital evidence from Michigan and Illinois to track down suspects.
2. Access the following story regarding the use of cell phone technology on the student companion Web site (http://criminaljustice.jbpub.com/CrimeSceneInvestigation). Explain how digital technology was used during the investigative process and how absent this technology, the suspect may not have been convicted.
3. Research the name Tanya Rider, of Seattle, on the student companion Web site (http://criminaljustice.jbpub.com/CrimeSceneInvestigation) and explain how cell phone technology, along with good police work, is credited with saving her life and changing the focus of her missing persons investigation.

References

Chan, S. (2009, May 12). Police used GPS illegally, court rules. *New York Times*. Retrieved August 19, 2009, from http://www.nytimes.com/2009/05/13/nyregion/13gps.html

Electronic Crime Partner Initiative. (2008, April). *Technical resources* (2nd ed.). Retrieved August 19, 2009, from http://ecpi.us.org/Technical resources.html

Foley, R. J. (2009, May 7). Wisconsin court upholds GPS tracking by police. *Chicago Tribune*. Retrieved August 19, 2009, from http://archives.chicagotribune.com/2009/may/07/news/chi-ap-wi-gps-police

National Institute of Justice (2008, April). *Electronic Crime Scene Investigation: A Guide for First Responders* (2nd ed.). Retrieved August 19, 2009, from http://www.ojp.usdoj.gov.nij

Singletary, K. (2008, February 15). Man is charged with stalking. He's accused of putting a GPS device in the car of his estranged wife. *Wisconsin State Journal*. Retrieved August 19, 2009, from http://www.madison.com/archives/read.php?ref=/wsj/2008/02/15/0802140389.php

U.S. Department of Justice, National Institute of Justice (NIJ). (2008). *Electronic crime scene investigation: A guide for first responders* (2nd ed.). Retrieved August 19, 2009, from http://www.ojp.usdoj.gov/nij/topics/law-enforcement/crime-scene-guides/welcome.htm

Specialized Investigations

Death Investigation

You can take him to the mortuary now.

Sir Arthur Conan Doyle (1859–1930)
Sherlock Holmes, *A Study in Scarlet* (1887)

KEY TERMS

- Adipocere
- Algor Mortis
- Antemortem
- Asphyxiation
- Autopsy
- Biological Profile
- Cadaveric Spasm
- Cause of Death
- Cone of Foam
- Death
- Decomposition
- Excusable Homicide
- Exsanguination
- Homicide
- Hyperthermia
- Hypothermia
- Incision
- Justifiable Homicide
- Laceration
- Livor Mortis
- Manner of Death
- Mummification
- Penetrating Gunshot Wound
- Perforating Gunshot Wound
- Perimortem
- Postmortem
- Postmortem Interval (PMI)
- Putrefaction
- Rigor Mortis
- Stippling
- Tache Noire
- Taphonomy
- Tardieu Spots
- Vitreous Draw

LEARNING OBJECTIVES

- Describe the various types of homicides.
- Gain a foundation in forensic pathology and differentiate between cause and manner of death.
- Recognize the various contributors to the determination of a postmortem interval.
- Understand the various types of trauma and how each is identified and investigated.
- Know the correct methods of evidence documentation, collection, and preservation associated with a death scene.

Introduction to Death Investigation

For the purposes of this text, and most criminal investigations, **death** is the irreversible cessation of circulatory and respiratory functions. No crime scene and subsequent investigation will require as much from crime scene personnel as will a death investigation. Personnel will be called upon to use all aspects of their training and education in an effort to attempt to determine whether or not the death was natural or unnatural. However, not every death scene is a homicide, and not every homicide is a murder. It is important to understand the differentiation between the various classifications of deaths.

Challenges in the Investigation

Death investigations have added difficulty due to a number of challenges that exist in comparison to other types of investigations. One of the largest is pressure by the media and the public. The public have a desire and a right to feel safe. It is up to the police to ensure that the community is safe and that the reason for the death is determined in an effort to alleviate community concerns for future safety.

Sometimes this is not as easy as it would appear, however. Although there might be a death, there also might be difficulty establishing that the death is unnatural or that a crime has been committed. If the death involves advanced decomposition of remains or is particularly gruesome, then it may be difficult to identify the victim. Lastly, as will be discussed further within this chapter, establishing the cause, manner, and time of death is not always as simplistic as is shown in modern forensic television and movie dramas, and in many cases is particularly difficult.

Homicide

Homicide is the killing of one person by another. A basic requirement in a homicide investigation is to establish whether death was caused by a criminal action. Murder and homicide are not synonymous. All murders are homicides (and criminal); however, not all homicides are murders (or criminal). Although the term homicide is usually associated with crime, not all homicides are crimes.

Excusable homicide is the unintentional, truly accidental killing of another person. It is the result of an act that under normal conditions would not cause death, or from an act committed with due caution that, because of negligence on the part of the victim, results in death (i.e., running in front of a car).

Justifiable homicide is the killing of a person under authority of the law. This includes killing in self-defense or in the defense of another person if the victim's actions and capability present imminent danger of serious injury or death. It also includes killing an enemy during wartime, capital punishment, and deaths caused by police officers while attempting to prevent a dangerous felon's escape or to recapture a dangerous felon who has escaped or is resisting arrest.

Suicide versus Homicide

More Americans die by suicide than by homicide. Research studies have shown that 85% of suicides are premeditated and that 90% of those who take their lives communicate their intentions to someone they know (Bennett & Hess, 2001). Three basic considerations to establish if a death might be a suicide include:

1. The presence of the weapon or means of death at the scene.
2. Injuries or wounds that are obviously self-inflicted or that could have been inflicted by the deceased.
3. The existence of a motive or intent on the part of the victim to take his or her own life.

All death investigations should be treated as homicide investigations, regardless of the initial reporting or theory that the death is a suicide. This ensures that the investigation does not become narrowly focused or only documents evidence that supports the theory of suicide.

Preliminary Investigation

Regardless of what is reported, at any scene the first priority is to give emergency aid to the victim if he or she is still alive or to determine that death has occurred. Signs of death include lack of breathing, lack of heartbeat, lack of flushing of the fingernail bed when pressure is applied to the nail and then released, and failure of the eyelids to close after being gently lifted. Another quite obvious sign of death is advanced decomposition and related insect activity. If there are remains with insects coming and going from the body, it does not require one to be a physician to determine that the person is in fact dead.

In cases of sudden, unexpected, suspicious or unnatural death (homicide, accident, or suicide), the coroner/medical examiner (ME) must be notified and her or his representative will be responsible for the body remains and the subsequent death investigation. The crime leading to the death, and the investigation concerned with such remains the responsibility of law enforcement. It is imperative that the agencies work together in an effort to ensure the proper investigatory outcome. It is the responsibility of the coroner/ME to establish the cause and manner of death. It is in these instances when forensic pathology becomes a part of the investigative process. It should be noted that there is a difference between a "coroner" and a "medical examiner." A coroner is an elected official, second to the sheriff, within a jurisdiction. This is typically a sworn law enforcement position. The position of coroner does not necessarily require that a person be a physician, whereas a medical examiner is a physician and is not elected. Sometimes the position and person are one and the same; however, the terms mean different things and have different requirements for the position. **Table 17.1** shows the state-by-state breakdown of medico-legal investigation type within the United States.

TABLE 17.1 U.S. State Medico-Legal Investigation by Type

Type	Number	State
Medical Examiner System	**22**	
State Medical Examiners	19	Alaska, Connecticut, Delaware, District of Columbia, Iowa, Maryland, Massachusetts, New Hampshire, New Jersey, New Mexico, Oklahoma, Oregon, Rhode Island, Tennessee, Utah, Vermont, West Virginia
District Medical Examiners	1	Florida
County Medical Examiners	2	Arizona, Michigan
Mixed Medical Examiner and Coroner System	**18**	
State Medical Examiner and County Coroners/ Medical Examiners	7	Alabama, Arkansas, Georgia, Kentucky, Mississippi, Montana, North Carolina
County Medical Examiners/Coroners	11	California, Hawaii, Illinois, Minnesota, Missouri, New York, Ohio, Pennsylvania, Texas, Washington, Wisconsin
Coroner System	**11**	
District Coroners	2	Kansas, Nevada
County Coroners	9	Colorado, Idaho, Indiana, Louisiana, Nebraska, Nevada, North Dakota, South Carolina, South Dakota, Wyoming

Source: Data from Hanzlick, R. (2006). *Death investigation: Systems and procedures*. Boca Raton, FL: CRC Press.

Forensic Pathology

Forensic pathology is the branch of medicine that applies the fundamentals and knowledge of the medical sciences to the problems in the field of law and those that relate to public health and safety. It involves the investigation of sudden, unnatural, unexplained, or violent deaths.

In their role as coroners and MEs, forensic pathologists are responsible for answering the following questions:

- Who is the victim?
- What injuries are present?
- When did the injuries occur?
- Why and how were the injuries produced?

The primary role of the medical examiner is to determine manner of death and cause of death.

Manner of Death

The **manner of death** refers to the circumstances under which the cause of death occurred. The manner of death may be classified as natural or unnatural. A death is classified as natural when it is caused by disease. Other deaths are classified as unnatural and are given the following categories: homicide, suicide, accident, or undetermined, based on the circumstances surrounding the incident causing death.

Cause of Death

The **cause of death** is the injury or disease responsible for the pathological and physiological disturbances that resulted in death. In other words, it is the medical reason for death. If a cause cannot be found through observation, an autopsy is normally performed to establish the cause of death.

The Forensic Autopsy

An **autopsy** is the medical dissection and examination of a body in order to determine the cause of death. Sometimes there will be religious implications of such a dissection that must be addressed by investigative personnel. Some religions believe that desecration of the body is improper and unacceptable from a faith standpoint. However, although perhaps strongly believed, religion does not trump a criminal investigation regarding the collection and preservation of evidence. This can be extremely difficult to address, and must be handled in a professional manner, explaining the necessity of a thorough investigation.

A forensic autopsy entails the removal of internal organs through incisions made in the chest, abdomen, and head. It is customary for a forensic pathologist to prepare a written report of each autopsy examination. It is considered a gross exam if it deals with what is seen by the unaided eye. An exam is microscopic if it involves the examination of tissue under a microscope. Most reports of a gross autopsy consist of discussions of external examination, medical treatment, evidence on the body, evidence of injuries, dissection technique, diagnoses, and toxicology reports.

Time of Death Determination

No problem in forensic medicine has been investigated as thoroughly as that of determining the time of death on the basis of postmortem findings. An effort to find the moment of death for a person dates back to as early as 1247 AD with the first known forensic handbook written by Sung Tz'u (Sachs, 2001). In his book entitled, *The Washing Away of Wrongs*, Tz'u noted that decomposition rates change between seasons, even stating that winter decomposition rates will be five times slower than that of summer, and occurs at a different rate in heavier people.

Repeated experience teaches the investigator to be wary of relying on any single observation for estimating the time of death. Factors that help in estimating the time of death include:

- Hypostasis/postmortem lividity/livor mortis
- Rigor mortis
- Algor mortis
- Vitreous draw
- Appearance of the eyes
- Stomach contents
- Stage of decomposition
- Evidence suggesting a change in the victim's normal routine

Many methods remain in use today to help investigators estimate the time of death, which is usually referred to as the **postmortem interval (PMI)** of a victim. PMI is the time that has elapsed since a person has died. While only an estimate, it assists in narrowing the interval between death and discovery, providing a timeline for investigators to begin the search for the suspect. The PMI is estimated through various scientific observations of the changes that occur to a body after death. The scientific observations are based on the biological processes of the body when it is both living and dead. The heart circulates oxygen-rich blood throughout the body but also has the important role of removing bacteria and waste from the body. With death, the heart stops pumping, gravity causes blood to settle, and the bacteria remain in the body. Although their host is deceased, bacteria continue to thrive. These physiological events all contribute to the beginning of the decomposition process.

Hypostasis/Postmortem Lividity/Livor Mortis

Livor mortis is another aid in PMI estimations when combined with both algor mortis and rigor mortis. Also known as postmortem lividity, **livor mortis** is the visible color change that occurs from the pooling of blood once the heart stops pumping. The onset of lividity begins within a half an hour after death. Until the point in which lividity is set, blanching may occur, and the lividity may shift if the victim's body position is changed. Blanching can be noticed when applying pressure to an area where lividity is present; if the lividity temporarily disappears upon the application of pressure, lividity is not yet set, but becomes set once the blood clots. The time frame in which lividity sets varies, but typically occurs between 8 and 12 hours. After 8 to 12 hours, the blood will congeal within the capillaries or diffuse into the surrounding tissues, both of which will result in a lack of blanching being present, and the inability for displacement of blood to occur. In some cases of advanced stages of lividity, the capillaries may burst and cause what appears to be small, pinpoint hemorrhages, which are termed **Tardieu spots**. This is especially common in the lower extremities in situations involving hanging.

Lividity appears purple in color, which is caused by the dying, deoxygenated blood. The color of lividity presents several problems. In some instances lividity may be mistaken for bruising or injuries that occurred prior to death, and in other instances may be undetectable

Hypostasis/Postmortem Lividity/Livor Mortis

- The medical condition that occurs after death and results in the settling of blood in areas of the body closest to the ground.
- The skin will appear as a dark blue or purple color in these areas.
- Onset is immediate and continues for up to 12 hours after death.
- The skin will not appear discolored in areas where the body is restricted by either clothing or an object pressing against the body, but will instead lack coloring due to blanching.
- Can be useful in determining if the victim's position was changed after death occurred.

due to the color of a victim's skin or blood loss. Once again, because of the disparity in the time frame and the difficulty in detecting it, lividity is simply another tool to be used in combination with algor mortis and rigor mortis to estimate a time frame for death. Together, these three methods can be used to help investigators estimate the time of death, but these at the most are only present for up to 60 hours (Bennett & Hess, 2001).

Rigor Mortis

Rigor mortis, or the stiffening of the body, is another factor considered when estimating the PMI. Rigor mortis begins within one to four hours after death. Ultimately, rigor mortis is the contraction of body muscles; it begins in the smaller muscle groups and progresses to the larger groups, and may be found first in the jaw and neck. It is the direct result of chemical changes that occur within the body upon death. Adenosine 5′-triphosphate (ATP) nucleotide in cells is the chemical "energy" that allows muscles to contract and relax. Following death, the body's supply of ATP is depleted and lactic acid is produced, which results in the contraction of muscles. This chemical reaction is responsible for the fixation of a body in its position of rigor (stiffened muscles).

Because ATP is affected by muscle movement, there is no standard for the amount present in the body. Strenuous exercise, excitement, or a struggle before death will affect the amount of ATP present, which in turn directly affects the onset of rigor mortis. This is because muscles that were used extensively just prior to death will contain more lactic acid, which expedites the rigor process.

In temperate climes, rigor mortis becomes completely set by approximately 12 hours after death. It will then remain in place for approximately 12 hours, at which time it will begin to dissipate, disappearing after approximately another 12 hours. Rigor mortis has many other factors that affect the time of onset and duration in a cadaver, including temperature and body size. Warmer temperatures may accelerate the progression of rigor mortis, while colder temperature may significantly slow down the progression; rigor mortis may never develop in obese people but it may be expedited in lean people. These disparities clearly demonstrate that PMI estimations must be combined with other means in order to find a more accurate time frame.

Cadaveric Spasm

In some instances of death there is evidence of immediate rigor known as **cadaveric spasm**. This characteristic is present with no prior period of flaccidity and no extended onset. It typically involves the victim's

Rigor Mortis

- The medical condition occurs after death and results in the shortening of muscle tissue and the stiffening of body parts in the position they were in when death occurred.
- Appears within the first 24 hours and disappears within 36 hours.
- This can be useful in determining if the victim's position was changed after death occurred.

hands clenched around an object, such as a weapon, debris from a lake floor, clothing, or another object. Cadaveric spasm is sometimes associated with events such as drowning or homicide events that involved considerable excitement or tension preceding death.

Algor Mortis

The temperature of the body as well as the environment combined with other environmental and biological factors affect the rate of decomposition. Temperature has been a tool for determining the PMI since the early studies of the Greeks and the Egyptians more than 2,000 years ago (Sachs, 2001). In 1868, Dr. Harry Rainy from the University of Glasgow conducted many studies on the change of body temperature after death. While his research was rudimentary, he did monitor the temperatures of the dying and dead in the hospital and morgue until family members claimed the body, and through careful analysis devised a mathematical formula for the rate of temperature loss that is still used today (Sachs, 2001).

The body's loss of heat, or **algor mortis**, is based upon simple physics: heat loss will occur until the body reaches the temperature of the surrounding environment (ambient temperature). Numerous environmental factors can significantly alter the rate of heat loss. Humidity, wind, ambient temperature, body temperature at the time of death, surface the body is on, body position, body size and composition (fat acts as an insulator), and clothing all can affect the rate of heat loss of a body. It has even been suggested that the amount of blood that has been lost from a body may drastically affect this rate, because essentially the blood will be at the same temperature as the body and the less blood remaining in the body is one less source of heat (Sachs, 2001).

Forensic studies conducted by Rainy incorporated Newton's Law of Cooling, that heat loss is directly proportionate to the difference from the environment. Rainy concluded that the body will lose approximately one-tenth of its remaining heat each hour, using his equation derived from analyses of rectal temperatures. His study results showed that temperatures differed drastically depending on the location from where it was taken, and that a deep rectal temperature, close to the organs, was less likely to be affected by environmental conditions such as wind and humidity (Sachs, 2001). He then calculated that the time of death could be determined by the difference between the rectal temperature at the time of death and the environmental temperature (D), the rectal temperature after one hour (tI), and the rate of change between the initial rectal temperature and tI (R). Rainey's equation is:

$$\frac{\log D - \log tI}{\log R} = \times \text{ (time since death)}$$

Much like Sung Tz'u, Rainy stated that this time could not be exact but should be within a range of four hours (Sachs, 2001). He found that sometimes the body would not lose heat for up to four hours, and in some cases that the body would actually show a rise in temperature. This may be attributed to the enzymes within the body still breaking down sugars, starches, and fats before ultimately self destructing and converting into lactic acid. This process causes biochemical heat that ultimately may increase the body's temperature. This action in addition to bacteria remaining within the body suddenly thriving could account for the increase in postmortem body temperature (Sachs, 2001).

Although there are what seems to be endless amounts of variables that affect the body temperature, algor mortis is still widely used throughout the United States as a beginning estimation of time of death. Algor mortis is best used as a predictor within the first 10 hours after death, where the heat is lost at approximately 1.5°F per hour. This is based on the assumption that the body was at the normal internal temperature of 98.6°F and that the environmental temperature is between 70°F and 75°F. However, with so many varying climates throughout

Algor Mortis

- Postmortem changes cause a body to lose heat. The process in which the body temperature continually cools after death until it reaches the ambient or room temperature is algor mortis.
- It is influenced by factors such as the location and size of the body, the victim's clothing, and weather conditions.
- As a general rule, beginning an hour after death, the body will lose heat at a rate of approximately 1° to 1.5°F per hour until reaching ambient temperature. However, this is only an estimate!

the United States, a touch test is often used as a predictor. If the body is warm, death is within a few hours; if a body is cold and clammy, death occurred between 18 and 24 hours ago (Geberth, 1996). This method also may be combined with rigor mortis to give a closer time frame: if the body is warm and flexible, death is within a few hours; if warm and stiff, death is within a few hours and half a day; cold and stiff, death is within half a day to two days; and finally, if the body is cold and not stiff, death occurred more than two days prior (Sachs, 2001).

■ Vitreous Draw

A relatively recent method of assisting with determination of PMI is to use a syringe to take a sample of ocular fluid (vitreous humor) from the eye to determine potassium levels. This is referred to as a **vitreous draw**. Studies have shown that following death, cells within the inner surface of the eyeball will begin to release potassium into the ocular fluid. Through the collection and analysis of the amount of potassium present at various intervals after death, the forensic pathologist can determine the rate at which potassium is released into the vitreous humor. This allows the time of death to be approximated.

■ Appearance of the Eyes

The eyes are often the first part of the body to exhibit the earliest signs of postmortem change. This can include corneal clouding due to the eyes remaining open following death. A thin film can often be observed within minutes of death, converting to complete cornea clouding being observed within two to three hours, postmortem. If, however, the individual's eyes are closed, the corneal film development may be delayed by several hours, and corneal cloudiness may not be present for over 24 hours.

If the individual dies in an arid environment, and his or her eyes are open at the time of death, the exposed area of the eyeball (sclera) may develop a brownish-black line known as **tache noire** (black spot).

Eyeball collapse, resulting from absence of intraocular fluid within the eyeball, typically will occur after 24 hours; however, it may take as long as four or five days following death.

■ Stomach Contents

The theme of variance continues when examining stomach contents. Stomach contents have been examined in a hope to determine the PMI. This may even include the contents throughout the gastrointestinal tract. Like the heart, digestion halts after death occurs and, therefore, the amount of food in the stomach combined with the knowledge of when the victim's last meal was eaten should be able to present a narrow time frame for death. Digestion occurs as a process and the presence of food or digested materials in the stomach and small intestines can give a time frame for death. Digestion rates may have extreme variance from victim to victim, however, due to metabolism, activity prior to death, or narcotics.

■ Decomposition

The postmortem breakdown of body tissues is referred to as **decomposition**. During this process, the components that make up the tissues of the body will begin to leak and break down. The bacteria and other microorganisms present within the body (mouth, digestive tract, etc.) thrive on the newly unprotected organic components of the body, resulting in **putrefaction** (postmortem changes produced as a result of actions by bacteria and microorganisms).

Putrefaction

These decompositional changes are entirely dependent upon environmental conditions and the health of the victim prior to death. From an environmental standpoint, changes that occur within temperate climates in a matter of days may be evident in a warm environment within only hours. Additional considerations such as the victim's proximity to a heat source also may exacerbate putrefaction, while exposure to cold will significantly retard the process. Proximity to a heat source is not the only concern with regards to positioning and location of a victim. The rate of putrefaction is also dependent upon whether or not the victim is exposed to air, buried, or in water. In general, putrefaction is more rapid in air than in water, and in water is more rapid than in soil. Approximately one week in air is the equivalent of two weeks in water and approximately seven to eight weeks in a soil environment (Becker, 2005, p. 453). Putrefaction is more accelerated in obese victims and significantly slower in infants and thin individuals.

Under normal conditions, decomposition will initially be visually present within 24 to 30 hours. Early forms will manifest themselves as bluish-green discoloration of the abdomen region, due to the thriving bacterial present within the colon and digestive tract. The area of greatest discoloration is typically the lower right area of the abdomen.

The next phase of putrefactive changes to the body is gaseous bloating. As bacteria proceed to breakdown body tissues and organic material within the body, methane is produced, which causes certain areas of the body to swell and bloat. This is especially true of the abdominal area but often is found in the breasts and scrotum as well.

Next, dark purple and green discoloration of the face is common, along with purging of decomposition fluids from the orifices of the body. The discoloration of the face typically will spread to the chest and extremities within 36 to 48 hours after death. This period is characterized by venous marbling, a breakdown of the walls of the veins and arteries of the body with the seepage of blood into the surrounding tissues (**Figure 17.1**).

Decomposition will continue to progress with the separation of the epidermal and dermal skin layers, resulting in what is commonly referred to as gloving (on the hands) or stocking slippage (on the legs or feet) (**Figure 17.2**). This is especially true for bodies that have been immersed in water for a considerable amount of time, or were involved in second-degree burns.

Insect Activity

Insects are a foreign contributor to the process of decomposition, and yet their contribution is an increasingly effective manner of estimating the PMI of a victim. Forensic entomology is the study of insects as they relate to a criminal investigation. Forensic entomology is a relatively new area of study in determining the PMI that has been gaining credibility since the 1980s. Entomology may prove useful in identifying previous locations of a body, but most importantly, how long a body

Figure 17.1 Decomposition Showing Bloating, Veinous Marbling, and Abdominal Discoloration.

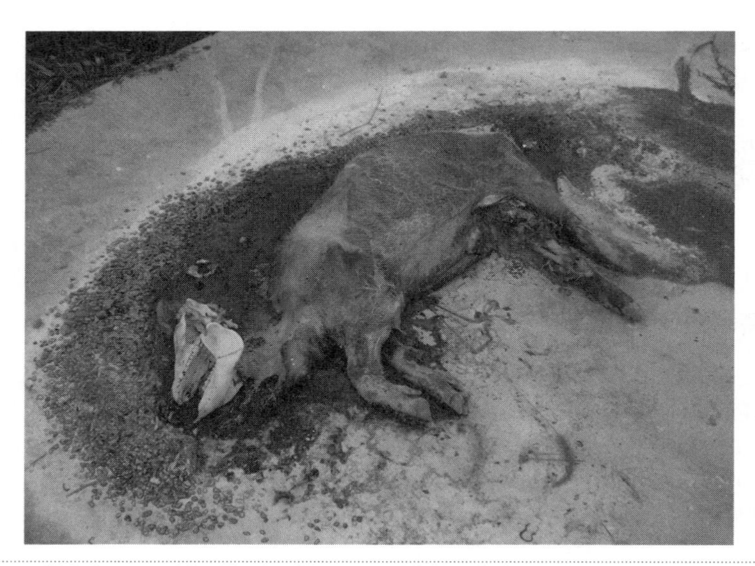

Figure 17.2 Decomposition Showing Slippage and Epidermis Separation.

has been dead. Through progression of insect eggs and presence on a body or even insect bite marks left postmortem, a PMI may start to be established. As a body continues to decompose, different odors attract different insects. Although the specific type of the insect may vary, insects within the same family tend to appear on a carcass at the same stage. The stages of decomposition harbor conditions that are good for different families of insects, which make the order of their appearance, regardless of the region, fairly predictable.

Insects found on a body are characterized into the following categories: (a) scavengers of the body, (b) predators and parasites of the scavengers and the body, (c) predators and parasites of only the scavengers, and (d) insects that use the body as a shelter. The first insects typically present on a body are flies, which will show up soon after death occurs and inhabit the body until the advanced decay stage. Soon after they arrive, the flies lay eggs on the body, typically in the corners of the eyes, mouth, and other mucous membrane areas, as well as in open wounds. Flies are still present through the advanced decay stage, but the fresh stage is the only stage in which the flies mate, which gives an important timeline towards the determination of a time of death for a carcass. A minimum time of death can be established once the eggs develop into maggots. The maggots will proceed to rapidly consume the soft tissues of the body and will concentrate on body openings and wounds or perforations to the body. Therefore, anytime a decomposing body is found with a high concentration of maggot activity in a certain area, it sometimes can be deduced that the area was most likely the site of a wound to the individual (defensive wounds on hands, gunshot wound [GSW] in chest, etc.). In cases of contact or close-range firearms injuries it is common for the maggots to consume the surrounding flesh but leave soot covered flesh or bone untouched.

The succession of insects continues as beetles and other predators, such as bees, arrive to prey on the maggots. This continues in a predictable manner, allowing forensic entomologists to determine the time of death of the victim by determining the insect present on the carcass and the current stage of development of the insect (**Figure 17.3**).

Many insect inhabitants occupy the carcass in overlapping periods with other insects, making the PMI estimation much easier. Although this method of PMI estimations has many benefits, insects are very sensitive to environmental changes as well as the manner of death. As with other areas of PMI estimation, temperature and humidity both affect the activity of insects on a body: while warmer temperatures accelerate insect activity, cooler temperatures retard insect activity, with activity even ceasing during periods of extreme cold.

Mummification

Mummification is the dehydration of soft tissues as a result of high temperatures, low humidity, and wind or other form of ventilation. The skin will appear brown, leather-like, and tight. The mummification process begins at the tips of the fingers and toes and progresses towards the hands

Figure 17.3 Insect Larvae Activity in a Cadaver.

and feet, face, and other extremities. Fingers found in a mummified state cannot be inked and fingerprinted. As discussed in Chapter 8, in such a case, the fingers first must be soaked in warm water so that the tissues will rehydrate, which allows them to be more pliable and closer to their original texture so that the ridge detail can be documented correctly.

In cases where mummification is allowed to become fully developed, a body will be preserved for a relatively long period of time. The rate at which mummification develops is a factor of environmental elements and conditions in the vicinity of the body, but often will require a postmortem interval of longer than three months to develop fully. This of course can be exacerbated by extreme heat and low humidity, for example, being present indoors during the winter months, with the furnace on, or if present in an arid, desert climate.

Adipocere

When a body is exposed to conditions of high humidity, and high temperatures, it will often exhibit signs of **adipocere**. Adipocere is the hydration and dehydrogenation of the body's fat, which results in an off-white, waxy, clay-like substance that in many cases preserves the body, and retards the decomposition process. It is especially common in the subcutaneous tissues of body extremities, the face, buttocks, breasts, and in individuals with a high percentage of body fat.

■ Evidence Suggesting a Change in the Victim's Normal Routine

Yet another indicator of postmortem interval involves evidence within a victim's home, place of business, car, etc. that suggests a an interruption or break in behavior, or aids in isolating specific dates. Examples include:

- Dated items within a fridge
- Newspapers or mail
- Answering machine/voicemail messages
- Computer activity (email)
- Receipts
- Calendars/date books

■ Taphonomy

In order to study human remains, knowledge about the human physical form and function must be combined with scientific input regarding postmortem changes. The combined information is essentially the postmortem history of the body, which is referred to as **taphonomy**. Examples of postmortem changes could include: normal decomposition, alteration and scattering by scavengers, and movement and modification by flowing water, freezing, or mummification.

The study of taphonomy requires input from forensic experts in a variety of fields. Essentially, all of the biological and geological sciences overlap in the study of taphonomy. Anthropologists, entomologists, botanists, biologists, geologists, archeologists, pathologists, and many others often are called to contribute to telling the story or giving the history of a body following death. Assembling a taphonomic history of a body involves accounting for information relating to the circumstances of the death event itself as well as the interval of time between death and discovery and collection of the body. This includes any modifications to the hard and soft tissues of a body by environment, activity, event, and scavengers. While it has already been discussed that it is a difficult process to collect accurate data and determine PMI, this difficulty is significantly compounded through the activities of weather, insects, rodents, and other animals. Small gnaw marks on bone by rodents sometimes can be mistaken for tool or weapon marks by an untrained eye. Cockroaches, ants, and other insects will sometimes leave postmortem bitemarks on a body that will be mistaken for defensive wounds or other perimortem-related trauma. This natural scavenging activity can add a greater degree of difficulty for the investigative process when the involved scavenger removes a bone, limb, or other portion of a body and drags it away to its burrow or den. Often these components will go undiscovered even after a thorough search.

One of the largest parts of a taphonomic history involves the interpretation of traumatic injuries to the body. It must be determined whether the injuries were made before death and were beginning the healing process, if they occurred at or near the time of death, or if they were injuries to the body after death.

■ Antemortem Trauma

Antemortem injuries are those sustained prior to death and that have healed or begun to heal. These can include injuries such as broken bones that have been cast or set, cuts that have been stitched or begun healing on their own, or other injury activities where evidence of healing is shown.

■ Postmortem Trauma

A body may be subjected to modification by an assortment of taphonomic methods after death. These injuries are referred to as occurring **postmortem**. These types of injuries can include carnivores, freezing and thawing, transport by flowing water, and many other events that may damage bone or soft tissue. Such postmortem damage is not related to the cause of death and must be differentiated from perimortem trauma. This is accomplished by noting the patterns of bone breakage and identifying modifications made by scavengers, plants, or other weather-related or geological processes.

Perimortem Trauma

Perimortem injuries are those that show evidence of having occurred at or near the time of death. These show no signs of healing, but do show signs that the body was still alive at time of injury (as evidenced by hemorrhaging, subcutaneous blood, etc.). This type of injury can occur with hard tissues (bones) as well. However, unlike soft tissue, bone will not exhibit vital reaction and requires several days of healing time in order to see evidence of healing. Therefore, bone damage that shows no signs of healing and for which no postmortem explanation exists, often is considered to be related to perimortem causes. There are several classifications of perimortem-related injuries. These include: mechanical trauma (sharp force injuries and blunt force injuries), thermal trauma, chemical trauma, and electrical trauma.

Sharp Force Injuries

A sharp force injury is a subcategory of mechanical trauma that refers to injuries received from sharp implements, such as knives, machetes, saws, and axes. The amount of mechanical force required for a sharpened instrument to exceed the tensile strength of tissue is significantly less than the force required with a blunt object. Sharp objects produce incised wounds, known as **incisions** (**Figures 17.4** and **17.5**). In deaths resulting from sharp trauma, the cause is typically due to **exsanguination**, death due to loss of blood/bleeding out.

Defensive Wounds

In cases involving sharp force injuries, it may be necessary to determine whether or not the person sustained defensive wounds during the altercation resulting in death. Persons who are beaten or stabbed to death typically will have defense wounds on the little finger side of the forearm and palm of the hand, sometimes resulting in the severing of fingers in sharp force injury instances. However, the absence of such injuries does not necessarily mean that the victim did not see his or her attacker or was suicidal. If the victim is restrained by physical or chemical means, then no defensive wounds will be present. One of the most common methods of chemical restraint is alcohol.

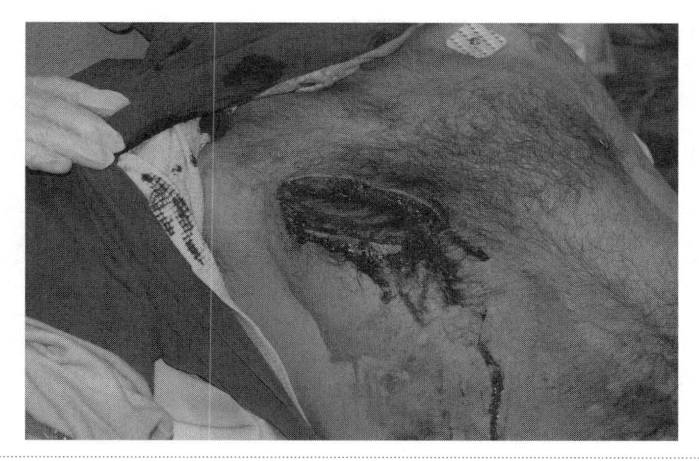

Figure 17.4 Example of an Incision Made from a Stabbing Incident Involving a Knife.

Blunt Force Injury

Another subcategory of mechanical trauma is blunt force injury. Blunt objects produce **lacerations**. Blunt traumas are further subdivided into nonfirearm and firearm groups. Typically, deaths due to blunt trauma are the result of significant damage to the brain or from laceration to the heart or aorta, which then leads to exsanguination.

Firearm Injuries

Injuries by firearm are the most common suicidal and homicidal wounds that are seen in the United States. They may be classified as close/near contact, intermediate range, or long range. Firearm injuries

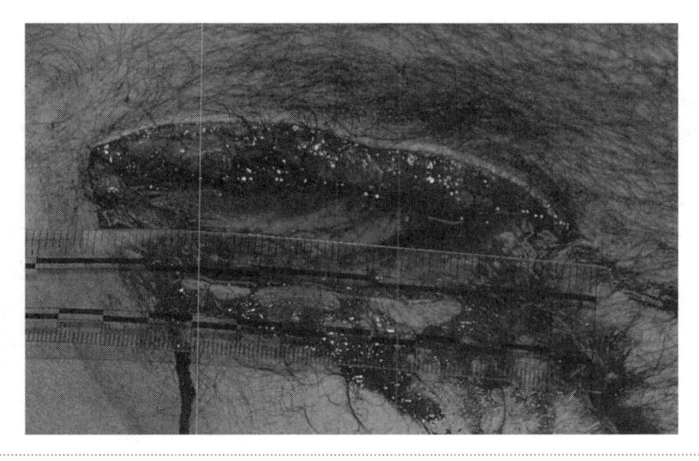

Figure 17.5 Scaled Photograph of the Incision in Figure 17.4.

Suicide versus Homicide Sharp Injuries

Suicide indicators:

- Hesitation wounds
- Wounds under clothing
- Weapon present, especially if tightly clutched
- Usually wounds at throat, wrists, or ankles
- Seldom disfigurement
- Body not moved

Homicide indicators:

- Defense wounds
- Wounds through clothing
- No weapon present
- Usually injuries to vital organs
- Disfigurement
- Body moved

can be further classified as penetrating or perforating. A **penetrating gunshot wound** (GSW) has an entrance wound but no exit and, therefore, a projectile should be recovered at autopsy for every penetrating GSW. A **perforating gunshot wound** has an entrance wound and an exit wound. This means that generally no projectile will be recovered by a pathologist at autopsy. This is important information to relay back to those responsible for processing the crime scene as a projectile should be recovered on scene for each identified exit wound.

Contact/Near Contact Injury. Firearms injuries that are made at contact or near contact range (less than 0.5 centimeters from the target) will exhibit signs of blackening and charring of the skin. Dependent upon the body location of the injury, the skin also may show stellate (star-like) lacerations that originate at and radiate away from the point of impact. These lacerations form because the gas blown into the wound results in the skin being torn apart. The gases associated with a firearm discharge also will cause red discoloration of the underlying tissue and hemoglobin present within a wound. This cherry red discoloration is similar to that found within the skin and organs in carbon monoxide–related deaths.

Intermediate Range Injury. As the distance increases between the barrel of a firearm to its target, the effect of the gas and associated unburned powder accompanying the event diminishes. Intermediate range injuries are considered to be those occurring at a distance of between 0.5 centimeters and approximately one to one and a half meters. At this distance, unburned powder that penetrates the skin produces a defect injury known as **stippling**. This injury occurs as the result of impact of burned and unburned particulates associated with the discharge of a firearm. They surround the bullet impact wound in a roughly circular pattern due to the fact that gunpowder is discharged in a conical pattern as it exits a firearm (as discussed in Chapter 13). This pattern is seen to enlarge as the muzzle to target distance is increased. Such stippling patterns are not typically observed at a range greater than 1 to 1.5 meters because the speed of the powder slows sufficiently so that it cannot penetrate the skin.

Distant Range Injury. A GSW that occurs at a muzzle-to-target distance greater than 1 to 1.5 meters will lack the characteristics exhibited by close contact and intermediate range injuries (i.e., soot damage and stippling). It is typical for a distant wound to exhibit a circular skin defect known as a marginal abrasion ring around the entrance wound edges. This is due to the stretching of the skin relating to the blunt force trauma.

Entrance and Exit Wounds. Conventional wisdom is that GSW exits are larger than entrance wounds; however, this is not always true. In most cases it is advisable to not make assumptions as to entrance or exit wounds at the scene of a crime or within a report. The autopsy process will determine which wounds are associated with an entrance or an exit of a projectile (**Figure 17.6**).

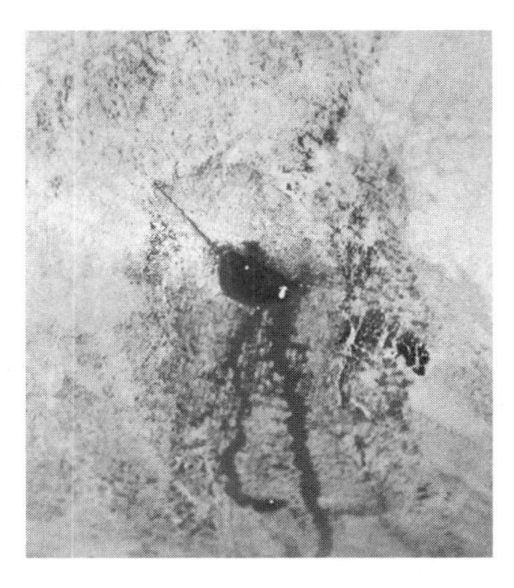

Figure 17.6 Gunshot Entrance Wound.

Suicide versus Homicide Gunshot Wounds

Suicide indicators:

- Gun held against skin
- Wound in mouth or in right temple if victim is right-handed and left temple if left-handed

Homicide indicators:

- Gun fired from more than a few inches away
- Angle or location that rules out self-infliction
- Shot through clothing
- No weapon present

Other Blunt Force Trauma Injuries

Another type of blunt force trauma injury encountered in crime scene investigations is that of a bumper fracture **(Figure 17.7)**. These types of blunt force injury are typically seen in motor vehicle versus pedestrian types of accidents, where the bumper of the automobile strikes the victim and causes lacerations and broken bones at the height of the vehicle. If the suspect car is found and the height of the bumper is measured off of the ground, it may be found to match the corresponding height of the injuries on the victim. If the measurement is less than the normal height of the bumper to ground distance, then that may be evidence that the vehicle was in the act of breaking upon striking the victim (and there should then be corresponding skid marks to corroborate that assumption). If the height of the injuries is longer than the height, then the vehicle may have been accelerating at the time of

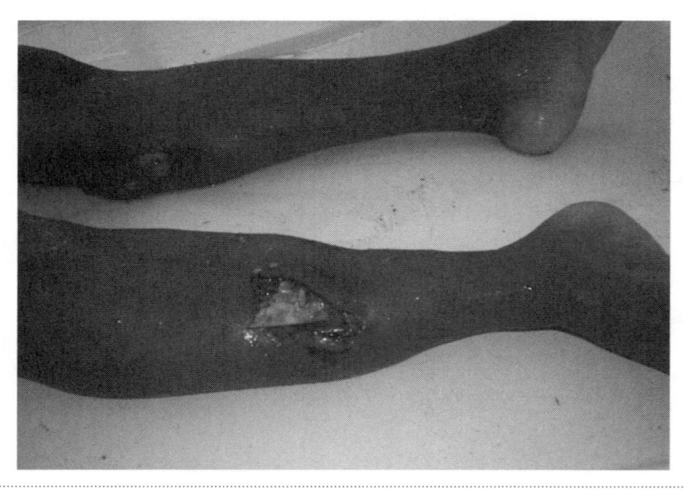

Figure 17.7 Example of Bumper Fractures.

impact. There is typically other evidence on scene to corroborate this and it is simply one way in which forensic pathology and investigation of injuries can aid the investigatory process.

■ Chemical Trauma Injuries

In approximately 50% of deaths, ethyl alcohol is a contributing factor to the death (James & Nordby, 2003). In most cases of alcohol-related deaths, it is alcohol-induced comas that result in the cessation of respiratory functions, resulting in death. However, as discussed in Chapter 14, another common chemical injury is that of carbon monoxide–related poisoning, resulting in death. Carbon monoxide–related deaths are characterized by the victim's cherry red skin.

■ Thermal Trauma Injuries

If an individual is exposed to excessive heat or cold the result can be death. **Hypothermia**, the lowering of the body's core temperature, results from exposure to excessive cold. **Hyperthermia**, a rising of the body's core temperature, is the result of exposure to excessive heat. Persons who die at the scenes of fires most commonly succumb from the inhalation of carbon monoxide and/or other products of combustion rather than thermal-related injuries.

■ Electrical Trauma Injuries

Electrical related deaths are quite rare, resulting in approximately 1,000 fatalities a year within the United States (Spitz, 1993). High voltage–related deaths are quite evident and present little difficulty for the examining pathologist. High voltage burns can result in poration, the flow of a current through tissues that creates holes in the membranes of the cells. This often leads to loss of limbs. High voltage burns are characterized by ferning marks on the body.

Whereas the passage of low voltage electricity through a person may cause cardiovascular-related difficulties, death can be due to the heart experiencing ventricular fibrillation that can lead to non-resuscitability within minutes. Low voltage may or may not produce electrical burns and will rarely exhibit signs of ferning.

■ Asphyxia-Related Deaths

Asphyxia is the interruption of oxygenation of the brain. There are a number of ways for this interruption to occur. These include, but are not limited to, drowning and strangulation.

Drowning

Drowning is death by asphyxiation from immersion in water or other liquid. There are several forensic ways to determine whether or not a body recovered from water died as a result of drowning or was dead prior to entering the water. At autopsy, the forensic pathologist can conduct a lung "float test." If a person died due to drowning, his or her lungs will be heavier and will sink when exposed to this test. There is also the possibility of conducting microscopic examination and analysis of bone marrow collected from the victim, to look for the presence of microorganisms known as diatoms, which are present within water and if ingested by the victim will find their way into the marrow. This can also prove or disprove drowning.

Manual Strangulation and Hanging

Asphyxial deaths associated with strangulation will often necessitate determining whether or not the event was self-induced or homicidal in nature. To determine this, the hyoid bone of the neck is typically examined at autopsy to see if fracture has occurred. Fracturing of the hyoid typically will be found only in cases of manual strangulation. Manual strangulation constricts the airway by compressing the neck or crushing the airway. Ligature strangulation, when from hanging or garroting, characteristically does not involve fracture of the hyoid. In addition to the lack of fracturing of the hyoid bone, there are ligature differences between hanging and strangulation-related deaths. Typically, the ligature marks made by hanging will be vertical, making a U-shape that extends under the victim's chin and around the back of the ears, whereas homicidal strangulation ligature marks typically will be horizontal marks extending from front to back, and may include an area near the back of the victim's neck where the ligature material was crossed, tied, or pressed against the victim to ensure the constriction of airflow. Hanging-related deaths typically involve the victim dying as a result of cutting off the blood supply (and relaying oxygen) to the brain through pressure placed on the carotid arteries, versus strangulation, which compacts or crushes the trachea, thus depleting the intake of oxygen and thereby causing death.

Homicide, Suicide, or Accidental?

In attempting to determine whether or not an asphyxial death was suicidal, accidental, or homicidal, the following is offered as suggestions. It must be remembered, however, that all deaths should be investigated as homicides until such point as the evidence determines otherwise.

- Most cases of choking, drowning, and smothering are accidental.
- Most cases of hanging are suicides.
- Most cases of strangulation are murder.

Deaths Due to Drug Overdose

In some cases of drug overdose and drowning, the victim will exhibit signs of froth emanating from the mouth and nostrils. This froth is sometimes referred to as a **cone of foam**. It is a result of severe pulmonary edema and may appear initially off-white but will advance to pinkish in color as the decomposition process advances.

Public Assistance in Death Investigations

Death investigations are complex and oftentimes confusing undertakings. In some cases, it may be necessary to seek the assistance of the public in attempting to gain information that can assist in determining the activities associated with a death or the identity of a recovered victim. In these cases, a **biological profile** is developed that is distributed to the media through law enforcement channels. A biological profile is assembled by studying the remains and noting characteristics of shape and size, which may allow an estimation of height, build, age, sex, ancestry, and any individualistic features such as tattoos, jewelry, medical apparatus, and clothing. Stature is estimated by measuring total body (or skeletal) length. Unique antemortem characteristics such as a healed bone fracture or an unusual dental configuration are also included. The goal of developing a profile is to describe the individual in such a way that law enforcement or acquaintances can narrow the range of possible identities.

This process is used extensively where prisoners of war (POW) and missing in action (MIA) service members are concerned. Hickam Air Force Base, Hawaii, is home to the Joint POW/MIA Accounting Command. According to the associated Web site (http://www.jpac.pacom.mil/index.php?page=mission_overview), the "mission of the Joint POW/MIA Accounting Command (JPAC) is to achieve the fullest possible accounting of all Americans missing as a result of the nation's past conflicts. The highest priority of the organization is the return of any living Americans that remain prisoners of war." This mission dates back to World War II. While most countries throughout the world bury their war dead where they fall or are found, the United States vows that it will do everything in its power to bring service members remains home (JPAC, 2009).

This promise results in joint-service teams traveling to burial and crash sites around the world. Once teams have recovered remains, and they are repatriated within the United States, the command's forensic laboratory attempts to identify them. The laboratory portion of JPAC, referred to as the Central Identification Laboratory (CIL), is

RIPPED FROM THE HEADLINES

Reality Television Meets Real-World CSI

In August of 2009, a man was searching through a rubbish bin in Buena Park, Orange County, California, looking for recyclables. What he instead discovered was the body of a young female's remains packed into a suitcase. This find would eventually launch an international investigation involving a reality television star and crime scenes in both the United States and Canada.

On August 14, the body of Jasmine Fiore, an aspiring actress and swimsuit model, was found within a suitcase. At the time, investigators were unaware of the identity of the remains. The remains were badly mutilated and it was discovered that the female's fingers and teeth had been removed, most likely to prevent identification. Identification was made by investigators through the serial numbers on the victim's breast implants, positively identifying the victim as Jasmine Fiore.

Ms. Fiore, 28, had married Ryan Jenkins, 32, a U.S. reality television star, in a quickie Las Vegas wedding ceremony earlier the same year. The investigation revealed that the couple checked into a San Diego hotel on August 13, and that Mr. Jenkins checked out alone the next morning. Mr. Jenkins immediately became a person of interest to the investigation. Authorities were initially unable to locate the whereabouts of Mr. Jenkins, although his boat was found in a marina not far from the U.S.–Canada border, just south of Vancouver. The body of Ryan Jenkins was eventually discovered in a motel room in British Columbia on August 23, where he had apparently killed himself.

the largest forensic anthropology laboratory in the world. The lab consists of state-of-the-art instrumental and analytical techniques to determine identity from the remains. DNA sampling and other cutting edge forensic equipment is advancing JPAC experts ability to make positive identifications that were once impossible.

According to JPAC, since its inception, on average, JPAC identifies about six MIAs each a month. To date, the U.S. government has identified over 1,300 individuals. At any given time, there are more than 1,000 active case files under investigation (JPAC, 2009).

Chapter Summary

Postmortem interval estimations are extremely important in death investigations. A time frame can be established in order to aide investigators with their search for the suspect. Numerous methods are used to narrow down the PMI and all are affected by environmental changes as well as many other variables. The PMI cannot currently be determined by the use of any one method, but a more accurate range may be determined through a combination of all the methods and consideration for environmental conditions.

PMI is not the only consideration when investigating a death. One must also consider the cause and associated manner of the death. This can include analysis of the circumstances surrounding the death as well as the injuries associated with the death. In looking at the history of the body after death, and the history of the person preceding his or her death, one will be able to best determine whether or not the death was natural or unnatural.

Review Questions

1. The irreversible cessation of circulatory and respiratory functions is known as ______________.
2. Explain the differences between excusable homicide and justifiable homicide.
3. Explain the differences between cause of death and manner of death.
4. What is blanching and how is it relevant to death investigations?
5. The postmortem breakdown of body tissues is referred to as ______________.
6. Explain the differences between antemortem, perimortem, and postmortem injuries.
7. Knowledge about the human physical form and function combined with scientific knowledge about postmortem changes as applied to the study of human remains is known as ______________.
8. Death due to a loss of blood is known as ______________.
9. ______________ is the interruption of oxygenation of the brain.
10. What is PMI and how is it done?
11. Which chemical is looked for in a vitreous draw?

References

Becker, R. F. (2005). *Criminal Investigation* (2nd ed., p. 453). Sudbury, MA: Jones and Bartlett Publishers.

Bennett, W. W., & Hess, K. M. (2001). *Criminal investigation* (6th ed.). Belmont, CA: Wadsworth.

Doyle, A. C. (1887). *A Study in Scarlet.* London, United Kingdom: Ward Lock & Co.

Geberth, V. J. (1996). *Practical homicide investigation: Tactics, procedures, and forensic techniques* (3rd ed.). Boca Raton, FL: CRC Press.

Hanzlick, R. (2006). *Death investigation: Systems and procedures*. Boca Raton, FL: CRC Press.

James, S. H., & Nordby J. J. (2003). *Forensic Science*. Boca Raton, FL: CRC Press.

Joint POW/MIA Accounting Command, United States Department of Defense. (2009). *Mission overview*. Retrieved September 11, 2009, from http://www.jpac.pacom.mil/index.php?page=mission_overview

Sachs, J. S. (2001). *Corpse: Nature, forensics, and the struggle to pinpoint time of death*. Cambridge, MA: Perseus Publishing.

Spitz, W. U. (1993). *Medicolegal investigation of death* (3rd ed.). Springfield, IL: Charles C. Thomas Publishing.

Special Scene Considerations

Law enforcement agencies must encourage their officers to see the benefits of exploring new methods of investigating underwater crime scenes and not rely solely on past policies and procedures.

Ronald F. Becker (2000)

LEARNING OBJECTIVES

- Understand the safety concerns that apply whenever processing an underwater crime scene.
- Understand the environmental concerns and challenges when documenting and processing an underwater crime scene.
- Describe black water diving including its dangers and what causes it.
- Know the proper documentation, collection, and preservation methods associated with underwater evidence and underwater crime scenes.
- Describe the duties of a DMORT and know why and when one is utilized.
- Know the proper documentation, collection, and preservation methods associated with both buried and scattered human remains.

KEY TERMS

Black Water Diving
Buoyancy Control Device (BCD)
Cadaver Dogs
Disaster Mortuary Operational Response Team (DMORT)
Probing
Underwater Search and Evidence Response Team (USERT)

Underwater Crime Scene Response

Underwater Investigative Teams

The number of people making use of recreational waterways across this country has dramatically increased in recent years. A corresponding increase in the number of accidents, drownings, and violent crimes occur in such settings. Criminals also have found these watery estuaries useful in attempting to conceal evidence of their crimes. These increased incidents have caused law enforcement agencies to become more involved in underwater recovery operations.

When an investigation leads to an underwater environment, it becomes an underwater criminal investigation. Its process should be as forensically thorough and systematic as any crime scene process conducted out of water. However, unlike the separation of duties often found in land-based investigations, underwater scenes typically call for the investigator to become the crime scene photographer, fingerprint specialist, and evidence recovery specialist. They also call for the investigator to be a skilled diver. This is an extremely specialized position requiring highly specialized training and equipment.

The first documented dive team dedicated to investigative purposes was the Miami-Dade County Underwater Recovery Unit, which was established in 1960. The Miami-Dade team was assembled due to the increasing number of water-related crime scenes occurring within the water laden area of Miami-Dade County (Becker, 2005).

Federal agencies also found reason for training and equipping such teams as a result of national and world situations. The FBI established its first **Underwater Search and Evidence Response Team (USERT)** in 1982 (FBI, 2009). There are currently four such teams in the United States and they are located in New York, Washington, Miami, and Los Angeles. The teams are managed from the FBI Laboratory in Quantico, Virginia. Each team is composed of twelve members; however, divers respond and fill in on other teams as necessary. The USERTS have been used in a number of newsworthy events that include: searching for evidence and remains in the explosion of TWA Flight 800; preventative diving to search for explosives under the spectator stands for the 1996 Atlanta Olympic Games; and also overseas in such terrorist attacks as that on the 2000 bombing of the USS Cole (Becker, 2005).

While the FBI maintains several well trained and extremely well equipped dive teams, many municipal, county, and state agencies across the country also have realized the necessity and benefit of implementing diver operational teams.

Why Is an Investigative Dive Team Necessary?

A common misconception is that evidence that has been submerged lacks the forensic value of evidence found top-side (**Figure 18.1**). There are numerous cases where submerged items have yielded identifiable blood evidence, fingerprints, hair and fiber evidence, and many other types of trace evidence that have aided in identifying the perpetrator of the criminal act. Some people would question why a body would have to be handled any differently in water than out of water. However, bodies should be placed in body bags when found underwater because correctly processing and packaging it can result in a variety of

Figure 18.1 Forensic Divers Prepare to Recover Crime Scene Evidence.

forensic evidence being saved. Bagging a body underwater minimizes damage and postmortem injuries to the body, maintains hair and fiber evidence, and keeps clothing intact.

Investigators must understand the value and necessity of properly collecting, marking, and recording the location of items recovered from a dive site. Often investigators, administrators, crime lab personnel, and prosecutors lack the knowledge and training about the underwater investigative process. Because of this deficit, they are unfamiliar with the forensic value and scientific methodology involved with underwater crime scene processing. For instance, if only a diver is aware of the evidentiary value of collecting control samples from the water and bottom material, then investigators may not understand or request the steps to be done. Collection of proper control and elimination samples for exclusion of background and marine debris is as important below the water as it is above. Because the ultimate goal of any underwater crime scene is not merely the salvaging and recovery of submerged items, but discovery of the truth and conviction of the perpetrators, dive teams must adapt to accomplish such tasks.

Where Is Investigative Diving Used?

The working environment for the underwater investigator is much different than those shown on television or in movies. A typical work environment could consist of diving in a quarry, dirty river, raw sewage facility, or any of a host of other less than hospitable environments. Investigative diving can occur in any environment, at any time of the day, and in any area involving a body of water. In the vast majority of these cases, the visibility confronting the forensic diver will be less than three to four feet. In some instances, the visibility will be so drastically reduced that a diver's gauges will not be readable or even visible. This is referred to as **black water diving**. Unlike in a land-based criminal investigation where one or two persons could effectively work some scenes, in a black water diving environment, the underwater investigator relies upon the coordinated efforts of the dive team to ensure both survival and success. This dive team includes both the surface personnel and fellow underwater investigators (**Figure 18.2**).

Who Should Comprise the Team?

Depending upon the location, there are volunteer teams, teams of fire and rescue personnel, law enforcement teams, and in some cases a collaboration of all of the above. All types are involved in attempting to fill the niche of "underwater investigator." The problem is that when dealing with a potential crime scene, the question shouldn't be "Who can help?" Rather, the question should be, "Who *should* help?" Except in very few instances (e.g., an arson investigation), fire departments do not investigate or become involved with criminal investigations on land. Therefore, why should they participate or why are they qualified to investigate crime scenes when the environment has changed to

Figure 18.2 Dive Team Members Processing an Underwater Crime Scene.

underwater? Law enforcement does not allow fire or rescue personnel to investigate a homicide occurring within a residence. So why then allow them to investigate a crime scene underwater?

It must be realized that in some instances, if fire departments do not train their personnel to respond to underwater scenes then nobody will. However, if a need has been documented and recognized, then law enforcement agencies should not ignore the responsibility and defer to fire or rescue personnel. The responsibility for investigating crime scenes belongs with the criminal justice community.

Team Design

There are many different theories and operational plans for the makeup of an underwater investigation team. However, at the bare minimum, the following personnel are required:

Team Leader/Commander

As in any operational response performed by law enforcement, it is crucial to have an incident commander for each scene. This duty would fall to the Team Leader/Commander. This individual is not in the water and is responsible for supervising the diving process and ensuring that the individuals involved are conducting their tasks in a safe and proper manner. This individual is also responsible for ensuring that the members acquire the equipment and training necessary to conduct their jobs, as well as documenting all training, medical, and other personnel issues.

Line Tenders

Due to environments of low visibility, fast moving water, and other hazardous situations, divers are typically tethered to a line that is tended by someone above the water. These individuals are the eyes and ears for the forensic diver. It is preferable to have a tender be a current or former diver so that he/she is familiar with the underwater environment. In some jurisdictions, underwater communication gear has been implemented that is similar to those worn by tactical teams. However, even in these cases, and certainly in those where such communication gear is not possessed, communication between the tender and a diver occurs through a process of predetermined line tugs. This communication can involve such matters as changes of direction, status checks, ascents, and emergencies. Line communication varies by team, but must be trained on and understood by all team members.

Divers

There is no set number of divers for an operational dive team, but it is recommended that at the very minimum there be two divers for any dive situation. The primary diver is the person who is in the water and actively searching or investigating the scene. The safety diver (sometimes referred to as a 100% diver) is either in the water or out of the water and fully geared up with the ability to enter the water to provide emergency backup or aid to the primary diver. In some situations, a backup diver (90% diver) may be used as the backup to the primary diver and as a safety diver should the safety diver find him or herself deployed.

The team set up is similar to the arrangement law enforcement and fire agencies typically utilize in dismantling clandestine laboratories. This is the case due to the similarities between the two environments. In both cases personnel are confronted with hazardous and inhospitable environments that require the use of respiratory and exposure equipment. Both situations are physically taxing on an individual and require that there be constant monitoring and documentation to avoid overexertion or other potentially fatal situations. It is for this reason that both situations involve the use of safety or standby personnel **(Figure 18.3)**.

Assembling, Equipping, and Training the Team

It is suggested that all members of an agency dive team be full-time, trained members of that agency for liability, training, and policy and procedure purposes. However, some smaller agencies with limited resources require volunteers to supplement their dive recovery operations. Other

Figure 18.3 Members of a Dive Team Conduct a Search for a Missing Person.

departments sometimes share equipment and members. This is often pointed to as being "cost-effective"; however, it can prove problematic when divers from one location fail to respond, or when the necessary equipment is in another location other than where it is needed. Some locations have solved these problems by establishing specialized regional dive teams. One team (perhaps fire department personnel) is trained in and specializes in rescue diving while another team (law enforcement) specializes in evidence search, recovery, and collection. This minimizes evidence handling and ensures that the necessary equipment is operationally available.

There must be minimum entry requirements established for team membership. Many times this minimum certification is a recreational certification such as the Professional Association of Dive Instructors (PADI), National Association of Underwater Instructors (NAUI), Scuba Schools International (SSI), and/or other recognized organizations. Often teams will write their own standard operational guidelines (SOGs) that list the minimum entry requirements and establish training and equipment requirements. As a point of caution in this world of lightning litigation, all established requirements for a dive team must be job-related. If the requirements are not shown to be specifically related to the job applied for or involved in, they could be challenged as being discriminatory based on age, race, gender, or disability.

In addition to being (at a minimum) recreationally certified, all applicants should be required to pass a thorough and appropriate medical physical to verify medical ability to perform tasks and to set a baseline for future medical exams. After joining the team, all team members should undergo annual dive physicals. Dive physicals should be administered also following any dive-related injury, major injury or illness, or any medical procedures that could interfere with the performance of duty.

The merits and benefits of fielding an underwater investigation team are many; however, with this comes the equally involved and elaborate budget and steps necessary to do so. This should not dissuade law enforcement from the assembly of such resources to ensure that each scene is processed in its entirety to the best of the agency's ability. All too often underwater crime scenes—and the evidence contained therein—are not properly documented or processed. This failure can result in a loss, contamination, or inadmissibility of the evidence. Proper underwater crime scene processing and documentation methods will ensure that the evidence is located, properly documented, preserved, and able to be used in subsequent litigation. It is important that the underwater investigation team be properly trained as well as take part in ongoing training specifically relating to matters of crime scene documentation and processing methods involving the underwater environment. Remember this author's adage: "What is not looked for will not be found. What is not found cannot be analyzed to uncover the truth."

Today's crime scene investigator is well versed in the methodology and logistics associated with crime scene processing and documentation. However, what if the crime scene in question was located underwater? Would the same practices and methodologies apply? Would the same equipment and personnel be appropriate?

Oftentimes, the standard operating procedures (SOPs) associated with processing crime scenes are forgotten or ignored when encountering an underwater crime scene. There are many reasons for this occurring, some of which include equipment, manpower, and/or environmental issues. More often it is a result of improperly trained or equipped personnel or a case of rushing and thinking that, "it won't really matter as long as we get the stuff." However, the underwater scene and its contents are equally important and subject to the same scrutiny and legal considerations as a land-based scene, so it should be processed in an equally thorough and competent manner.

Personnel Requirements and Training

The first obvious hurdle is that typically an underwater scene will require the investigator to submerge him or herself in order that the scene be documented. This usually requires that the investigator be a certified diver in order to effectively document an in-water event. With an exception being very shallow water photography that can be accomplished by either wading or taking photographs from a boat or other floating object, the individual tasked with the duty must be a competent swimmer and underwater diver as well as a skilled photographer.

Whenever a person enters an environment for which his or her body was not meant to live within, there are hazards, risks, and restrictions involved. The individual assigned the duty of underwater photography must be both physically and mentally capable of venturing into and working within the underwater environment.

Several organizations can certify an individual as a diver. Some are dedicated to recreational divers and a more civilian concentrated population (PADI, NAUI, SSI), and others are specifically related to public safety professionals (Dive Rescue International, Miami-Dade). Each has a purpose and a niche; however, the important matter is that an underwater investigator be certified, comfortable, and competent in order that he or she can effectively perform the assigned duties.

Processing and Documenting an Underwater Scene

When encountered with an underwater crime scene, the investigation team must take into account that there are multiple scenes and levels that must be accounted for. Just as in an above-water scene, multiple methods of documentation must occur. These remain the same as for land-based operations. They are typically: still photography, videography, sketching/mapping, and the written report.

Documenting the Surface Crime Scene

When documenting the scene through these four methods, the investigation team must remember that there are two types of scenes that must be documented. The first is the surface scene. This includes any water access points such as piers, shorelines, or waterfronts, as well as the surface of the water. All of the aforementioned must be thoroughly searched, photographed, and located items of evidence noted on a sketch.

Documenting the Underwater Crime Scene

The other type of scene is the submerged scene. The submerged scene has the added difficulty of depth as well as the aforementioned environmental issues and visibility issues that compound the problem. This is why it is imperative that only those individuals who have received proper training and certification in such matters be utilized to conduct underwater search and recovery operations. The submerged scene includes the objects located at depth as well as the level at which they were found. It is a three-dimensional scene, often with very little in the way of fixed markers or items to aide in reference. Just as with the surface scene, this area must be thoroughly searched. Items of evidence must be photographed, marked, and their positions and depths noted on a sketch. They must then be collected and preserved.

Searching for and Marking Evidence

In addition to proper personnel, underwater investigative teams also must possess the necessary equipment to conduct proper underwater crime scene processing. One item of value where searches are concerned is sonar. It can help to locate items of evidence, as well as identify potential hazards.

A number of different search techniques are utilized for underwater investigation, dependent upon the item being searched for, condition of the water, and location of the body of water. Briefly, some of these methods are described.

Once an item of evidence is located, it is suggested that the position of the item be marked. This is especially important in areas of low visibility. The are several suggested methods for doing this, but the most common method is to "float" a marker buoy, which consists of an inflated buoy attached to a line affixed to the item of evidence or to a tie down in close proximity to the item. This will serve several purposes. First, it allows the investigator to follow the line down directly to the item. Second, it allows surface personnel to visually gauge the evidence location and assist in crime scene sketching. It is important to remember not to mark too many items of evidence in an area with marker buoys while divers are in the water, as it becomes increasingly hazardous that divers will become entangled. The scene will have to be processed in sections. GPS devices are also available that can be used to more easily and safely mark items of evidence. In either case, it is important that the underwater investigator take note of the depth at which the item was discovered. It is suggested that the diver or dive team assemble a diver's slate (i.e., underwater note pad) made specifically for the purpose of annotating underwater scene information.

Photographing in the Underwater Environment

It is not enough to be a skilled top-side photographer and think that one will be equally as successful employing that knowledge and technique sublevel. Several issues must be addressed in order to effectively and accurately capture the underwater crime scene.

Underwater Equipment

The first obstacle is equipment. A photographer is not typically able to employ the same photographic equipment that he or she uses for land-based crime scenes. Special marine or underwater cameras must be used or underwater housings utilized to encapsulate the equipment. This requires the photographer to have a working knowledge of the use and maintenance of such equipment. However, having the correct equipment is not enough. The photographer must next overcome the difficult environmental issues involved with underwater photography.

Environmental Issues

The first major environmental issue encountered by the underwater photographer is distortion. Water refracts light rays differently underwater. Unlike above the water, underwater light refraction causes objects to be magnified. For this reason images and distances are distorted. In fact, objects underwater will appear approximately 33% larger. Thus, it is extremely important to include a scaled object for reference. Although the scale will also be magnified, it will be magnified in direct relation to the object in question, thereby enabling a viewer to interpret the true size and dimension of the subject matter.

CASE IN POINT

Underwater Documentation of a Bank Robbery

Two men allegedly held up a local branch bank with a semi-automatic assault rifle and a handgun, fired one shot, doing no harm, and made off with approximately $110,000. The suspects, both wearing ski masks, were seen leaving the bank by an off-duty firefighter who called dispatch on his cell phone to report the violation. The firefighter followed the two men in his vehicle as they left the bank premises in a vehicle. Due to the firefighter's descriptions and real-time updates, police officers were able to intercept the vehicle prior to it leaving the area. The suspects became involved in a high speed pursuit, covering approximately 13 miles, and in which three law enforcement agencies participated.

At one point, while driving over a highway viaduct that passed over an area lake, several objects were seen to be thrown out of the passenger window of the suspect vehicle. The suspects were observed to split up in the area of a subdivision, when the passenger exited the vehicle, while the vehicle was moving. The driver continued on, and the passenger was pursued by police while he broke into and ran through several houses in an attempt to elude them. The passenger was seen to be carrying a semi-automatic rifle during the foot pursuit. The suspect on foot managed to evade officers and was not found until several hours later, after calling his girlfriend from a payphone to come pick him up. The suspect did not have a firearm in his possession at the time of his arrest. The vehicle pursuit finally terminated when the suspect vehicle disregarded a traffic control device and collided with another passenger vehicle, resulting in the occupant's death and the incapacitation of the suspect driver.

Upon investigation of the incident, crime scene investigators found one spent 40 cal. shell casing in the lobby of the bank. One firearm, an AR-15 assault rifle, was located in the living room of a home in which the passenger was seen entering. There was no additional firearm located in the vicinity of the foot pursuit, nor in the vehicle of the suspect driver. The assault rifle that was located was missing the magazine. The bag containing the money was found within the vehicle.

The money found in the possession of the driver was what would be considered a "smoking gun" in most cases. However, because the possession of semi-automatic weapons and their use in a federal offense carried additional repercussions, it was imperative that the firearm evidence be located.

The firearm located on the living room floor was photographed and collected for evidence. Pursuing officers recalled that the passenger had possibly thrown something out of his window while the vehicle traversed the viaduct crossing over the lake. Investigators were sent to the location to determine if there had been an attempt to destroy or ditch evidence. When investigators arrived, they soon recognized that it would not be possible to conduct a thorough investigation without divers. The dive team was mobilized and responded to the location. The dive team included a crime scene technician whose duty was the documentation of the scene and the collection of any discovered evidence.

(continues)

The crime scene technician had recently been issued a Sony DSC-P200, 7.2MP, point-and-shoot camera, with compatible underwater housing and strobe configuration for use in underwater environments. The agency's administration realized that there could be a potential need for underwater documentation, and wanted to address the need, but did not want to expend inordinate amounts of money on a dedicated underwater system. They rationalized that their decision was meeting the problem half way.

The crime scene technician was tasked with documenting both the submerged and surface scene. He photographed the suspected entry area of the unknown objects from numerous angles, including from the position of the viaduct from which they were believed to have been thrown. The technician also photographed the presence of the dive team while conducting the search and all equipment associated with the effort.

The dive team entered the water and subsequently found one high capacity magazine for a .223 rifle, the same caliber as the assault rifle that had been located in the living room **(Figure 18.4)**. The team also discovered a magazine for a 40 caliber semi-automatic handgun, the same caliber as the expended cartridge found in the bank. Nothing else was located that was believed to be associated with the alleged crime. The magazines were photographed in situ by the crime scene technician, utilizing the underwater system assigned to him.

Visibility in the dive area was less than two feet. Photographs were difficult. An extreme amount of environmental debris had been disturbed as a result of the search performed. This resulted in severe turbidity and added to photographic backscatter. The photographer elected to shoot the scene using available light, thereby avoiding backscatter. A fast shutter speed was used due to movement of both the diver and the water surrounding the subject matter. The items were photographed both with and without a scale of reference. Coloration was yellow/green due to the severe algae growth in the marine environment. A color correction filter was not used. Underwater video was also taken utilizing the same underwater photographic equipment.

The underwater photographs were a large part of this investigation and showed an attempt to disguise evidence associated with criminal activity. It also helped tie the expended round in the bank to the individuals, because the round did not match the firearm located in the living room. The round, however, did match one of the magazines found and photographed in the lake, although the related handgun was never located.

Despite the assistance of the public and numerous police searchers, the semi-automatic handgun was never located. There were several points within the vehicle pursuit when the vehicle was lost from law enforcements' line-of-sight, and the firearm was possibly discarded during one of those moments. The evidence located, however, was sufficient to convict both men of federal charges for bank robbery, possession of illegal firearms, use of a firearm to commit a felony, and also first degree murder for the killing of the driver of the passenger vehicle.

Figure 18.4 Underwater Photograph of a Rifle Magazine.

Note the distortion of the item caused by light refraction.

Coloration and Lighting

Another major problem is subject coloration (**Figure 18.5**). As depth increases, light rays of red, orange, and yellow are filtered out by the water. Eventually, the diver is left with only blue and green rays. Even in the clearest of waters, only blue and green wave lengths typically penetrate at a depth of over 30 feet. This depth is significantly reduced if the water is polluted or murky. One way in which the photographer can overcome this loss of color is to shoot the photograph facing upward towards the surface, thereby utilizing the most natural light penetration available.

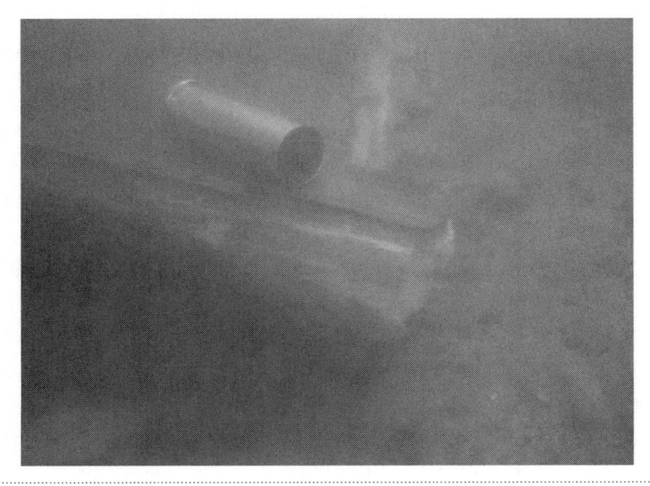

Figure 18.5 Underwater Photograph of a Cartridge Casing Located at a Crime Scene.

Because natural light is quickly absorbed or scattered by water, artificial light is often essential. The addition of a strobe, thereby adding electronic flash lighting to the scene, is useful for two reasons: to illuminate the subject matter, and to obtain the true color of underwater objects and surroundings. However, water is significantly more dense than air and an electronic flash typically will not penetrate or light objects further away than 6 to 8 feet, depending on the strength of the strobe employed.

Color correction filters can be attached to the lens in instances where overall photos are being taken and a strobe will not illuminate the subject matter; however, a strobe must not be used for this purpose and it must be reflected on a diver's photo log. Remember, the important thing about crime scene photography is that it is a true and accurate portrayal of the scene as it was when the photograph was taken. Technically, the scene was viewed by the photographer in unnatural colors of shades of green and blue, not in the "true" colors compensated for by the filter. However, the filter does make the photo more realistic as to what the subject matter would or should look like were it not for the environmental abnormalities associated with being underwater.

In addition to distortion, coloration, and lighting issues, often the environment itself will create difficulty for the underwater photographer. Silt, sediment, algae, and pollution can create what are called black water diving conditions. These are conditions where visibility can be less than 1 or 2 feet. It is especially important for the photographer to have good diving skills to ensure proper buoyancy so as not to disturb the environment and affect photo quality.

Suggestions for Adapting to Underwater Environmental Issues

To overcome the difficulties encountered when photographing an underwater scene, a few basic methods can be employed.

- *If possible, stay shallow.* This will reduce the color loss from light reaching the subject matter. However, if the crime scene is deep and photographs must be taken at a greater depth, the use of a strobe or color correction filter must be employed.
- *Whenever possible, use a strobe (electronic flash).* This will replace the light that is lost underwater.
- *Stay close to your subject.* Because of underwater distortion, coloration issues, and environmental haze, it is wise to keep the distance between the subject and camera as close as possible.
- *Maintain proper buoyancy.* If a diver-photographer is able to maintain his or her proper attitude, it will reduce or eliminate distortion and obliteration caused by stirring up environmental elements.

Evidence Recovery

Items of evidence recovered from water will require special handling and packaging. Just as in a surface level investigation, the ultimate success of an underwater investigation will be determined by the proper location, preservation, and documentation of the evidence. In addition to the obvious underwater hazards and inconveniences that affect search and documentation efforts, bottom suction is another factor that affects evidence recovery.

This phenomenon typically occurs in areas containing silty or soft mud bottoms. Because of the laws of physics, raising an object that is in suction contact with the bottom requires greater lifting force than on land. However, an attempt to exert too much pressure on the item to gain its release can result in a runaway item that could injure the investigator or be lost into the current. For this reason, heavy lift bags are typically used to lift objects underwater. A lift bag is slowly inflated with one of the divers' underwater air sources, and when inflated, will help to raise the object safely. Most of these lift bags contain over-pressurization valves that release air from the lift bag as the bag ascends and aids in controlling the item's ascent to the surface (**Figure 18.6**).

Depending upon the item, sometimes it is suggested to package the item underwater, thereby ensuring the collection of any trace material adhering to the item. Often it is suggested that the item be packaged in

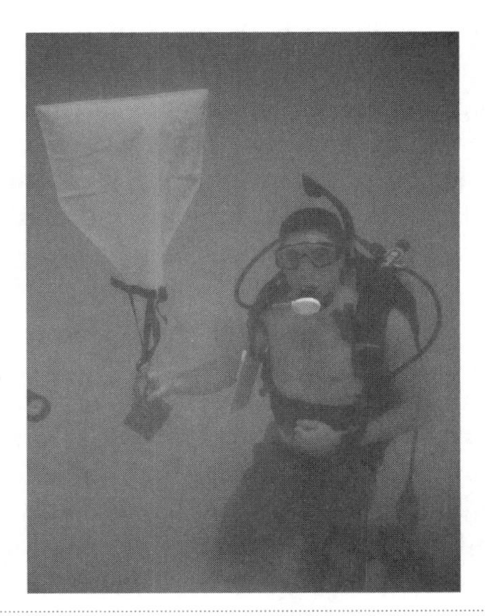

Figure 18.6 A Diver Demonstrates the Use of a Lift Bag to Recover Underwater Evidence.

the water in which it was found to slow environmental deterioration possibly caused by removing the item from water. It is extremely important that an individual familiar with the collection and preservation of underwater evidence be utilized for such matters.

Not Everything Is Submerged

Many of those involved with investigating an underwater crime scene will lose sight of the fact that not all evidence associated with the event is submerged. In fact, a great deal of the time investigative personnel will encounter black water or other environmental conditions that lend underwater photography useless. However, even if the environmental conditions do not allow for photographs to be taken, there is plenty of surface level material that should still be documented. The recommended items to remember to photograph include:

- Object or subject entry point or route
- Object or subject exit point or route
- Overall photographs of geographical area adjacent to the water
- Photographs of personnel and equipment utilized in the operation (for liability reasons)
- Photographs of staged equipment and personnel not utilized in the operation (for liability reasons)

Underwater Investigation Safety

Obviously, the goal of any criminal investigation—to include the photographic documentation of such an event—is to determine whether or not a crime has been committed, and if so, to document and collect evidence to identify and prosecute the guilty party. However, during this process the safety of all personnel involved in the process is paramount (**Figure 18.7**). Underwater photography is no exception to this rule. The underwater environment is inherently dangerous. It is foreign to humans and hard on equipment. The additional weight and bulk of the required equipment result in fatigue, entanglement, and other safety issues. The photographer should ensure that all of his or her equipment is connected properly and does not present an entanglement threat. A neck strap should never be used. Instead, equipment should be affixed to the wrist, dry suit, or **Buoyancy Control Device** (**BCD**), which is the jacket-like piece of equipment used to keep the diver neutrally buoyant. It is extremely important to learn dive-related skills from a certified professional and to participate in ongoing education and training related to the underwater environment to ensure both the safety and competency of the underwater investigator.

Figure 18.7 A Diver Makes Use of a Safety Line to Search a Scene.

Buried Remains

Although buried remains account for relatively few domestic forensic cases, proper scene processing methodology of such bears mention in light of recent world and domestic events. There are many agencies and individuals within the United States who are donating their time and talents both domestically (Hurricane Katrina; September 11, 2001 terror attacks) and internationally (Thailand's tsunami, mass graves of Kosovo) to assist with the processing of mass disasters and mass gravesites. Typically, these individuals are part of what is referred to as a **Disaster Mortuary Operational Response Team (DMORT).** DMORT is a program of the U.S. Department of Health and Human Services that responds only when requested by local law enforcement agencies. The goal of DMORT is to assist local authorities during a mass fatality incident that is beyond the scope of local agency resources and abilities. Sometimes these instances will involve buried remains associated with human rights abuses, war crimes, and genocide. In any event, whether a mass event or an isolated case, buried remains present the problem of being difficult to locate.

Locating Buried Remains

There are a variety of techniques available to assist law enforcement with locating the buried remains of an individual or individuals. Witness statements typically form a starting point, and more sophisticated methods of probing, trenching, photography, **cadaver dogs** (trained to seek out the scent of decomposition), instrumental sensing, and ocular clues help to define a more specific area. Often a clandestine gravesite area will exhibit signs such as the discoloration of vegetation, primary and secondary areas of depression, and cracked soil that are clues to the investigation team of potential buried remains at the location (Ramey Burns, 2006). (**Figure 18.8**).

The preferred strategy is to progress from nonintrusive methods towards more destructive activities. Nonintrusive methods include remote sensing technologies such as airborne near-infrared imaging. Intrusive methods include any activities that disturb a site. Examples include digging, probing, disrupting vegetation, and compacting the soil.

Turning to Specialized Organizations

As a rule, it is best that buried remains be recovered by individuals familiar with archeological techniques. One such agency is NecroSearch International, located in Fort Collins, Colorado. NecroSearch is a private, volunteer organization whose mission is to both train and assist law enforcement agencies in the search for and/or recovery of human remains and evidence. Due to the fact that NecroSearch International is a private organization, it can only respond to assist at a crime scene when asked to do so by the law enforcement agency in charge of a case, so that the scene and related evidence will not be compromised. (For more information on NecroSearch International, visit http://www.necrosearch.org/)

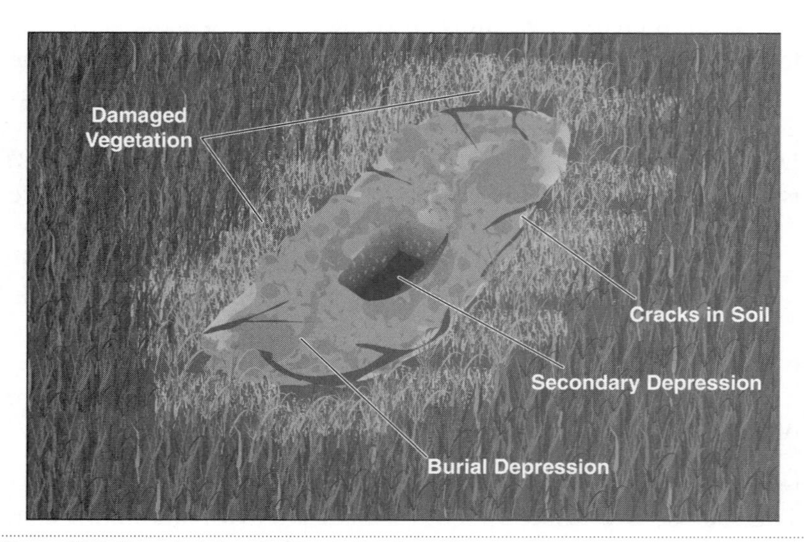

Figure 18.8 Characteristics of a Clandestine Gravesite.

Source: Erica Lawler, University of Wisconsin–Platteville.

Effect of Burial Depth

Research has shown that buried bodies typically decompose at a slower rate than do bodies that are discovered on the surface. Of course, this is mostly dependent upon the amount of time that a body lay exposed prior to burial. If insects are given a chance to lay their eggs, insect propagation and development can continue along with the subsequent decomposition and degradation of a body after burial.

Depth of burial has an effect on both decomposition and discovery. Temperatures within deeper graves (deeper than 2–3 feet) remain relatively stable, whereas shallower graves are more susceptible to the fluctuations of related surface temperatures. The soil density of a gravesite also has a bearing on the decomposition process, to include the tampering of the remains by scavengers. Except for extremely shallow graves or those in noncompacted soil, burial of remains offers protection of the remains from scavenging.

Processing Locations Involving Buried Remains

The area should be properly photographed before any work is begun at the site. Then the area to be searched and excavated should be organized into a grid in order to best preserve the information present within it. If available, a metal detector may be employed for surface investigation. If metallic objects are located they should be marked with flags. The area should also be examined for insect and larvae activity that may be associated with the body. If located, these should be collected and preserved.

A location should be set aside for a "screening area" in which surface materials and material eventually removed from the gravesite can be systematically searched through to locate small items (broken bones, fibers, insects, bullets, shell casings, etc.) sometimes missed in the excavation process. This should be located in an area that has already been searched and is in close proximity to the grave.

The next step in the process is to set up a staging area for equipment and the grave excavation process. This area should extend several yards, in all directions, surrounding the suspected gravesite. Vegetation and debris not associated with the burial should then be cut down and removed and screened. It may be advisable to once again make use of a metal detector after the vegetation and debris has been removed. Any discovered items should be documented and preserved.

In many instances relating to buried remains, it is helpful to have an idea as to the position of the body prior to beginning the excavation process. This can sometimes be determined through visual cues such as secondary depressions, and also from the use of ground penetrating radar or other instrumentation, based upon mass and density. The periphery of a gravesite can be determined through the use of the same instrumentation, in a nonpenetrating manner, or through **probing**, which is much more invasive and risks damaging potential evidence, but is very easy, inexpensive, and efficient. Probing involves pushing a metal rod into the ground and marking where the rod enters more easily than its surroundings. This is measuring for soil density, and can also give some idea as to grave depth. Once the periphery has been determined, it is identified through the use of evidence flags or driving stakes into the ground (**Figure 18.9**).

Once the excavation process has begun, it should proceed slowly, deliberately, and systematically. The work should involve small instruments, rather than taking a shovel or backhoe approach. Small trowels and brushes are the tools of choice for this undertaking, in order to minimize the loss or destruction of evidence.

In cases where the body is relatively fresh or still in the soft tissue decomposition process, it will be necessary to minimize damage to the body. In these cases, the exposed skin, especially the hands, feet, and head, should be bagged and/or covered in order to minimize environmental exposure and best preserve trace evidence.

All material removed from the grave area should be screened in an effort to discover additional evidence that might have eluded investigators upon the initial excavation and removal process. For this step,

Figure 18.9 Evidence Flags Denote the Periphery of a Suspected Clandestine Gravesite.

the material is sifted through screens of various widths, in order to systematically process and discover evidence of various sizes (**Figure 18.10**). If the material removed is mud, then water can be used to wash the material through the screen grates. Those individuals responsible for screening material should be experienced in the identification of small bones and other forensically significant material. The bones of small children, teeth, broken hyoid bones, and others are quite small and can go undetected by the untrained eye.

Throughout the excavation process, at regular intervals and when encountering items of importance, photographs and/or video should be taken. This will ensure proper documentation of the event as well as serve as a record for the measures undertaken in an effort to properly process the evidence associated with the case (**Figure 18.11**).

When the remains have been completely exposed, they should be removed to a body bag for transportation and examination to the medical examiner's office. Personnel must also remember to inspect the surface areas of the original grave excavation for the presence of tool marks or footwear impressions that could help in identifying a suspect. After the body has been removed, the floor of the grave should be excavated and screened down several additional inches in order to ensure that any artifacts that may have settled (teeth, bullets, etc.) are recovered.

Figure 18.10 Crime Scene Personnel Process Material Associated with a Clandestine Gravesite.

Figure 18.11 Crime Scene Personnel Document a Clandestine Gravesite.

Sketching and Mapping the Scene

As was discussed in Chapter 6, there are a number of methods to choose from when sketching and mapping the scene of a clandestine gravesite. The important thing to remember is that with a burial, a minimum of two sketches will be necessary. The documentation process should include an overhead view and an elevation view at the very minimum (**Figure 18.12**).

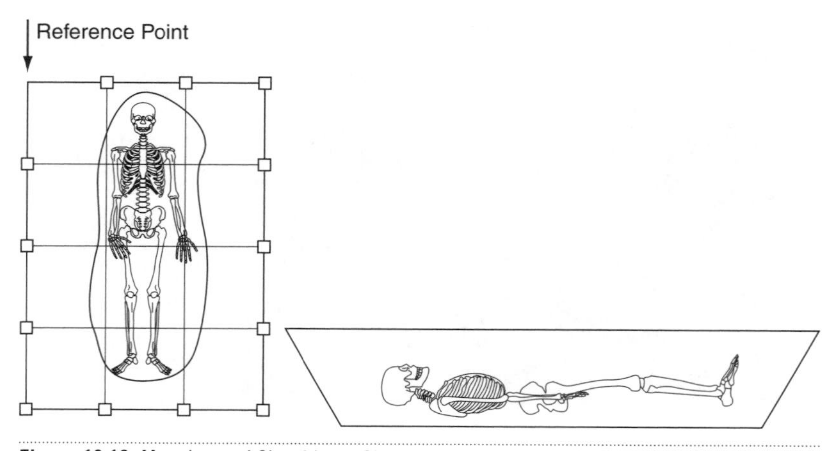

Figure 18.12 Mapping and Sketching a Clandestine Gravesite.

Source: Courtesy of Ellie Bruchez, University of Wisconsin–Platteville.

Chapter Summary

While each investigation and each crime scene presents its own challenges there are some that call upon a crime scene investigator to utilize very specialized training. Two of the most challenging crime scenes encountered by crime scene personnel are those involving underwater crime scenes and the investigation of buried remains. Each requires additional skills above and beyond the "routine" crime scene processing methodology. These situations present additional difficulties and hazards to crime scene personnel. The work is hard and logistically cumbersome. However, when employed, the proper processing of these difficult scenes can be the turning point in a case.

All too often, underwater crime scenes and scenes involving buried remains are not properly documented or processed. This failure can result in a loss, contamination, or inadmissibility of the evidence. Proper scene processing and documentation methods will ensure that the evidence is located, properly documented, preserved, and able to be used in subsequent litigation. It is important that the investigation team be properly trained and take part in ongoing training that is specifically related to matters of crime scene documentation and processing methods involving the underwater environment and clandestine gravesites. What is not looked for will not be found. What is not found cannot be analyzed to uncover the truth.

Review Questions

1. What are the two benefits of using floating marker buoys?
2. DMORT is the acronym for ______________.
3. Which federal agency established USERTS and how is it used?
4. A common misconception is that submerged evidence lacks the ______________ of evidence found topside.
5. Another name for the safety diver is the ______________.
6. The two types of scenes need to be processed in an underwater investigation are ______________ and ______________.
7. Approximately how much larger than actual size will objects underwater appear?
8. Silt, sediment, algae, and pollution can create what are called ______________ conditions.
9. The jacket-like piece of equipment used to keep the diver neutrally buoyant is called a ______________.
10. Pushing a metal rod into the ground and marking where the rod enters more easily than its surroundings in order to determine the location of burial is known as ______________.

References

Becker, R. F. (2000). Myths of underwater recovery operations. *The FBI Law Enforcement Bulletin*, *69*, 1–5.

Becker, R. F. (2005). *Underwater forensic investigation*. Upper Saddle River, NJ: Prentice Hall.

Caplan, J. (2001). *Documenting individual identity: The development of state practices in the modern world*. Princeton, NJ: Princeton University Press.

Federal Bureau of Investigation. (2009). Retrieved from http://www.fbi.gov.

International Association of Identification (IAI). (2008). Retrieved August 21, 2009, from http://www.theiai.org

Ramey Burns, K. R. (2006). *Forensic Anthropology Training Manual* (2nd Ed.). Upper Saddle River, NJ: Prentice Hall.

Crime Scene Investigations Historical Timeline

Date	Historically Significant Event *Period*
	Ancient
Prehistoric	Evidence of fingerprints in early paintings and rock carvings found throughout the human-populated world.
	1000s
c. 1000	An attorney within the Roman courts proved that bloody palm prints were intended to frame a blind man in his mother's murder.
	1100s
1100s	Establishment of the *Crowners* (First Coroner's Office) in England.
	1200s
1248	The Chinese book entitled, *Xi Yuan Ji Lu* (*Collected Cases of Injustice Rectified*) described how to distinguish drowning due to strangulation. This was the first known record of medical knowledge being applied to crime solution.
	1500s
1543	Andreas Vesalius publishes *De Humani Corporis Fabrica*, a seven volume book on the structure of the human body. This essentially defined the science of anatomy.
1554	Ambrose Paré's research relating to gunshot wounds published. His official autopsy studies were an effort to describe how GSWs were inflicted and to prove that they were not themselves poisonous.
	1700s
1784	In England, John Toms was convicted of the crime of murder based upon the first documented use of physical matching evidence. In the case, a torn edge of newspaper was found within a firearm that in turn matched a remaining piece discovered within his pocket.

Date	Historically Significant Event *Period*
	1800s
1813	In France, Mathiew Orfila was credited with the development of the science of toxicology and with the first recorded use of a microscope to analyze blood and semen stains.
1835	Henry Goddard was the first credited with the analysis of ballistic evidence to catch a murderer. His analysis involved visible flaw comparison in a bullet that was then traced back to the mold that produced it.
1845	The New York Police Department was established.
1849	In Chicago, Allen Pinkerton was appointed as the city's first detective. This was noteworthy because it was the first known separation of investigations and police patrol.
1857	The New York Police Department established an investigative division and developed a photo gallery of criminals.
1858	In India, British civil servant William James Herschel used thumbprints to identify Indian workers who signed official contracts. This was the first official use of the observed unique patterns of an individual's fingerprint.
1865	The United States Secret Service was created in response to an increase in finding counterfeit currency.
1870	The United States Department of Justice was created.
1879	In France, Alphonse Bertillon developed the first scientific method of identification, known as anthropometry.
1880	In Tokyo, Scottish physician Henry Faulds published a paper in the peer-reviewed scientific journal, *Nature*, suggesting that fingerprints at a crime scene could be used to identify a criminal. In one of the first recorded cases involving fingerprints, Faulds used his theory to eliminate suspicion of an innocent man and identify the criminal in a burglary in Tokyo.
1887	The first Sherlock Holmes detective story, *A Study in Scarlet*, was written by Sir Arthur Conan Doyle. Doyle referred to the not yet developed science of serology within the text.
1893	In Germany, Hans Gross published *Handbuch fur Untersuchungsrichter als System der Kriminalistik (Criminal Investigation)*, the first known treatise on criminal investigation. This text provided the first descriptions of the importance of reconstructing a crime scene and the use of physical evidence in crime-solving. Gross is often credited with creation of the term criminalistics.
	1900s
1901	Austrian biologist and physician, Dr. Karl Landsteiner discovered human blood grouping relating to whole blood.
1903	At Leavenworth Federal Penitentiary, Kansas, Will and William West were indistinguishable by visual and anthropometric measurements. This showed the weakness of the system of anthropometry and

Date	Historically Significant Event *Period*
1903 (cont.)	strengthened the science of fingerprints after the men were later (1905) easily differentiated by examining their fingerprints.
1907	Chief of Police, August Vollmer, in Berkley, CA, instituted a formal procedure for the handling of physical evidence after a failure to indict a suspect.
1910	Edmond Locard established the first police crime laboratory in Lyon, France.
1915	In Italy, Dr. Leon Lattes developed a method to determine blood grouping of a dried bloodstain.
1920	In France, Locard published *L'enquete criminelle et les methods scientifique*, in which he presented his cross transfer exchange principle, stating "every contact leaves a trace."
1923	August Vollmer established the first forensic laboratory in the United States at the Los Angeles Police Department.
1929	In Chicago, the "Scientific Crime Laboratory of Chicago" was established by Colonel Calvin Goddard, its first director, at the Law School of Northwestern University, after his work on the St. Valentine's Day Massacre.
1932	The Federal Bureau of Investigation (FBI) established its crime laboratory.
1937	Dr. Paul Leland Kirk established the first university criminalistics program at the University of California-Berkeley.
1950	The American Academy of Forensic Science (AAFS) was formed, and began publication of the *Journal of Forensic Science* (JFS).
1953	Dr. Kirk published *Crime Investigation: Physical Evidence and the Police Laboratory*, which brought increased scientific rigor into criminal investigations nationwide.
1966	The Supreme Court ruled in *Miranda v. Arizona* that the accused has the right to remain silent and that statements made by defendants while in law enforcement custody may not be used against them unless the police have advised them of this right. This case had a significant impact on the value of physical evidence.
1980's	Automated Fingerprint Identification System (AFIS) was established. This helped to dramatically increase fingerprint comparison rates.
1987	DNA profiling was introduced for the first time in a United States criminal court to convict Tommy Lee Andrews of a series of sexual assaults in Orlando, Florida.
1999	The FBI upgraded and integrated its computerized AFIS system, dubbing it the Integrated Automated Fingerprint Identification System. This system allowed for paperless submission, storage, and search capabilities.
1999	The ATF and FBI combine to organize the National Integrated Ballistics Information Network (NIBIN). This system allowed for the paperless comparison, submission, storage, and search of firearms and ballistic evidence.

Sources

Block, E. B. (1979). *Science vs. crime: The evolution of the police lab*. San Francisco, CA: Cragmont Publications.

Dillon, D. (1977). *A history of criminalistics in the United States 1850–1950*. Doctoral Thesis. Berkeley, CA: University of California, Berkeley.

Else, W. M., & Garrow, J. M. (1934). The detection of crime. *The Police Journal*, xiv–xv.

Gaensslen, R. E. (1983). *Sourcebook in forensic serology*. Washington, DC: U.S. Government Printing Office.

Gerber, S. M., Saferstein, R. (1997). *More chemistry and crime: From Marsh arsenic test to DNA profile*. Washington, DC: American Chemical Society.

German, E. (1999). *Cyanoacrylate (Superglue) Discovery Timeline*. Retrieved August 23, 2009, from http://onin.com/fp/cyanoho.html

German, E. (1999). *The History of Fingerprints*. Retrieved August 23, 2009, from http://onin.com/fp/fphistory.html

Kind, S., & Overman, M. (1972). *Science against crime*. New York: Aldus Book Limited, Doubleday.

Morland, N. (1950). *An outline of scientific criminology*. New York: Philosophical Library.

Olsen, R. D., Sr. (2008). A fingerprint fable: The Will and William West case. (Initially published in: [1987]. *Identification News*, which became *Journal of Forensic Identification*, *37*, 11.) Retrieved August 23, 2009, from http://www.gaston.k12.nc.us/schools/northgaston/Marr%20Class%20Webpage/Forensic%20Science/PDF/Will%20and%20William%20West.pdf

Thorwald, J. (1966). *Crime and science*. Translation, Winston R & Winston C. New York: Harcourt, Brace & World, Inc.

Forensic Databases

■ Integrated Ballistic Identification System (IBIS)

Maintained by the U.S. Bureau of Alcohol, Tobacco, Firearms and Explosives' National Integrated Ballistic Information Network, this forensic database contains bullet and cartridge casings that have been retrieved from crime scenes and test-fires of guns found at a crime scene or on a suspect. A limitation of this database is that there must be a suspected gun in order to make a comparison. Because the database contains information on bullets and casings—and not on specific guns—a test-fire bullet from a gun must be compared to a bullet found at a crime scene, for example, to determine whether a bullet came from a specific gun. A comparison microscope is used to perform a manual examination. For more information, see http://www.atf.gov.

■ Paint Data Query (PDQ)

Maintained by the Royal Canadian Mounted Police (RCMP), PDQ contains the chemical compositions of paint from most domestic and foreign car manufacturers and the majority of vehicles marketed in North America after 1973. The PDQ software is free to agencies that supply a minimum of 60 paint samples per year. The database information comes from the street (more than 60% from body shops and junkyards) and from manufacturers. In 1998, RCMP entered into agreements with the German Forensic Institute and the Japanese National Police Agency, which resulted in 1,500 samples being added to the database each year. Not all manufacturers, however, are willing to divulge the chemical composition of paint used on their vehicles. If a particular sample has not been entered into the database from the street, it would not be possible to obtain a match.

Each paint layer is examined to determine the spectra and chemical composition. The chemical components and proportions are coded into the database. These known samples are compared against a paint sample from a crime scene or a suspect's vehicle to search the make, model, and year of manufacture of a vehicle involved in a hit-and-run or other criminal activity. For more information, see http://www.rcmp-grc.gc.ca.

■ TreadMark

The number of shoe prints at a crime scene can be so large that the process of impression recovery becomes very time-consuming. TreadMark™ is a commercial product that uses four parameters—pattern, size, damage, and wear—to identify individual outsole impressions. These are then compared with shoe print data from two sources: suspects in custody and crime scenes. A match could yield the name, date of birth, criminal record number, places of interest, and similar offenses for possible suspects.

Impressions from a crime scene are obtained using the current recovery methods of photograph, gel lift, dust lift, and adhesive lift. These are input directly into the analytical system by high-resolution digital imaging. The same procedure is used with an impression of a suspect's shoe print: It is photographed using a high-resolution digital camera, and these impressions (along with the offender's details) are input into the analytical system, where the operator can measure, analyze, and compare crime-scene and suspect images. Both image sources can be searched within themselves and against each other, allowing such images to be transmitted to other users. For more information, see http://www.csiequipment.com/systems.aspx.

■ SoleMate

This commercial database contains information—manufacturer, date of market release, an image or offset print of the sole, and pictorial images of the uppers—for more than 12,000 sports, work, and casual shoes. Sold on DVD, the product is updated and distributed to subscribers every three months. One limitation is that different manufacturers often use the same sole unit and, therefore, it may be difficult to determine the exact make and model of a shoe. The software links such records, however, so that all footwear that might match a crime-scene print can be considered.

The pattern of an unidentified shoe print is assigned a set of codes to isolate basic features, such as circles, diamonds, zigzags, curves, and blocks. Options, with variations, are presented pictorially, which allows an investigator to code features that best match the shoe print. These codes form the database search, with results presented in descending order of pattern correlation. For more information, contact Foster & Freeman USA Inc., at 888-445-5048 or http://www.fosterfreeman.com.

■ TreadMate

The pattern of an unidentified tire mark is assigned a set of codes for pattern features, such as waves, lines, diamonds, zigzags, curves, and blocks, which then form the basis of the database search. Results are presented in descending order of correlation. For more information, contact Foster & Freeman USA Inc., at 888-445-5048 or http://www.fosterfreeman.com.

■ Forensic Information System for Handwriting (FISH)

Maintained by the U.S. Secret Service, this database enables document examiners to scan and digitize text writings such as threatening correspondence. A document examiner scans and digitizes an extended body of handwriting, which is then plotted as arithmetic and geometric values. Searches are made on images in the database, producing a list of probable "hits." The questioned writings, along with the closest hits, are then submitted to the Document Examination Section for confirmation. For more information, see http://www.secretservice.gov/forensics.shtml

■ International Ink Library

The collection—maintained jointly by the U.S. Secret Service and the Internal Revenue Service—includes more than 9,500 inks, dating from the 1920s. Every year, pen and ink manufacturers are asked to submit their new ink formulations, which are chemically tested and added to the reference collection. Open-market purchases of pens and inks ensure that the library is as comprehensive as possible.

Samples are chemically analyzed and compared with library specimens. This may identify the type and brand of writing instrument, which can be used to determine the earliest possible date that a document could have been produced. If the sample matches an ink on file, a notation is made in the database. The U.S. Secret Service generally provides assistance to law enforcement on a case-by-case basis. For more information, contact 202-406-5708.

■ Ident-A-Drug

The Therapeutic Research Center, a private company, publishes a computer program and book to help identify drugs in tablet or capsule form. To make an identification, sufficient information about the unknown drug must be available. Data used for comparison purposes contain codes that are imprinted on tablets and capsules, information on color and shape, the national drug code (NDC #), and drug class. Schedule information is shown if the drug is a narcotic or in one of the U.S. Drug Enforcement Administration schedules. For more information, see http://www.therapeuticresearch.com.

■ Ignitable Liquids Reference Collection (ILRC)

Maintained by the National Center for Forensic Science, this database and associated liquid repository allows a laboratory to isolate an ignitable liquid of interest for inclusion in an in-house reference collection. Designed for screening purposes only, it parallels—but does not replace—American Standard Testing Materials requirements for an in-house reference collection. A laboratory does not need to adopt the ILRC classification system to use this database.

Users enter the name of the liquid into the searchable database. The database can also be organized by classification of the liquid for quick reference. Users can then purchase samples of the liquid. Commercial samples are obtained directly from manufacturers and distributors. The products are then repackaged for distribution using the product name and sent to forensic science laboratories. For more information, see http://ilrc.ucf.edu.

■ Integrated Automated Fingerprint Identification System (IAFIS)

This FBI-maintained database contains:

- Fingerprints acquired after arrest at the city, county, State, and Federal levels.
- Fingerprints acquired through background checks for employment, licensing, and other noncriminal justice purposes (if authorized by State or Federal law).
- Latent prints found at crime scenes.

Although IAFIS offers electronic search and storage capabilities, it has some limitations. The database contains the fingerprints of only a small percentage of the population. Moreover, to make a comparison, the latent print must be of sufficient quality to identify certain individual characteristics. For example, the cores and deltas must be present in the print to determine the orientation of the print.

The database receives data electronically, in hard copy, or in a machine readable data format. IAFIS accepts, stores, and distributes photographs, including the results of remote 10-print and latent searches. These are returned electronically to the requesting agencies with a list of potential matching candidates and their corresponding fingerprints for comparison and identification. For more information, see http://www.fbi.gov/hq/cjisd/iafis.htm or contact the FBI's Criminal Justice Information Services Division at 304-625-2000.

■ Combined DNA Index System (CODIS)

This FBI-run database blends forensic science and computer technology into a tool for solving violent crimes. CODIS enables Federal, State, and local crime labs to exchange and compare DNA profiles electronically, thereby linking crimes to each other and to convicted offenders. CODIS uses two indexes: (1) the Convicted Offender Index, which contains profiles of convicted offenders, and (2) the Forensic Index, which contains profiles from crime-scene evidence.

Searches are performed to find a match between a sample of biologic evidence and an offender profile. Matches made between the Forensic and Offender Indexes provide investigators with the identity of a suspect. DNA analysts in the laboratories share matching profiles, then contact each other http://www.fbi.gov/hq/lab/html/codis1.htm.

Crime Scene Equipment

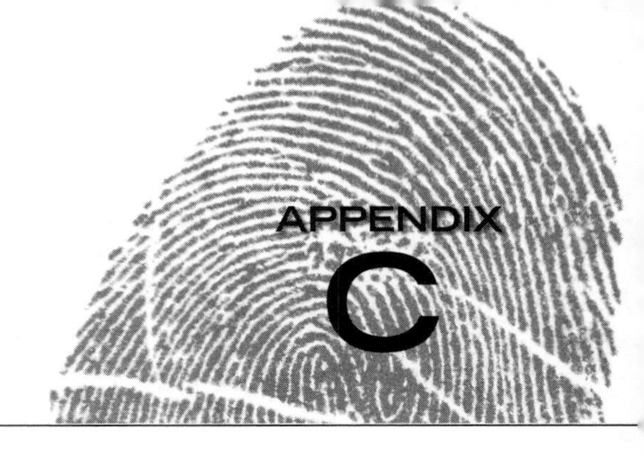

TABLE C.1 Equipment Needed by the Initial Responding Officer(s)

Essential	Optional
Consent/search forms	Audiotape recorder
Crime scene barricade tape	Camera with flash and extra film
First-aid kit	Chalk
Flares	Directional marker/compass
Flashlight and extra batteries	Disinfectant
Paper bags	Maps
Personal protective equipment (PPE)	Plastic bags
	Reflective vest
	Tape measure
	Tarps to protect evidence from the weather
	Traffic cones
	Waterless hand wash (towelette with germicide)
	Wireless phone

* Essential items should be in police vehicles or readily available to the initial responding officer(s).

Source: Courtesy of the National Institute of Justice. (2004, June). *Crime Scene Investigation: A Reference for Law Enforcement Training*, U.S. DOJ, Office of Justice Programs. NCJ200160.

TABLE C.2 Equipment Needed by the Crime Scene Investigator/ Evidence Technician

Essential*	Optional
Bindle paper	Audiotape recorder
Biohazard bags	Bloodstain pattern examination kit
Body fluid collection kit	Business cards
Camera (35 mm) with flash/film tripod	Chalk
Casting materials	Chemical enhancement supplies
Consent/search forms	Entomology (insect) collection kit
Crime scene barricade tape	Extension cords
Cutting instruments (knives, box cutter, scalpel, scissors)	Flares
Directional marker/compass	Forensic light source (alternate light source, UV lamp/laser, goggles)
Disinfectant	Generator
Evidence collection containers	Gunshot residue kit
Evidence identifiers	Laser trajectory kit
Evidence seals/tape	Maps
First-aid kit	Marking paint/snow wax
Flashlight and extra batteries	Metal detector
High-intensity lights	Mirror
Latent print kit	Phone listing (important numbers)
Magnifying glass	Privacy screens
Measuring devices	Protrusion rod set
Permanent markers	Reflective vest
Personal protective equipment (PPE)	Refrigeration or cooling unit
Photographic scale (ruler)	Respirators with filters
Presumptive blood test supplies	Roll of string
Sketch paper	Rubber bands
Tool kit	Sexual assault evidence collection kit (victim and suspect)
Tweezers/forceps	Shoe print lifting equipment
	Templates (scene and human)
	Thermometer
	Traffic cones
	Trajectory rods
	Video recorder
	Wireless phone

* Essential items should be in police vehicles or readily available to the crime scene investigator/evidence technician.

Examples of Evidence Collection Kits

Blood Collection

Bindle
Coin envelopes
Disposable scalpels
Distilled water
Ethanol
Evidence identifiers
Latex gloves
Photographic ruler (ABFO scales)
Presumptive chemicals
Sterile gauze
Sterile swabs
Test tubes/test tube rack

Bloodstain Pattern Documentation

ABFO scales
Calculator
Laser pointer
Permanent markers
Protractor
String
Tape

Excavation

Cones/markers
Evidence identifiers
Metal detectors
Paint brushes

Shovels/trowels
Sifting screens
String
Weights
Wooden/metal stakes

Fingerprint

Black and white film
Brushes
Chemical enhancement supplies
Cyanoacrylate (super glue) wand packets
Flashlight
Forensic light source
Lift cards
Lift tape
Measurement scales
One-to-one camera
Powders

Impression

Bowls/mixing containers
Boxes
Dental stone (die stone)
Evidence identifiers
Measurement scales
Permanent markers
Snow print wax
Water

Pattern Print Lifter

Chemical enhancement supplies
Electrostatic dust lifter
Gel lifter
Wide format lift tape

Tool Marks

Casting materials

Trace Evidence Collection

Acetate sheet protectors
Bindle paper
Clear tape/adhesive lift
Flashlight (oblique lighting)
Forceps/tweezers
Glass vials
Slides and slide mailers
Trace evidence vacuum with disposable collection filters

Trajectory

Calculator
Canned smoke
Dummy
Laser
Mirror
Protractor
String
Trajectory rods

Source: Courtesy of the National Institute of Justice. (2004, June). *Crime Scene Investigation: A Reference for Law Enforcement Training*, U.S. DOJ, Office of Justice Programs. NCJ200160.

Processing the Scene of Mass Fatality Incidents

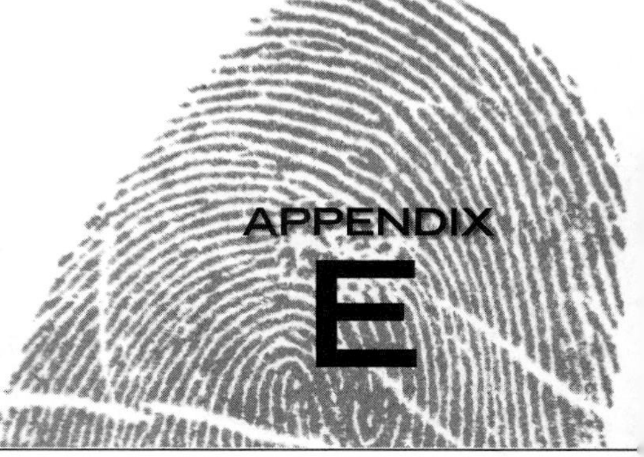

■ Initial Considerations

The complete and accurate identification of remains and evidentiary processing begins at the scene of the mass fatality incident. In most circumstances, the medical examiner/coroner has the ultimate responsibility for the recovery and identification of the deceased. The remains/ evidence processing teams have to assume that any mass fatality scene could be a crime scene. They are expected to carefully document every piece of physical evidence recovered from the scene. The scene should be large enough to ensure its protection from public access until all agencies have agreed to release the scene. Although teams can discard information later, scene processing always involves the physical destruction of the actual scene, and additional information may not be recoverable after the scene has been processed and released. Efficient information recovery proceeds from the least intrusive to the more intrusive (e.g., taking photographs is allowed only after teams locate, flag, and sequentially number the remains). Although protocols may change in the middle of an event depending on the scope and extent of the incident, documenting every aspect of the remains and the evidence processing operation will ensure the preservation of information.

Before processing the scene, the incident command, in consultation with the medical examiner/coroner, is expected to:

A. Identify team leaders responsible for remains/evidence processing.

B. Determine the size and composition of the remains/evidence processing teams (usually a function of the team leaders), which may include:

1. Medical examiner/coroner
2. Forensic anthropologist
3. Odontologist

4. Police crime scene investigator
5. Forensic photographer
6. Evidence technician
7. Scribe/notetaker

C. Integrate the remains/evidence processing teams according to existing interagency jurisdiction and chain of command. The scope and extent of the mass fatality incident determines the number of agencies involved.

D. Establish and/or verify control over access to the scene.

E. Establish communication among transport vehicles, the incident command, and the morgue.

F. Establish an on-scene remains-processing station.

G. Consider the recovery of remains and personal effects as evidence and preserve the chain of custody throughout the recovery operation.

Effective organization and composition of the remains/evidence processing teams ensures the proper collection and preservation of remains, personal effects, and evidence.

■ Establish a Chain of Custody

Establishing and maintaining a chain of custody verifies the integrity of the evidence. The remains/evidence processing teams are expected to maintain the chain of custody throughout the recovery process. Throughout the investigation, those responsible for preserving the chain of custody are expected to:

A. Document the time of arrival and departure of other personnel at the scene.

B. Establish a standard numbering system at the scene that relates back to the location of the remains/evidence. Ensure that the numbering system is:

1. Internally consistent and cross-referenced with other agencies
2. Expandable
3. Simple to interpret
4. Capable of indicating where the remains, personal effects, and evidence were recovered

5. Capable of tracking remains, personal effects, and evidence throughout the investigation
6. Related to subsequent individual results without error
7. Integrated into all protocols and reports

C. Document the collection of evidence by recording its location at the scene and time of collection.

D. Document all transfers of custody (including the name of the recipient and the date and manner of transfer).

Maintaining the chain of custody by properly documenting, collecting, and preserving the evidence ensures its integrity throughout the investigation.

Scene Imaging and Mapping Principle

The remains/evidence processing teams can use a grid system to divide the scene into manageable units to show the location and context of items (i.e., their positions relative to other items) at the scene. A grid system may need to be three-dimensional.

The remains/evidence processing teams are expected to:

A. Record overall views of the scene (e.g., wide-angle, aerial, 360-degree) with a designated photographer to relate items spatially within the scene and relative to the surrounding area. A combination of still photography, videotaping, and other techniques is most effective. Remember to:
   1. Consider muting the audio portion of any video recording unless there is narration.
   2. Minimize the presence of scene personnel in photographs/videos.
   3. Maintain photo and video logs.

B. Identify boundaries and fixed landmarks (e.g., a utility pole, building corners, or global positioning system [GPS]-located points).

C. Establish a primary point of reference for the scene.

D. Divide the scene into identifiable sectors and create a checkerboard.

E. Use accurate measuring devices.

The remains/evidence processing teams are responsible for establishing an accurate, logical mapping system for the scene.

Document the Location of Remains, Personal Effects, and Evidence

The remains/evidence processing teams are expected to include documentation in the permanent record of the scene. Photographic documentation creates a permanent record of the scene that supplements the written incident reports. The teams are expected to complete this documentation, including location information, before the removal or disturbance of any items. Videotaping may serve as an additional record but not as a replacement for still photography.

The remains/evidence processing teams are expected to:

- **A.** Photograph individual items (midrange and close) with an identifier (i.e., a grid identifier and/or individual item number) and scale. Consider including a directional compass arrow that points north.
- **B.** Attach identifying numbers and flag all remains, personal effects, and evidence in the grid:
  1. Use a waterproof ink marker.
  2. Ensure that numbers on the flags correspond with those on the remains and are also clearly discernible in the photograph.
- **C.** Ensure that the systematic on-scene documentation of all remains, personal effects, and evidence includes:
  1. The sequential numbering system at the scene
  2. Recovery location information
  3. Notes that may help with personal identification or scene reconstruction (e.g., generic descriptors, such as a foot or shoe)
  4. Documentation of the evidence collector (e.g., the collector's unique identifier and the date and time of recovery)
- **D.** Conduct the systematic removal of remains, personal effects, and evidence:
  1. Using a permanent marker, mark the outside of the primary bag or container and tag with the identifying number, the collector's unique identifier, and the date and time of collection.
  2. Place the same identifying number on the inside of the body bag or other bag or container.

3. Do not remove any personal effects on or with the remains. Transport all personal effects on or with the remains to the morgue.
4. When necessary, wrap the head before moving it to protect cranial/facial fragments and teeth.

E. After removing the remains, photograph the areas from which evidence was recovered to document whether anything was under the remains.

F. After the remains/evidence processing teams have cleared the area and before releasing the scene for public access, conduct a final shoulder-to-shoulder sweep search to locate any additional items.

The remains/evidence processing teams must properly document the collection of all remains, personal effects, and evidence before removing them from the scene.

On-Scene Staging Area

The remains/evidence processing teams should use the on-scene staging area for checking documentation, maintaining the chain of custody, and conducting potential triage functions. At this area, the remains/evidence processing teams can add notes to aid personal identification at the morgue (e.g., comments about tattoos, marks, and scars) and identify contents of body bags (e.g., watches, body parts). The remains/evidence processing teams are responsible for closing and locking body bags at this point.

The remains/evidence processing teams are expected to:

A. Establish a staging area proximate to the incident scene that provides maximum security from public and media scrutiny and access (including a no-fly zone over the site).

B. Remand evidence that is not required to accompany the remains to the mortuary to the custody of the appropriate agency.

C. Maintain the chain of custody of body bags:

1. Maintain a log of the body bags that are transported from the staging area to the morgue.
2. Record drivers' names and the license numbers of vehicles.
3. Record dates and times that all vehicles leave for the morgue.

- **D.** Maintain equipment and supplies at the staging area. Inventory resources may include:
    1. A large tent
    2. Body/storage bags
    3. Litters, gurneys, and stretchers for remains transport
    4. Refrigeration vehicles
    5. Emergency lighting
    6. Sawhorses with plywood boards for makeshift examination tables
    7. Tarpaulins or other screening materials to create visual barriers
    8. Decontamination control
    9. Inventory control system
    10. Equipment storage
    11. Personal protective equipment
- **E.** Notify the morgue when transport of remains will begin.

The remains/evidence processing teams are expected to maintain a secure triage area for initial examination of remains and other evidence and to ensure secure transport to the morgue. Strongly consider placing forensic identification specialists at the staging area, as the initial evaluations at this point will dictate the efficiency of subsequent morgue operations.

Source: Courtesy of the National Institute of Justice (NIJ). (2005, June). *Mass Fatality Incidents: A Guide for Human Forensic Identification.* Retrieved August 23, 2009, from http://www.ncjrs.gov/pdffiles1/nij/199758.pdf

Professional Organizations Pertaining to Crime Scene Investigation

American Academy of Forensic Science
http://www.aafs.org/

American Board of Criminalistics
http://www.criminalistics.com/

American Board of Forensic Document Examiners
http://www.abfde.org/

American Board of Medico-legal Death Investigators
http://www.slu.edu/organizations/abmdi/

American College of Forensic Examiners
http://www.acfei.com/

Association for Crime Scene Reconstruction
http://www.acsr.org/

Association of Firearm and Tool Mark Examiners
http://www.afte.org/

Association of Forensic Document Examiners
http://www.afde.org/

Evidence Photographers International Council
http://www.epic-photo.org/

International Association of Arson Investigators
http://www.firearson.com/

International Association of Bloodstain Pattern Analysis
http://www.iabpa.org/

International Association of Computer Investigative Specialists
http://www.iacis.com/

International Association of Forensic Nurses
http://www.forensicnurse.org/

International Association of Identification
http://www.theiai.org/

International Crime Scene Investigators Association
http://www.icsia.org/

National Association of Document Examiners
http://documentexaminers.org/

Drugs of Abuse: Uses and Effects

TABLE G.1 Drugs of Abuse: Uses and Effects

Drugs	CSA Schedules	Trade or Other Names	Medical Uses	Dependence			Duration (Hours)	Usual Method	Possible Effects	Effects of Overdose	Withdrawal Syndrome
				Physical	Psycho-logical	Tolerance					
NARCOTICS											
Heroin	Substance I	Diamorphine, Horse, Smack, Black tar, *Chiva, Negra (black tar)*	None in U.S., Analgesic, Antitussive	High	High	Yes	3–4	Injected, snorted, smoked	Euphoria, drowsiness, respiratory depression, constricted pupils, nausea	Slow and shallow breathing, clammy skin, convulsions, coma, possible death	Watery eyes, runny nose, yawning, loss of appetite, irritability, tremors, panic, cramps, nausea, chills and sweating
Morphine	Substance II	MS-Contin, Roxanol, Oramorph SR, MSIR	Analgesic	High	High	Yes	3–12	Oral, injected			
Hydrocodone	Substance II, Products III, V	Hydrocodone w/ Acetaminophen, Vicodin, Vicoprofen, Tussionex, Lortab	Analgesic, Antitussive	High	High	Yes	3–6	Oral			
Hydromorphone	Substance II	Dilaudid	Analgesic	High	High	Yes	3–4	Oral, injected			
Oxycodone	Substance II	Roxicet, Oxycodone w/ Acetaminophen, OxyContin, Endocet, Percocet, Percodan	Analgesic	High	High	Yes	3–12	Oral			
Codeine	Substance II, Products III, V	Acetaminophen, Guaifenesin or Promethazine w/ Codeine, Florinal, Floricet or Tylenol w/ Codeine	Analgesic, Antitussive	Moderate	Moderate	Yes	3–4	Oral, injected			

TABLE G.1 Drugs of Abuse: Uses and Effects (continued)

Drugs	CSA Schedules	Trade or Other Names	Medical Uses	Dependence: Physical	Dependence: Psycho-logical	Tolerance	Duration (Hours)	Usual Method	Possible Effects	Effects of Overdose	Withdrawal Syndrome
NARCOTICS (continued)											
Other Narcotics	Substance II, III, IV	Fentanyl, Demerol, Methadone, Darvon, Stadol, Talwin, Paregoric, Buprenex	Analgesic, Antidiarrheal, Antitussive	High–Low	High–Low	Yes	Variable	Oral, injected, snorted, smoked			
DEPRESSANTS											
gamma Hydroxybutyric Acid	Substance I, Product III	GHB, Liquid Ecstasy, Liquid X, Sodium Oxybate, Xyrem®	None in U.S., Anesthetic	Moderate	Moderate	Yes	3–6	Oral	Slurred speech, disorientation, drunken behavior without odor of alcohol, impaired memory of events, interacts with alcohol	Shallow respiration, clammy skin, dilated pupils, weak and rapid pulse, coma, possible death	Anxiety, insomnia, tremors, delirium, convulsions, possible death
Benzo-diazepines	Substance IV	Valium, Xanax, Halcion, Ativan, Restoril, Rohypnol (Roofies, R-2), Klonopin	Antianxiety, Sedative, Anti-convulsant, Hypnotic, Muscle Relaxant	Moderate	Moderate	Yes	1–8	Oral, injected			
Other Depressants	Substance I, II, III, IV	Ambien, Sonata, Meprobamate, Chloral Hydrate, Barbiturates, Methaqualone (Quaalude)	Antianxiety, Sedative, Hypnotic	Moderate	Moderate	Yes	2–6	Oral			

TABLE G.1 Drugs of Abuse: Uses and Effects (continued)

Drugs	CSA Schedules	Trade or Other Names	Medical Uses	Dependence			Duration (Hours)	Usual Method	Possible Effects	Effects of Overdose	Withdrawal Syndrome
				Physical	Psycho-logical	Tolerance					
STIMULANTS											
Cocaine	Substance II	Coke, Flake, Snow, Crack, *Coca, Blanca, Perico, Nieve*, Soda	Local anesthetic	Possible	High	Yes	1–2	Snorted, smoked, injected	Increased alertness, excitation, euphoria, increased pulse rate & blood pressure, insomnia, loss of appetite	Agitation, increased body temperature, hallucinations, convulsions, possible death	Apathy, long periods of sleep, irritability, depression, disorientation
Amphetamine/ Meth-amphetamine	Substance II	Crank, Ice, Cristal, Krystal Meth, Speed, Adderall, Dexedrine, Desoxyn	Attention deficit/ hyperactivity disorder, narcolepsy, weight control	Possible	High	Yes	2–4	Oral, injected, smoked			
Methylphenidate	Substance II	Ritalin (Illy's), Concerta, Focalin, Metadate	Attention deficit/ hyperactivity disorder	Possible	High	Yes	2–4	Oral, injected, snorted, smoked			
Other Stimulants	Substance III, IV	Adipex P, Ionamin, Prelu-2, Didrex, Provigil	Vaso-constriction	Possible	Moderate	Yes	2–4	Oral			
HALLUCINOGENS											
MDMA and Analogs	Substance I	(Ecstasy, XTC, Adam), MDA (Love Drug), MDEA (Eve), MBDB	None	None	Moderate	Yes	4–6	Oral, snorted, smoked	Heightened senses, teeth grinding and dehydration	Increased body temperature, electrolyte imbalance, cardiac arrest	Muscle aches, drowsiness, depression, acne

TABLE G.1 Drugs of Abuse: Uses and Effects (continued)

Drugs	CSA Schedules	Trade or Other Names	Medical Uses	Dependence			Duration (Hours)	Usual Method	Possible Effects	Effects of Overdose	Withdrawal Syndrome
				Physical	Psycho-logical	Tolerance					
HALLUCINOGENS (continued)											
LSD	Substance I	Acid, Microdot, Sunshine, Boomers	None	None	Unknown	Yes	8–12	Oral	Illusions and hallucinations, altered perception of time and distance	(LSD) Longer, more intense "trip" episodes	None
Phencyclidine and Analogs	Substance I, II, III	PCP, Angel Dust, Hog, Loveboat, Ketamine (Special K), PCE, PCPy, TCP	Anesthetic (Ketamine)	Possible	High	Yes	1–12	Smoked, oral, injected, snorted		Unable to direct movement, feel pain, or remember	Drug seeking behavior *Not regulated
Other Hallucinogens	Substance I	Psilocybe mushrooms, Mescaline, Peyote Cactus, Ayahausca, DMT, Dextro-methorphan* (DXM)	None	None	None	Possible	4–8	Oral			
CANNABIS											
Marijuana	Substance I	Pot, Grass, Sinsemilla, Blunts, *Mota*, *Yerba*, *Grifa*	None	Unknown	Moderate	Yes	2–4	Smoked, oral	Euphoria, relaxed inhibitions, increased appetite, disorientation	Fatigue, paranoia, possible psychosis	Occasional reports of insomnia, hyperactivity, decreased appetite
Tetrahydro-cannabinol	Substance I, Product III	THC, Marinol	Antinauseant, Appetite stimulant	Yes	Moderate	Yes	2–4	Smoked, oral			
Hashish and Hashish Oil	Substance I	Hash, Hash oil	None	Unknown	Moderate	Yes	2–4	Smoked, oral			

TABLE G.1 Drugs of Abuse: Uses and Effects (continued)

Drugs	CSA Schedules	Trade or Other Names	Medical Uses	Dependence: Physical	Dependence: Psycho-logical	Dependence: Tolerance	Duration (Hours)	Usual Method	Possible Effects	Effects of Overdose	Withdrawal Syndrome
ANABOLIC STEROIDS											
Testosterone	Substance III	Depo Testosterone, Sustanon, Sten, Cypt	Hypo gonadism	Unknown	Unknown	Unknown	14–28 days	Injected	Virilization, edema, testicular atrophy, gyneco-mastia, acne, aggressive behavior	Unknown	Possible depression
Other Anabolic Steroids	Substance III	Parabolan, Winstrol, Equipose, Anadrol, Dianabol, Primabolin-Depo, D-Ball	Anemia, Breast cancer	Unknown	Yes	Unknown	Variable	Oral, injected			
INHALANTS											
Amyl and Butyl Nitrite		Pearls, Poppers, Rush, Locker Room	Angina (Amyl)	Unknown	Unknown	No	1	Inhaled	Flushing, hypotension, headache	Methemo-globinemia	Agitation
Nitrous Oxide		Laughing gas, balloons, Whippets	Anesthetic	Unknown	Low	No	0.5	Inhaled	Impaired memory, slurred speech, drunken behavior, slow onset vitamin deficiency, organ damage	Vomiting, respiratory depression, loss of consciousness, possible death	Trembling, anxiety, insomnia, vitamin deficiency, confusion, hallucinations, convulsions
Other Inhalants		Adhesives, spray paint, hair spray, dry cleaning fluid, spot remover, lighter fluid	None	Unknown	High	No	0.5–2	Inhaled			
ALCOHOL		Beer, wine, liquor	None	High	High	Yes	1–3	Oral			

Source: Courtesy of the U.S. Drug Enforcement Administration. (2004, June). *Drugs of Abuse/Uses/Effects*. Retrieved August 23, 2009, from http://www.usdoj.gov/dea/pubs/abuse/chart.htmGlossary

Glossary

Accelerant: Any substance used to accelerate (and sometimes direct) the spread of a fire.

Accreditation: An endorsement of a forensic laboratory's policies and procedures by law enforcement professional organizations or industry; a necessary component of establishing credibility within the court system. To qualify for accreditation, a crime laboratory must meet minimum requirements set forth by the certifying authority.

ACE-V: Acronym for Analysis, Comparison, Evaluation, and Verification. A method devised for the scientific comparison of prints to either identify a print (via individualization, as having originated from the same source) or exclude impressions as having no common origin.

Adipocere: The hydration and dehydrogenation of the body's fat, which results in an off-white, waxy, clay-like substance that in many cases preserves the body, and retards the decomposition process; commonly found in the subcutaneous tissues of body extremities, the face, buttocks, breasts, and in individuals with a high percentage of body fat.

Algor Mortis: Postmortem cooling of the body. Heat loss will occur until the body reaches the temperature of the surrounding environment (ambient temperature).

Alternate Light Sources: Light-emitting devices supplied with colored filters that filter the source light so that the potential evidence can be viewed with light of a narrow wavelength range, rather than at the usual "white light" viewing range.

American Academy of Forensic Sciences (AAFS): A multidisciplinary professional organization that provides leadership to advance science and its application to the legal system. The objectives of the Academy are to promote education, foster research, improve practice, and encourage collaboration in the forensic sciences.

Antemortem: Prior to death.

Anthropometry: A series of 11 body measurements of the bony parts of the body, and an in-depth description of marks (scars, moles, warts, tattoos, etc.) on the surface of the body; developed by Alphonse Bertillon.

Area of Convergence: A two-dimensional point, derived from analyzing bloodstain directionality. It is used to ascertain the point from where an event occurred (i.e., impact) that led to the subsequent dispersal of the blood. This is only a two-dimensional explanation (X and Y axis) and does not determine how far away from the area that the blood event originated; instead it provides an area in which to determine such information.

Area of Origin: (1) Fire. The large track of space or area where a fire would have started. It can be located where the fire was able to grow and develop. (2) Blood Pattern Analysis. By establishing the impact angles of representative bloodstains and projecting their trajectories back to a common axis (Z), extended at 90 degrees from the area of convergence, an approximate location of where the blood source was when it was impacted may be established.

Arson: The willful and malicious burning of a person's property.

Asphyxiation: The interruption of oxygenation of the brain (e.g., drowning, strangulation, etc.).

Associative Evidence: Evidence that can be attributed to, or associated with, a particular person, place, or thing, thus establishing inferred connectivity.

Automated Fingerprint Identification System (AFIS): An automatic pattern recognition system, for the identification of fingerprints that consists of three fundamental stages: data acquisition, feature extraction, and decision-making.

Autopsy: The medical dissection and examination of a body in order to determine the cause of death; entails the removal of internal organs through incisions made in the chest, abdomen, and head.

Backlogged: A case where the analysis has not been completed within 30 days of the item being submitted to the lab.

Ballistics: The study of a projectile (most likely from a firearm) in motion.

Barrier Protection: Involves creating a barrier between the personnel and their surroundings to ensure that they are not contaminated by the scene and that they themselves do not contaminate the scene and evidence therein.

Biological Profile: A description of an individual in such a way that law enforcement or acquaintances can narrow the range of possible identities. The profile is assembled by studying the remains of an individual, and noting characteristics of shape and size, which may allow an estimation of height, build, age, sex, ancestry, and any individualistic features such as tattoos, jewelry, medical apparatus, and clothing.

Black Water Diving: Conditions where silt, sediment, algae, and pollution create underwater visibility that is typically less than one foot.

Bloodborne Pathogen: Microorganisms found within the blood that can cause infection and disease, and may be transported in other biological fluids.

Bloodstain Pattern Analysis (BPA): The science of examining and interpreting blood present at a bloodshed event in order to determine what events occurred, in what order, and who possibly left the stains.

Bore: Interior of a firearm barrel.

Breech Lock: Component of a firearm that supports the base of a cartridge within the chamber. The face of this component (breech face) may leave identifiable marks upon the base of the shell casing.

Buoyancy Control Device (BCD): The jacket-like piece of equipment used to keep a diver neutrally buoyant.

Cadaver Dogs: Canines that are trained to seek out the scent of decomposition; used in area searches for a victim(s) or for clandestine gravesites.

Cadaveric Spasm: Immediate stiffening of a dead body, with no prior period of flaccidity and no extended onset. It will typically involve the victim's hands clenched around an object, such as a weapon, debris from a lake floor, clothing, or another object, and sometimes is associated with events (i.e., drowning or homicide) that involved considerable excitement or tension preceding death.

Caliber: Diameter of the bore of a rifled firearm, measured between opposing lands; usually expressed in hundredths of an inch or millimeters.

Cast Off: Bloodstain patterns created when blood is released from an object through the influence of centrifugal acceleration, cessation, or stop-action.

Cause of Death: The injury or disease responsible for the pathological and physiological disturbances that resulted in death; the medical reason for death.

Certification: A voluntary process of peer review by which a practitioner is recognized as having attained the professional qualifications necessary to practice in one or more disciplines of criminalistics.

Chain of Custody: A log or other record of who collected and subsequently handled evidence of a crime. Accurate accounting strengthens the court case; if evidence is found to be illegally collected, whether intentionally or unintentionally, then it will be found to be inadmissible in court.

Circumstantial Evidence: A series of facts that, although not the fact at issue, through inference tends to prove a fact at issue.

Classification: A formula given to a complete set of 10 fingers as they appear on a fingerprint card generally based on pattern type, ridge count, or ridge tracing. The FBI National Crime Information Center—Fingerprint Classification (NCIC-FPC) and Henry System are used to classify prints.

Close-Up Photographs: A type of image that allows the viewer to see all evident detail on a particular item of evidence. This frame is filled with the evidence itself; photos are taken with and without a scale; and can be called comparison, examination, or macro photographs.

Code of Ethics: A contract signed by an employee who agrees to function by the terms of employment or membership. This code lists what the department or organization believes are acceptable behaviors, professional expectations, and values to which employees should adhere. Failure to comply with the code can result in job dismissal or removal of membership and certification.

Combined DNA Index System (CODIS): An electronic database of DNA profiles administered through the Federal Bureau of Investigation (FBI), where federal, state, and local crime labs can share and compare DNA profiles to match DNA from crime scenes with convicted offenders and other crime scenes.

Combustible: Capable of burning, generally in air under normal conditions of ambient temperature and pressure.

Command Post: Established to coordinate on-scene activities and efforts; also called incident command under the incident command and control structure. From this location, all supervisory decisions are made and the crime scene is managed.

Comparison Microscope: Two compound microscopes connected by an optical bridge; allows two specimens to be viewed side-by-side.

Complimentary Base Pairing: The unique way that the two strands of the DNA double-helix formation are bonded together. Bases in the necleotides (adenine [A], cytosine [C], guanine [G], and thymine [T]) on each strand align in specific paired combinations, where G always pairs with C, and A always pairs with T (see *Deoxyribonucleic Acid*).

Compression Evidence: Marks left when an instrument is in some way pushed or forced into a material capable of picking up an impression of the tool.

Computer Network: Two or more computers linked by data cables or by wireless connections that share or are capable of sharing resources and data.

Concentric Fractures: Cracks or breaks that appear to make a typically broken series of concentric circles around an impact point.

Conchoidal Fractures: Stress marks in glass that are shaped like arches, located perpendicularly to one side of the glass surface, and curved nearly parallel to the opposite glass surface.

Conduction: The transfer of heat by direct contact.

Cone of Foam: In some cases of drug overdose and drowning, the victim will exhibit exudates in the form of froth emanating from the mouth and nostrils; a result of severe pulmonary edema; may appear initially off-white but will advance to pinkish in color as the decomposition process advances.

Contraband: An item that is found to be illegally possessed or for some legal reasons is illegal to possess.

Controlled Substance: A substance (typically a drug) whose possession or use is regulated by the government; Title 21 of the United States Code (21 USC) defines the substances.

Controlled Substances Act (CSA): Title II and Title III of the Comprehensive Drug Abuse Prevention and Control Act of 1970 regulate the manufacture and distribution of drugs and other substances placed by the Drug Enforcement Agency (DEA) or the Department of Health and Human Services (DHHS) into five schedules based upon medical use, potential for abuse, and safety or potential for dependence.

Convection: Transfer of heat caused by changes in density of liquids and gases; the most common method of heat transfer; when liquids or gases are heated they become less dense and will expand and rise.

Cortex: The region of hair between the medulla and cuticle in the hair shaft, which contains the pigment cells responsible for imparting hair color characteristics.

Credentials: A certificate, letter, experience, or anything that provides authentication for a claim or that qualifies somebody to do something. In the context of forensic and crime scene–related work, credentials as an expert will be established by the court through questioning pertaining to the witness' education, training, and experience.

Crime: An act or the commission of an act that is forbidden by a public law and that makes the offender liable to punishment by that law.

Crime Scene: Anywhere that evidence may be located that will help explain the events that occurred.

Crime Scene Investigation (CSI): The systematic process of documenting, collecting, preserving, and interpreting physical evidence associated with an alleged crime scene, in an effort to determine the truth relating to the event in question.

Crime Scene Sketch: A permanent mapped record of the size and distance relationships of the crime scene and the physical evidence within it; serves to clarify the special information present within the photographs and video documentation, because it allows the viewer to easily gauge distances and dimensions.

Criminalistics: The application of science through the analysis of physical evidence within the enforcement of law.

Criminalistics Light-Imaging Unit (CLU): A multispectrum imaging system that uses various colors of light to view the substance or structure being examined; can locate body fluids at crime scenes under normal lighting conditions. By using a strobe lamp, signal processing, and improved optics, CLU rejects surrounding light and thereby improves both the sensitivity and specificity of the area being viewed. CLU is five times more sensitive than current fluorescing methods.

Cross Contamination: The movement or transfer of material between two objects during the investigative processing efforts; should be avoided when possible.

CSI Effect: A phenomenon whereby forensic drama television has created unreasonable expectations in the public, thereby increasing the prosecution's burden of proof, while presenting an air of infallibility with regards to forensic science; also has educated the public and increased the overall interest in the area of forensic science and crime scene work.

Cuticle: The outermost layer of hair; contains the scaly protective layer that covers the shaft of the hair. Each species has identifiable cuticle characteristics.

Cyanoacrylate Ester Fuming: A technique that stabilizes latent prints using super glue. Super glue is induced to fume and the fumes interact with latent fingerprint residue by polymerizing them, yielding a stable friction ridge impression off-white in color (see *Latent Prints*).

Daubert Standard: From the case *Daubert v. Merrell Dow Pharmaceuticals*, 509 U.S. 579 (1993), the court held that federal trial judges are the gatekeepers of scientific evidence. Under this standard, trial judges must evaluate whether testimony is both relevant and reliable, resulting in a two-pronged test of admissibility.

Death: The irreversible cessation of circulatory and respiratory functions.

Decomposition: Postmortem breakdown of body tissues.

Deoxyribonucleic Acid (DNA): The nucleic acid that contains the genetic instructions used for the growth, development, and programmed death for cells of all organisms and some viruses; is a double-helix structure.

Depressant: Type of psychoactive drugs that temporarily reduce or incapacitate a specific area of the body or mind; also referred to as downers and sedatives.

Digital Evidence: Information and data of value to an investigation that is stored on, received, or transmitted by an electronic device.

Direct Evidence: A type of evidence that proves a fact without the necessity of an inference or a presumption.

Direct Flame Contact: A combination of two of the basic methods of heat transfer. As hot gases from the flame rise into contact with additional fuel, the heat is transferred to the fuel by convection and radiation until the additional fuel begins to vaporize, so that the vapors are ignited by the flames.

Directionality: In a crime scene, the direction that blood traveled that is determined by the bloodstains' shape.

Disaster Mortuary Operational Response Team (DMORT): A program of the U.S. Department of Health and Human Services that responds only when requested by local law enforcement agencies. Its goal is to assist local authorities during a mass fatality incident that is beyond the scope and abilities of local agencies.

Drug: Any chemical substance, other than food, which is intended for use in the diagnosis, treatment, cure, mitigation, or prevention of disease or symptoms.

Electrostatic Lifting Device (ELD): A type of machine that operates by electrically charging a lifting film that has been placed over a surface bearing a dust print impression. During operation, the electrostatically-charged film is drawn down to the surface, and the dust particles in the impression are attracted to the lifting film. The construction of the lifting film allows it to store the electrostatic charge and thus retain the dust particles after the power supply has been disconnected. The impression-bearing film is then viewed with an oblique light source to search for any impressions that may have been recovered.

Ethics: The study of moral standards and how they affect conduct.

Evidence: Anything that can help to prove or disprove that a crime was or was not committed, and by whom.

Excusable Homicide: Unintentional, truly accidental killing of another person; the result of an act that under normal conditions would not cause death, or from an act committed with due caution that, because of negligence on the part of the victim, results in death.

Expert Witness: A person who is called to answer questions on the stand in a court of law in order to provide specialized information relevant to the case being tried.

Expirated patterns: Images created when blood is blown out of the nose, mouth, or wound as a result of air pressure and/or airflow; often numerous, relatively small stain sizes are displayed that may vary in shape.

Exsanguination: Death due to loss of blood or bleeding out.

Extractor: A component found within the chamber area of a firearm (nonrevolver) that is responsible for ejecting the spent shell casing after it has been fired. This hook-like object will often leave a compression or striated mark within the ejector groove at the base of a spent shell casing that is potentially identifiable to a particular firearm (also called an extractor).

Federal Rules of Evidence (FRE): Adopted in 1975, these rules govern the introduction of evidence within proceedings, both civil and criminal, in U.S. Federal Courts. While they did not specifically apply to suits brought within state courts, the rules of many states have been closely modeled upon the provisions found within the FRE.

Final Sketch: A finished rendition of the rough sketch usually prepared for courtroom presentation; often will not show all of the measurements and distances originally recorded on the rough sketch but only significant items and structures. It may be either inked or a computer model, in a manner that is not able to be modified. The sketch is clutter-free and should accurately depict all pertinent items of evidence, typically through the use of an accompanying legend (see *Rough Sketch*).

Firearm: Any device that can fire a projectile or projectiles as a result of an explosive or propellant charge.

Fire Load: The total number of British Thermal Units (Btu's) that might evolve during a fire in a building or area under consideration, and the rate at which the heat evolves.

Firing Pin: The component in a firearm that strikes the base of the cartridge and causes the initial incendiary event leading to expulsion of the bullet from the barrel. These may leave marks upon a cartridge casing that may be identifiable to a particular firearm or brand of firearm.

First Responder: The police officer, fire fighters, and/or emergency medical personnel who is dispatched or arrives at the potential crime scene first.

Flame Point: Temperature at which the fuel will continue to produce sufficient vapors to sustain a continuous flame.

Flammable: Capable of burning with a flame.

Flammable Limit: The mixture of fuel vapors and oxygen, expressed as a concentration (percentage) of fuel vapors in air, which will result in flammability.

Flashover: When sufficient heat is generated to cause simultaneous ignition of all fuels in the confined area.

Flash Point: The lowest temperature at which a solid or liquid material produces sufficient vapors to burn under laboratory conditions.

Forensic Crime Laboratory: A scientific laboratory (with at least one full-time natural scientist) where physical evidence in criminal matters is examined; staff provides reports and opinion testimony with respect to such physical evidence in courts of law. These laboratories provide services for all levels various levels of government.

Forensic Science Educational Program Accreditation Committee (FEPAC): A professional working group established by the American Academy of Forensic Science (AAFS) to maintain and enhance the quality of forensic science education through a formal evaluation and accreditation of college-level academic programs.

Fruit of the Poisonous Tree: If evidence is found to be illegally collected (whether intentionally or unintentionally), the evidence will be found to be inadmissible in court.

Frye Test: The Federal Court of Appeals ruling (*Frye v. U.S.* 293 F. 1013 [D.C. Cir. 1923)]; held that evidence could be admitted in court only if "the thing from which the deduction is made" is "sufficiently established to have gained general acceptance in the particular field in which it belongs."

Gauge: The number of spherical lead balls that have the diameter of the interior of the barrel of the firearm that add up to weigh one pound.

Grooves: The low-lying portions between the lands within a rifled firearm bore.

Hallucinogen: A type of drug or substance that, taken in nontoxic dosages, produce changes in perception, thought, and mood.

Handgun: A type of weapon designed to be held in and fired with one hand; two primary subcategories are pistols (semi-automatic and automatic) and revolvers.

Hearsay: Unfounded information that is heard from other people. It must be corroborated by other sources to be admissible in court.

Henry System: Developed by Sir Edward Henry, a system of print classification used for well over a century. The system was built around the individual's whorl patterns in a fingerprint (primary classification) that were subdivided into five categories depending upon the type and size of the patterns.

High Explosive: A type of material designed to detonate and yield a near instantaneous release of energy; chemically detonates at a speed greater than 3,300 feet per second.

Homicide: The killing of one person by another.

Hyperthermia: Rising of the body's core temperature.

Hypothermia: Lowering of the body's core temperature.

Incision: A type of wound caused by sharp force trauma.

In situ: When items have not been moved, altered, or otherwise molested, and can be documented in their originally discovered location and condition.

Instrumentality: A device, system, or its associated hardware that has a significant role in the commission of a crime.

Integrated Ballistic Identification System (IBIS): A national imaging system developed by the FBI and ATF that digitally records images from fired bullets and cartridge cases used in crime scenes and test fires from recovered firearms for comparison with those used in unsolved crimes.

Investigate: To make a systematic examination, or to conduct an official inquiry.

Iterative Process: Where the investigator continually re-checks and re-analyzes the crime scene to assure that processing is done properly; continues until the results are negative, meaning that nothing further is required, and nothing has been overlooked.

Justifiable Homicide: The killing of a person under authority of the law. Includes killing in self-defense, killing an enemy during wartime, capital punishment, and deaths caused by police officers while attempting to prevent a dangerous felon's escape or to recapture a dangerous felon who has escaped or is resisting arrest.

Known Evidence: Any type of evidence that originates from a known, acknowledged, or accounted for source that is to be compared to an unknown or questioned material.

Laceration: A type of wound caused by blunt force trauma; subdivided into firearm and non-firearm types.

Lands: The raised portion between the grooves within a rifled firearm bore.

Latent Prints: Finger, hand, or other body part prints that require additional processing to be rendered visible and suitable for comparison.

Legend: A note of explanation inserted outside of the sketch area that relates to a specific item, symbol, or information contained within the graphical representation of a sketch (see *Rough Sketch*).

Ligature: An item used to bind, incapacitate, or kill; often leaves patterned impression evidence that can be matched later to the item that was used; the ligature or binding impressions may be either striated or compression.

Livor Mortis: The visible color change that occurs from the pooling of blood once the heart stops pumping; also called hypostasis or postmortem lividity.

Locard's Exchange Principle: Whenever two objects come in contact with one another, a cross-transfer of evidence occurs.

Low Explosive: A type of material that burns or detonates at a speed lower than 3,300 feet per second; they typically involve pyrotechnics that create smoke, light, heat, and sound.

Luminol: A type of presumptive search technique for blood that results in chemiluminescence as a result of the chemical reaction occurring between the reagent and the biological stain. It is fast acting and must be documented in darkness, with photography in an expeditious manner.

Manner of Death: Circumstances under which the cause of death occurred; is classified as natural or unnatural. A death is classified as natural when it is caused by disease. Other deaths are classified as unnatural, including: homicide, suicide, accident, or undetermined, based on the circumstances surrounding the incident causing death.

Mapping: Measurements and drawings associated with a crime scene. The basic methods utilized for crime scene sketching and mapping are: (1) baseline, (2) rectangular coordinates, (3) triangulation, and (4) polar/grid coordinates (see *Rough Sketch*).

Mechanical Loss: When, through the efforts of saving a life, evidence is lost. This is an accepted loss of physical evidence, and is easily articulated in the court.

Medulla: The innermost region of hair in the hair shaft. In humans it is amorphous and lacks visible cellular material, whereas the medulla of other species is often seen as cellular in nature, with some displaying characteristics similar to a bead of pearls.

Midrange Photographs: Their purpose is to frame the item of evidence with an easily recognized landmark, which visually establishes the position of the evidence in the scene in relationship to the item's surroundings.

Minutia: Tiny variations and irregularities within fingerprint ridges that are identifiable; also known as ridge characteristics.

Miranda Ruling: From the 1966 case of *Miranda v. Arizona*, mandated greater emphasis on the collection, preservation, and analysis of physical evidence. Established the Constitutional rights of an accused individual to be advised they can have legal counsel and can remain silent. Eliminated confessions by suspects and placed the emphasis on physical evidence relating to the crime in question.

Mitochondrial DNA (mtDNA): Found outside of the cell nucleus and is inherited solely from the mother. Each cell contains one nucleus and hundreds to thousands of mitochondria. Therefore, there are hundreds to thousands of mtDNA copies in a human cell compared to just one set of nuclear DNA in that same cell.

Morals: Although similar to values, morals are more widely reaching, socially accepted rules of conduct. A person is judged by others by societal definitions of immorality/morality.

Multilevel Containment: A crime scene preservation method utilizing several tiers of perimeters; the most effective for ensuring evidence integrity while also allowing a workable scene structure.

Mummification: The dehydration of soft tissues as a result of high temperatures, low humidity, and wind or other form of ventilation. Skin will appear brown, leather-like, and tight. The mummification process begins at the tips of the fingers and toes and progresses towards the hands and feet, face, and other extremities.

Narcotic: Derived from the Greek word for stupor, originally referred to a variety of substances that dulled the senses and relieved pain. Currently refers to opium, opium derivatives, and their semi-synthetic substitutes; here the term refers to drugs that produce morphine-like effects.

National Criminal Identification Center—Fingerprint Classification (NCIC-FPC): A system of fingerprint classification developed by the FBI that assigns a 20 character string of letters and numbers to a person's fingerprints. The database is used to compare existing prints with those from unsolved crimes.

National Integrated Ballistic Information Network (NIBIN): A networked computer database of fired cartridge casing and bullet images used by crime laboratories; developed to solve open cases by allowing firearms examiners to compare existing evidence with fired bullets, cartridge casings, shotgun casings, and firearms recovered in other jurisdictions.

Negative Impression: A type of pattern that an item, such as footwear, has left in a dusty area, essentially removing the dust that was present.

Ninhydrin: A chemical used to detect ammonia or amino acids within print residue. It reacts with these amino acids and forms a bluish-purple color; most useful on porous surfaces (e.g., paper and raw wood) and is primarily used in document processing efforts.

Overall Photographs: Taken with a wide angle lens or in a fashion that allows the viewer to see a large area in the scene. This documents the condition and layout of the scene as found and helps to eliminate issues of subsequent contamination (e.g., tracked blood, movement of items). Typically they are shot from the four corners of a crime scene looking inward.

Passive Bloodstain: A type of blood evidence that is not related to a specific violent action within the context of the scene. They are the aftermath associated with this violence, as the resulting bloodshed begins to move and cure due to environmental and gravitational forces.

Patent Prints: Finger or other body prints that require no processing to be recognizable and may be suitable for comparison.

Penetrating Gunshot Wound: A type of firearm projectile injury with an entrance but no exit, thus allowing a projectile be recovered at autopsy.

Perforating Gunshot Wound: A type of firearm projectile injury with an entrance wound and an exit wound. Generally no projectile will be recovered by a pathologist at autopsy; those responsible for processing the crime scene should try to recover a projectile for each identified exit wound.

Perimeter: The outer confines of a crime scene; involves some sort of delineation (e.g., plastic tape, rope, etc.) or physical boundary as to what area is considered inside of the supposed crime scene, and what is external to the area of investigation.

Perimortem: At or near the time of death.

Perjury: A lie told within a court of law by someone who has taken an oath to tell the truth.

Personal Protective Equipment (PPE): A type of barrier protection that typically consists of gloves, Tyvek suits, shoe covers, eye protection, and respiratory equipment; designed to protect crime scene personnel from the hazards inherent at the scene.

Photo Log: A permanent written record of all information pertaining to photodocumentation. Department policy often dictates what is included in a photo log; however, the following information should be included: date, time, case number, agency name, photo equipment used, numerical ordering of each photo taken, brief description of photo taken, direction facing for each photo taken, approximate distance from subject matter in each photo taken, and shutter speed, aperture setting, and ISO if taken on film (digital cameras automatically record such photo information).

Photo Placard: A handwritten or agency-developed sheet or board that lists pertinent case information for the photographs to follow. Taking a shot of the placard as the first photo on a roll of film or digital card ensures that personnel know which photographs pertain to a particular case and identifies the photographer.

Physical Evidence: Any type of evidence that has an objective existence; anything with size, shape, and dimension. Examples are gases, fingerprints, glass, paint, hair, blood, soil, and drugs. Also called real evidence.

Plant: A pool of flammable liquids or pile of combustibles (newspapers, rags, etc.) that is used to heat up a fire at a select location; designed to produce hot heat in the specific area the arsonist wishes to cause great damage; also called a booster.

Plastic Prints: A type of finger or other body print having a distinct three-dimensional appearance; these often do no require further processing.

Point of Origin: The location where a fire actually started; the place where it began.

Points of Comparison: Matching ridge characteristics between two compared prints. In the United States, prints should match at 12 points of comparison before an identification can be considered as positive. Current training in fingerprint comparison stresses that the quality of the print and the quality of the comparison are more important than placing emphasis on a numerical match.

Polymer: Hundreds of thousands (even millions) of the same type of monomer molecules link together into long linear (polymer) chains and sometimes structures. Properties can vary from sticky, pliable, to solid, depending upon the atoms in the monomers that repeat to form the polymer chain.

Postmortem: After death.

Postmortem Interval (PMI): Time elapsed since a person died; estimated through various scientific observations of the biochemical changes that occur to a body after death.

Preliminary Scene Survey: An overview of the entire crime scene to identify any threats to the scene's integrity and ensure protection of physical evidence. The primary purpose is to carefully asses the scene for logistical and safety considerations.

Primary Scene: The first encountered location where evidence was located, which often is the location where dispatch sends the officer, or from which a witness called in a complaint.

Primary Transfer: Occurs when a fiber from a fabric is transferred directly onto a victim's clothing.

Primer: In a firearm, the metal cup located within the center of the base of a cartridge that contains a small amount of incendiary compound that, when crushed by the firing pin, sets off the initial incendiary event. In rimfire weapons (e.g., a .22 caliber), the entire base of the cartridge serves as a primer.

Probability: Frequency with which an event will occur; also known as the odds of occurrence.

Probing: An invasive search technique that attempts to locate a gravesite periphery through pushing a metal rod into the ground and marking where the rod enters more easily than its surroundings; soil density is measured and this also gives some idea as to grave depth.

Product Rule: When the frequency of independently occurring variables is multiplied together to obtain an overall frequency of occurrence for the event or item.

Proficiency Testing: A measure for determining whether lab workers as individuals and the lab as a whole are operating at an industry established standard.

Projected Pattern: Images produced by blood released under pressure (i.e., arterial spurting). These patterns generally result from volumes of blood larger than those that produce passive drop stains or other dynamic patterns (i.e., impact patterns).

Protocols: The steps and processes that are undertaken by the laboratory to ensure that the correct tests are performed accurately.

Puddling: A type of pattern found when a flammable liquid is poured onto a floor to make a pool. When that pool is burning, only the vapors burn, not the liquid; the pool burns from the perimeter in towards the center, leaving heaver burning on the edges than the center of the pool.

Putrefaction: Postmortem changes produced as a result of biochemical actions by bacteria and microorganisms.

Pyrolysis: Transformation of a compound into one or more other substances by heat alone.

Quality Assurance (QA): A method to ensure and verify quality control. A laboratory's QA assessment measures are necessary to oversee, verify, and document the performance of the laboratory.

Quality Control: Measures to ensure that analysis results meet a specified standard of quality.

Questioned Evidence: Any type of evidence or material of an unknown, unacknowledged, or unaccounted for source.

Radial Fractures: Breaks or cracks originating from the point of impact and moving away from that point, in a radiating pattern.

Radiation: The transfer of heat from one source to another (e.g., heat waves from the sun) that generally is not visible to the naked eye.

Rifle: A type of weapon designed to be held in two hands while being fired from the shoulder.

Rifled Firearm: A type of weapon that contains rifling within the bore of the firearm's barrel.

Rifling: The spiral grooves that are formed in the bore of a firearm that are designed to impart spin upon a projectile as it passes through the barrel. The rifling is made up of grooves and lands; a spinning projectile improves its accuracy to impact its target.

Rigor Mortis: Stiffening of the body postmortem. It involves the contraction of body muscles, beginning in the smaller muscle groups and progressing to the larger groups and is a result of chemical changes that occur within the body upon death.

Rough Sketch: A hand-drawn image developed while on-scene, typically during the crime scene assessment/preliminary scene evaluation phase, to assist with development of a strategic plan for processing. The sketch is not done to scale, can be drawn with any implement (crayon, chalk, pencil, pen, etc.), and is crude in its artistry. As work progresses at the crime scene, the sketch will include not only the crude crime scene layout, but also be used to record measurements of items and structures, and distances between items.

3R Rule: *R*adial fractures form a *R*ight angle at the *R*everse side to which force was applied.

Scheduling: Placement of drugs or substances into one of five schedules according to the Controlled Substances Act. This is based upon the substance's medical use, potential for abuse, and safety or liability for psychological and/or physical dependence.

Scientific Investigation Method: An iterative process incorporating the fundamental principles behind the scientific method, which guides the investigator in ensuring a thorough and systematic investigatory methodology.

Scientific Method: Utilizes principles and procedures in the systematic pursuit of knowledge involving the recognition and formulation of a problem, the collection of data through observation and experiment, and the formulation and testing of a defined hypothesis.

Scientific Working Group on Friction Ridge Analysis, Study and Technology (SWGFAST): A working group established in 1997 in response to a number of inconsistencies and controversies relating to fingerprint identification and technological advancement; operates through Federal Bureau of Investigation sponsorship with a mission to assist the latent print community in providing the best service and product to the criminal justice system.

Secondary Device: A second explosive bomb placed at a scene to detonate after the original explosion. Secondary devices are typically targeted at emergency responders and investigators who respond to a bombing.

Secondary Scene: Other crime scenes that are later identified as being associated with a primary scene.

Secondary Transfer: Occurs when already transferred fibers on the clothing of a suspect transfer to the clothing of a victim.

Sexual Assault Evidence Collection Kit (SAE kit): Tools and instructions to assist the crime scene investigator and attending medical professional to properly collect and document the specimens required by the laboratory; can be used to collect appropriate samples from both male and female sexual assault victims and suspects.

Shoeprint Image Capture and Retrieval (SICAR): An automated shoeprint identification system, developed in England by Foster and Freeman Ltd., which incorporates multiple databases to search known and unknown footwear files for comparison against footwear specimens found in crime scenes.

Small Particle Reagent (SPR): A suspension of molybdenum sulfide grains in water and a detergent solution. The grains adhere to the fatty components of a latent print deposit, and assist with the visualization of latent print evidence.

Smoothbore Firearm: Type of firearm weapon with no rifling present within the bore.

Spatter: A random distribution of bloodstains that vary in size and may be produced by a variety of mechanisms. The pattern is created when sufficient force is available to overcome the surface tension of the blood.

Spoliation: The intentional or negligent altering of evidence.

Stellate Defect: An irregular, blown-out entrance wound associated with a gunshot. This type of wound is caused by the propellant gases separating the soft tissue from the bone and creating a temporary pocket of hot gas between the bone and the muzzle of the weapon.

Stimulants: A type of drug or substance (also called uppers) that can reverse the mental and physical effects of fatigue.

Stippling: Type of injury resulting from an impact of burned and unburned particulates associated with the discharge of a firearm. They surround the bullet impact wound in a roughly circular pattern due to the fact that gunpowder is discharged in a conical pattern as it exits a firearm.

Striation Evidence: Marks produced by a combination of pressure and sliding contact by a tool that result in microscopic patterns imparted to the surface onto which the tool was worked.

Swath: The effective area a searcher can cover while conducting a search.

Swipe: Occurs when a bloodied surface rubs across a nonbloodied surface.

Tache Noire: If an individual dies in an arid environment and his or her eyes are open at the time of death, the exposed area of the eyeball (sclera) may develop a brownish-black line.

Taphonomy: The study of postmortem changes to the body. Examples include: normal decomposition; alteration and scattering by scavengers; and movement and modification by flowing water, freezing, or mummification.

Terminal Velocity: An object falling through the air will increase its speed of descent until the force of air resistance that opposes the item is equal to the force of the downward gravitational pull.

Testimonial Evidence: Vocal statements most commonly made while the speaker is under oath, typically in response to questioning; also may be made by witnesses, victims, or suspects during the course of the investigation, while not under oath.

Trace Evidence: Any evidence that is small in size, such as hairs, fibers, paint, glass, and soil, which would require microscopic analysis in order to identify it.

Trailers: Arrangement of a combustible or flammable material (solids, liquids, or combinations of both) to ensure a fire is carried from one location to another. A trailer is regularly placed from a point of exit to the area the arsonist wants to be the area of origin; also used to connect multiple plants within a structure.

Transfer Evidence: A type of evidence that is passed from one item to another, typically as a result of contact or action. Careful analysis of this evidence can associate the questioned evidence with a known source.

Transfer Stain: Generally indistinct stains that can be of virtually any size or shape. The shape of a transfer pattern may retain some of the physical characteristics of the object that created it. In this way, the shape of a transfer pattern suggests the object that created it through the recognizable patent image.

Transient Evidence: Physical evidence present at the crime scene that is either fragile or at great risk for loss, alteration, or destruction if not properly identified, documented, collected, and preserved as soon as possible.

Trier of Fact: A judge or magistrate in a trial by the court, or a jury of one's peers in a trial by jury, whose duty it is to weigh the evidence presented and determine guilt or innocence.

Underwater Search and Evidence Response Team (USERTS): First formed in 1982, four USERT teams of 12 members each (located in New York, Washington, Miami, and Los Angeles) are managed by the FBI in Quantico, Virginia. USERTS can perform surface and underwater crime scene investigations in a variety of wet environments nationwide; also provide guidance, advice, and training to the public service community.

Values: Beliefs of an individual or group, for or against something in which they have some emotional investment; rules by which an individual makes decisions as to what is right or wrong.

Vapor density: The weight of a volume of a given gas to an equal volume of dry air, where air is given a value of 1.0.

Viscosity: Mutually attracted to one another; in this text, pertains to the properties of blood. The more viscous a fluid, the more slowly it flows; blood is approximately six times more viscous than water.

Vitreous Draw: A method of assisting with determination of the time of death in a postmortem investigation, in which a syringe is used to take a sample of ocular fluid (vitreous humor) from the eye to determine potassium levels.

Void Pattern: Created when blood is blown out of the nose, mouth, or wound as a result of air pressure and/or airflow; often displays numerous, relatively small stain sizes that may vary in shape. Also, the blank space against a surface inside of a spatter or smear (in a bloodstain) or char (in fire) where an object or person has been removed after the event occurred.

Voir dire: Preliminary examination of a witness or juror to determine his or her competency to give or hear evidence.

Wipe: Occurs when a non-bloodied surface moves through or across a stationary one.

Photography Credits

Chapter Openers © McMac/ShutterStock, Inc.

Case Studies Icon © juliengrondin/ShutterStock, Inc.

Chapter 5
5.1 © Visions of America, LLC/Alamy Images; **5.2 – 5.7** Courtesy of Joe LeFevre, University of Wisconsin–Platteville

Chapter 6
6.4 © D. Willoughby/Custom Medical Stock Photo

Chapter 7
7.1 Courtesy of the Massachusetts State Police

Chapter 8
8.1 Courtesy of the Federal Bureau of Investigation; **8.12** Courtesy of the Sudbury Police Department, Sudbury, MA

Chapter 10
10.1 © Jan Kliciak/ShutterStock, Inc.

Chapter 12
12.8 © Wellcome Trust Library/Custom Medical Stock Photo

Chapter 13
13.13 Courtesy of Savage Range Systems, Inc.

Chapter 15
15.1 © Scott Rothstein/ShutterStock, Inc.; **15.2** © Gary Paul Lewis/ ShutterStock, Inc.; **15.3** Courtesy of Orange County Police Department, Florida; **15.5** © Mitchell Brothers 21st Century Film Group/ ShutterStock, Inc.; **15.6** © Shawano County Sheriff's Department via

The Shawano Leader/AP Photos; **15.7** Courtesy of DEA; **15.8** © 2007 NBC Universal, Inc./AP Photos

Chapter 16

16.1 © Alex Kotlov/ShutterStock, Inc.; **16.2** © Ivan Montero Martinez/ShutterStock, Inc.; **16.3** © M_G/ShutterStock, Inc.; **16.4** © spaxiax/ShutterStock, Inc.

Chapter 17

17.6 © Chuck Stewart, MD.

Unless otherwise indicated, all photographs and illustrations are under copyright of Jones and Bartlett Publishers, LLC, or have been provided by the author.

Index

Italicized page locators indicate a photo/figure; tables are noted with a *t*.

B

C

E

F

G

H

I

N

O

Q

R

T

U

V